Polymer-modified Liquid Crystals

Soft Matter Series

Series editors:
Hans-Jürgen Butt, *Max Planck Institute for Polymer Research, Germany*
Ian W. Hamley, *University of Reading, UK*
Howard A. Stone, *Princeton University, USA*

Titles in this series:
1: Functional Molecular Gels
2: Hydrogels in Cell-based Therapies
3: Particle-stabilized Emulsions and Colloids: Formation and Applications
4: Fluid–Structure Interactions in Low-Reynolds-Number Flows
5: Non-wettable Surfaces: Theory, Preparation and Applications
6: Wormlike Micelles: Advances in Systems, Characterisation and Applications
7: Electrospinning: From Basic Research to Commercialization
8: Polymer-modified Liquid Crystals

How to obtain future titles on publication:
A standing order plan is available for this series. A standing order will bring delivery of each new volume immediately on publication.

For further information please contact:
Book Sales Department, Royal Society of Chemistry, Thomas Graham House, Science Park, Milton Road, Cambridge, CB4 0WF, UK
Telephone: +44 (0)1223 420066, Fax: +44 (0)1223 420247
Email: booksales@rsc.org
Visit our website at www.rsc.org/books

Polymer-modified Liquid Crystals

Edited by

Ingo Dierking
University of Manchester, UK
Email: ingo.dierking@manchester.ac.uk

ROYAL SOCIETY OF CHEMISTRY

Soft Matter Series No. 8

Print ISBN: 978-1-78262-982-5
PDF ISBN: 978-1-78801-332-1
EPUB ISBN: 978-1-78801-667-4
Print ISSN: 2048-7681
Electronic ISSN: 2048-769X

A catalogue record for this book is available from the British Library

The Royal Society of Chemistry is a charity, registered in England and Wales, Number 207890, and a company incorporated in England by Royal Charter (Registered No. RC000524), registered office: Burlington House, Piccadilly, London W1J 0BA, UK, Telephone: +44 (0) 20 7437 8656.

For further information see our web site at www.rsc.org

Printed in the United Kingdom by CPI Group (UK) Ltd, Croydon, CR0 4YY, UK

Preface

The use of polymers to modify liquid crystals and their properties can be done either by forming a continuous polymer matrix with inclusions of liquid crystalline droplets, or, by forming a bicontinuous system of a polymer network, which is dispersed in a liquid crystal host phase. These soft matter systems, polymer dispersed liquid crystals (PDLC) and polymer stabilized liquid crystals (PSLC), can be found on either side of the phase diagram. As a topic of research these fields are now relatively mature, so that it seems adequate to summarize their fabrication and properties in a single volume text. First applications, such as privacy windows, optic and electro-optic devices, optical filters, switchable THz devices and the like are available, or at least at a demonstrator stage. We have a relatively well-developed understanding of the relations between polymerization conditions, polymer morphology and electro-optic behaviour and can exploit these systems and materials for generating phase stability, as in the drastic widening of the range of frustrated Blue Phases, or for structural templating, as in the visualization of liquid crystal director fields.

The present book aims to provide an overview of the whole field of polymer-modified liquid crystals, from a brief introduction into the liquid crystalline state, *via* phase separation processes and photoreactive monomers (with a special thanks to Richard Mandle), to polymerization mechanisms. This is followed by a summary of PDLCs and PSLCs in general, and polymer stabilized nematics, cholesterics, ferroelectric and antiferroelectric phases in particular. Given that the most recent prospective applications lie in the field of polymer stabilized Blue Phase displays, special attention is devoted to this topic, before other frustrated phases, discotics and combinations with nanoparticle doping are considered. We hope that we thus cover most aspects of polymer-modified liquid crystals and their relevant applications to provide an introduction into the field for postgraduate

Soft Matter Series No. 8
Polymer-modified Liquid Crystals
Edited by Ingo Dierking

Published by the Royal Society of Chemistry, www.rsc.org

students and non-specialists, and generally for soft matter researchers who are the target group of the RSC soft matter series. Hopefully, the book will also capture the interest of those members of the liquid crystal community who are in fact already familiar with the topic.

At this point I would like to sincerely thank all the colleagues who have contributed their valuable expertise to this book. It would not have been possible to produce this volume without their dedication and willingness to take on the whole range of topics which are assembled in this text. Thanks so much. Finally, I am indebted to all the involved staff from the Royal Society of Chemistry for their continuous support and guidance throughout the completion of this project.

Ingo Dierking

Contents

Soft Matter Series No. 8
Polymer-modified Liquid Crystals
Edited by Ingo Dierking

Published by the Royal Society of Chemistry, www.rsc.org

CHAPTER 1

Introduction

INGO DIERKING

School of Physics and Astronomy, University of Manchester, Oxford Road, Manchester M13 9PL, UK
Email: Ingo.Dierking@manchester.ac.uk

1.1 What are Liquid Crystals?

Liquid crystals are one of the fundamental forms of soft condensed matter,[1–3] besides polymers and colloids, although there is a substantial overlap between all of the soft matter materials. They are generally considered to be materials which exhibit flow behaviour like a viscous liquid, and at the same time some anisotropy of their physical parameters, as is often observed for crystals.[4–6] The latter implies some kind of molecular ordering through self-organization. Liquid crystals are thus anisotropic liquids or partially ordered fluids, combining order and mobility, and are often referred to as the 4th state of matter. And indeed, the liquid crystalline phases are separated from the isotropic liquid and the crystalline solid through thermodynamic phase transitions, while their phase diagram can schematically be summarized as in Figure 1.1.

Two general classes of liquid crystalline materials are often distinguished, although their borderline becomes more and more diffuse. These are *thermotropic*[7–10] and the *lyotropic*[11,12] liquid crystals. The former are single component systems or mixtures, which exhibit their liquid crystalline behaviour on change of temperature, while the latter are mainly composed of surfactant molecules or shape-anisotropic particles on the nanoscale, which are dispersed in an isotropic, often polar, liquid, and form on change of

Soft Matter Series No. 8
Polymer-modified Liquid Crystals
Edited by Ingo Dierking

Published by the Royal Society of Chemistry, www.rsc.org

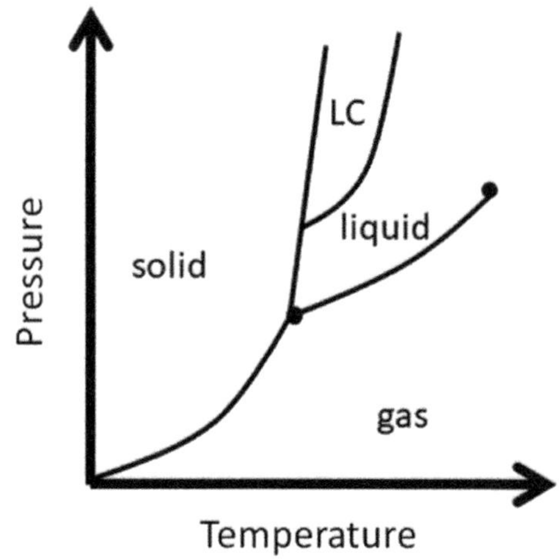

Figure 1.1 Thermodynamic phase diagram with the inclusion of the liquid crystalline state, beside the three well-known states of matter, gases, liquids and solids.

concentration. Materials that show both thermotropic and lyotropic liquid crystalline behaviour are called *amphotropic*.

It should be mentioned that most of the liquid crystalline behaviour can also be observed for macromolecular or polymeric materials.[13] There are for example main-chain and side-group liquid crystal polymers, combined polymers and rubber-like cross-linked systems, called elastomers.[14–16] These systems have been discussed in recent monographs and will not be covered again in any detail in this publication, because they do not strictly represent polymer-modified liquid crystals, but rather polymers exhibiting liquid crystalline behaviour. Polymer-modified liquid crystals are broadly considered in two groups, (i) liquid crystal droplets dispersed in a continuous polymer matrix, so called polymer disperse liquid crystals (PDLC), and (ii) polymer stabilized liquid crystals (PSLC), where a continuous polymer network is formed in a continuous liquid crystal phase, thus representing a bicontinuous system. PDLCs[17] are located at the high polymer concentration part of the phase diagram, generally larger than 70%, while PSLC[18] are formed at the opposite side of the phase diagram, at low polymer concentrations, generally smaller than 10%.

Thermotropic liquid crystals are further distinguished by the molecular shape of their constituent mesogens. The most common are *calamitic* and *discotic*[19] mesogens, which exhibit rod-like or disc-like molecules. Molecular shapes that have more recently attracted increasing interest are *bowlic*,[20] *sanidic*,[21] and *bent*-core[22] molecules, thus bow-shaped, brick- or lath-shaped, and banana-shaped molecules, respectively. A general feature of thermotropic mesogens is a relatively rigid, polarizable core, with flexible chains on the side. A few illustrative examples of some well-known mesogens, together with their schematic representations, are given in Figure 1.2. A further

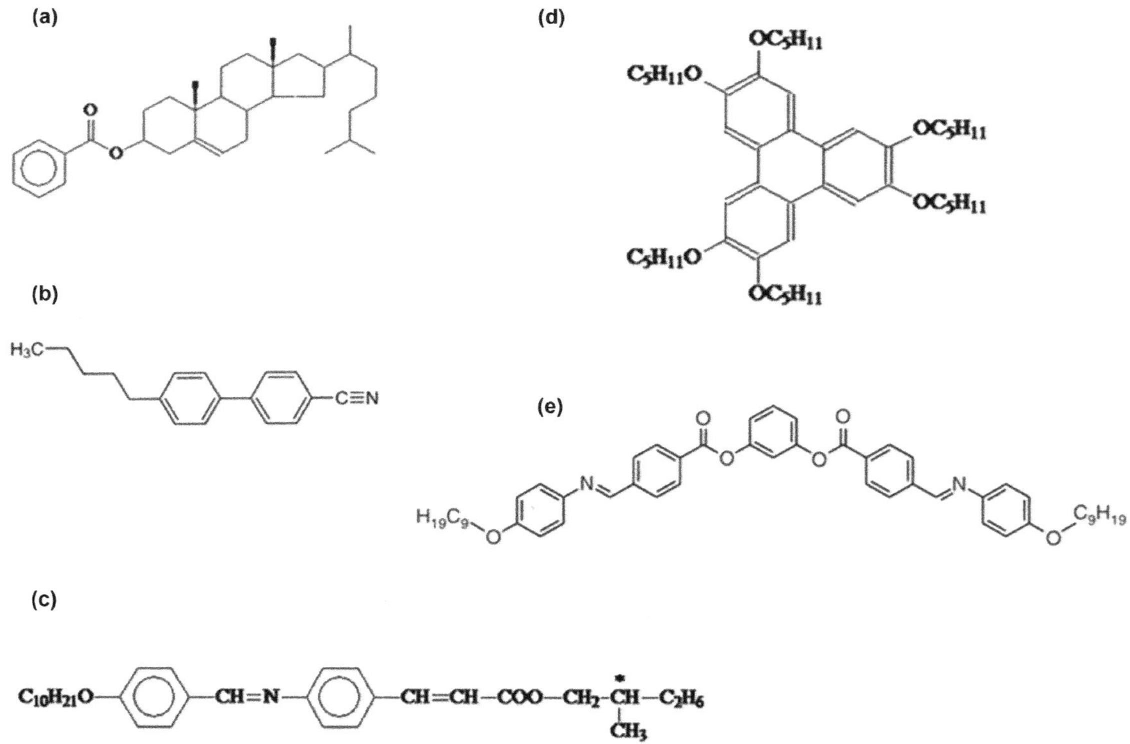

Figure 1.2 An illustrative selection of some well-known thermotropic, mesogenic molecules of different molecular shape (a) calamitic cholesteryl-benzoate, the first liquid crystal to be discovered, by the Austrian botanist Friedrich Reinitzer in 1888, (b) 4-cyano-4′-pentylbiphenyl known as 5CB, the first single component room-temperature liquid crystal, synthesized by George Gray, which greatly propelled the development of the display industry, (c) D-2-methylbutyl 4-[4-*n*-decyloxybenzylideneamino]-cinnamate, known as DOBAMBC, the first ferroelectric liquid crystal, (d) HAT, a typical discotic and (e) NOBOW, a common bent-core mesogen.

common feature of all liquid crystal forming molecules is a pronounced shape-anisotropy with aspect ratios generally larger than about 1 : 5.

Lyotropic liquid crystals on the other hand are mainly formed by *amphiphilic* molecules, *i.e.* those with a hydrophilic head-group and one or more hydrophobic tails. These self-organize into micelles, membranes or vesicles, which can then assemble into liquid crystalline structures.[23] A schematic view with some molecular structures is depicted in Figure 1.3. A different route to produce lyotropic liquid crystals is *via* the use of shape-anisotropic colloidal particles, dispersed in an isotropic, often polar, liquid.[24] These can be materials as diverse as inorganic and mineral liquid crystals, nanorods and nanowires, biological rod- or cylinder-shaped objects, such as tobacco mosaic and other viruses, DNA and RNA, cellulose nanocrystals, or carbon nanotubes, all representing rod-like colloids, but also graphene oxide or clays, which are of the plate-like shape. Over certain colloid concentration regimes, all of these materials form liquid crystal phases.

1.2 Liquid Crystal Phases

1.2.1 Thermotropic Phases

A large variety of liquid crystal phases has been described to date. They all differ in their qualitative degree of order and are delimited by thermodynamic phase transitions, which are most often of 1st order, but can also be 2nd order. First order transitions are those which exhibit discontinuous changes of physical properties and the order parameter at the transition, and which are connected to a transition enthalpy ΔH, while second order

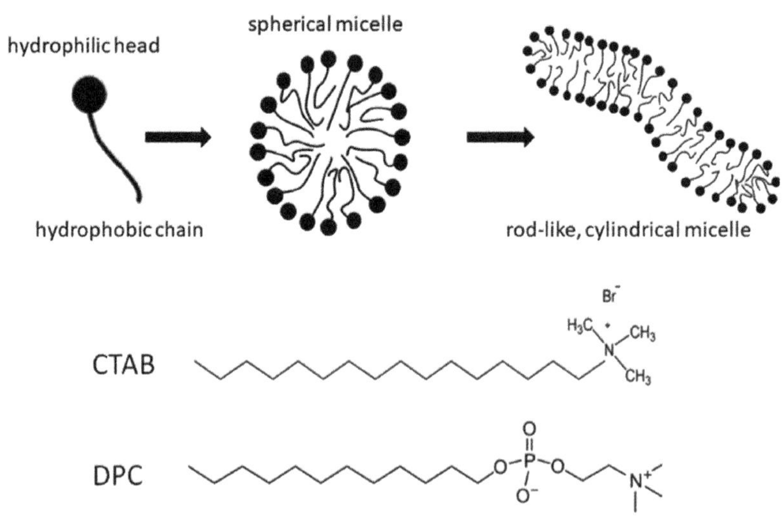

Figure 1.3 Schematic illustration of the formation of micelles from amphiphilic molecules, together with some often-employed molecular structures.

transitions are continuous and exhibit only a jump, Δc_p, in the heat capacity with $\Delta H = 0$.

To describe this self-organized ordering and the different phases observed, we may, without loss of generality, start with the isotropic liquid phase of a thermotropic, calamitic liquid crystal, observed at elevated temperatures (Iso.), and cool the material. Eventually, we will reach the first liquid crystal phase, which is often the so called *nematic* phase, N, from the Greek word for "thread". The schematic structure and a typical appearance seen using polarized microscopy are shown in Figure 1.4(a). The nematic phase solely exhibits orientational order of the long molecular axis, whose

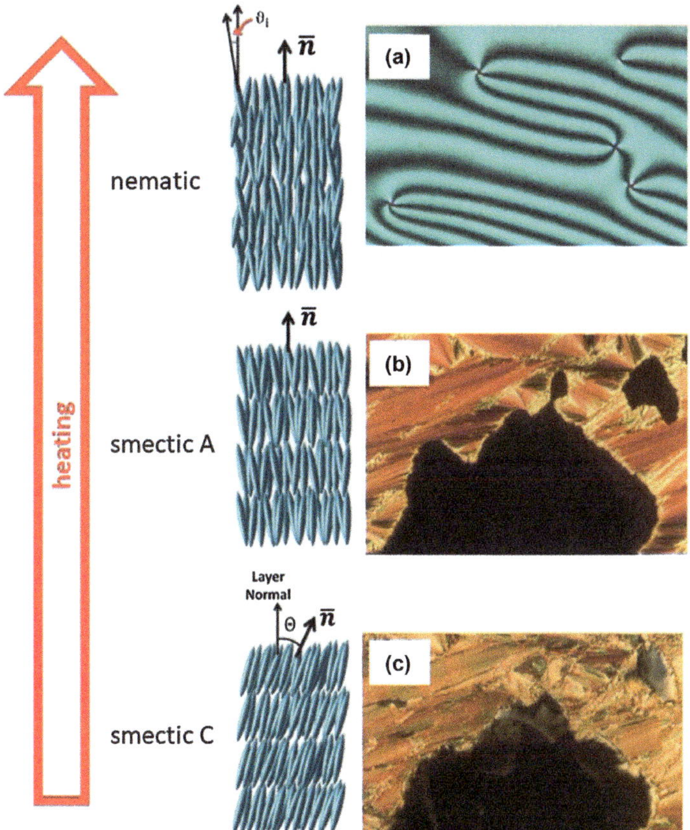

Figure 1.4 Schematic illustration of some of the most common liquid crystal phases, together with typical textures as they are observed by polarizing microscopy; (a) nematic phase with solely orientational order of the long molecular axes and an often-observed Schlieren texture, (b) additional one-dimensional positional order of the fluid SmA phase, which exhibits a fan-shaped texture, and (c) fluid SmC phase with a broken fan-shaped texture (same region as (b)), due to the collective tilt of the director with respect to the smectic layer normal.

average is denoted as the director **n**, a pseudo-vector of length unity, which in most cases has the property of $\mathbf{n} = -\mathbf{n}$ known as head-tail symmetry. The molecular centres of mass of the molecules in the nematic phase are isotropically distributed. The nematic phase is the least ordered of the liquid crystal phases and the one with the highest symmetry. The orientational order of the nematic phase is in its easiest form described by an order parameter S:

$$S = \frac{1}{2}\langle 3\cos^2\vartheta_i - 1\rangle \tag{1.1}$$

where ϑ_i is the angle between the long molecular axis and the director of an individual molecule, and the brackets denote the temporal and spatial average over an ensemble of molecules. One can easily see that $S = 0$ for the isotropic phase, and $S = 1$ for perfect orientational order. In practice the order parameter is approximately $S \approx 0.4$ at the Iso.–N transition and increases to about $S \approx 0.7$–0.8 with decreasing temperature.

Further cooling then often results in the formation of the fluid *smectic* phases, SmA, with the director **n** pointing along the smectic layer normal **k**, and SmC, where the director is tilted by a temperature dependent angle θ with respect to **k**. Both phases are illustrated in Figure 1.4(b) and (c), respectively. In addition to the orientational order of the nematic phase, the fluid phases also exhibit a one-dimensional positional order of the molecular centres of mass. They can thus be seen as a one dimensional crystal or a two dimensional liquid. Nevertheless, it should be stressed that the positional order is not perfect and should be seen rather like a sinusoidal density modulation. It is also worthwhile pointing out that the SmA–SmC transition is often found to be a second order, thus continuous, transition. The order parameter that can most suitably be used to describe this transition is the tilt angle θ, which is $\theta = 0$ in the SmA phase and has a finite value $\theta = \theta(\mathrm{T})$ at lower temperatures. For demonstration, Figure 1.5 depicts the temperature dependence of the tilt angle for a first order N*–SmC* and a second order SmA*–SmC* transition.

On still further cooling, a number of other phases are observed, which we do not want to discuss in much detail, as they are not of direct relevance to this publication. Below the fluid smectic phases, the hexagonal phases are located, which can exhibit a localized hexagonal order of the molecular centres of mass, but without extending across layers. This short range hexagonal order is sometimes also called bond-orientational-order. The molecules can be orthogonal to the smectic layer plane, SmB, or tilted either towards the apex of the hexagon, SmI, or its side, called SmF. Still further phases at even lower temperatures exhibit long range positional order and are generally classified as soft crystal phases. These can also be hexagonal (B, J, G), orthorhombic (E), or monoclinic (K, H) and are actually rather difficult to distinguish from the crystalline state.

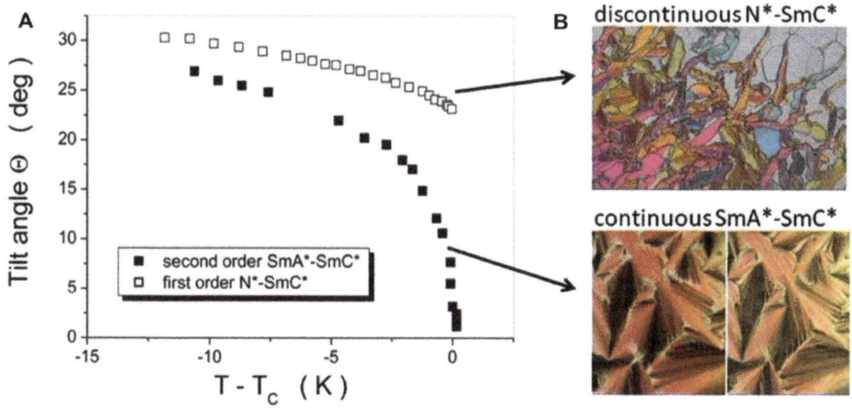

Figure 1.5 (A) Temperature dependence of the tilt angle Θ in the SmC phase after a discontinuous first order transition from N, and a continuous second order transition from SmA. (B) Textures illustrating the discontinuous and continuous behaviour of the two transitions N*–SmC* and SmA*–SmC*, respectively.

For completeness, it should be pointed out that sometimes phases can only be observed on cooling, not on heating. This behaviour is called *monotropic*. In general, liquid crystal phases become increasingly ordered and thus exhibit less and less symmetry. There are a few compounds which defy this general behaviour, having phase sequences where a less ordered phase appears below a more highly ordered one. This phenomenon is called *re-entrant* behaviour, and has for example been reported for phase sequences involving re-entrant nematic phases, but also re-entrant isotropic phases. Lastly, we would like to stress that the nomenclature of the different phases does not provide information about the degree of order of the involved phases. It is purely historical.

1.2.2 Lyotropic Phases

We will here concentrate on systems with amphiphilic molecules. As mentioned above, lyotropic liquid crystals are formed by increasing the concentration of the amphiphile beyond that of the critical micelle concentration (cmc), where micelles of different shapes (spherical, cylindrical) are formed, until these micelles interact and form continuous phases. The liquid crystal phase with the lowest amphiphile concentration is generally the hexagonal phase followed by the lamellar phase and, at even higher concentration, the inverted phases. The cubic lyotropic phase can occur at different positions within the phase diagram. Figure 1.6 illustrates a very simple and schematic phase diagram of a lyotropic liquid crystal with the structure of the phases indicated. Real life phase diagrams can be considerably more complicated, as different variables of state, concentration,

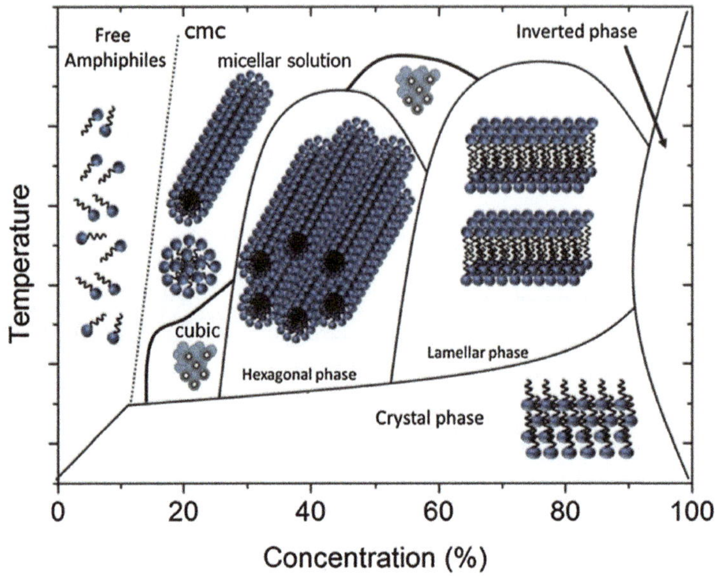

Figure 1.6 Schematic phase diagram of a simple lyotropic liquid crystal, with phase structures indicated. The cubic phase can occur in different regions of the phase diagram, which may also change dependent on which solvent is being used.

temperature and pressure may be involved. In addition, the diagram does depend on the solvent employed.

1.3 Chirality and Chiral Liquid Crystals

Chirality[25–27] is a fascinating aspect of geometry in its widest sense, and can be encountered in all aspects of physics, chemistry, biology, or mathematics, to name just a few of the subjects that it affects. Chirality manifests itself through the lack of mirror symmetry. It is an inherent property of many systems in nature like the helical macromolecules of DNA or polypeptides, shells, snails and plants, growing in a helical fashion and is also seen in the well-known molecular asymmetry of many molecules, known as stereochemistry. Probably one of the best-known examples is that of limonene, where the R-enantiomer exhibits a pronounced smell of orange, while the S-enantiomer smells of turpentine with a lemon note. The two enantiomers have different configurations of the chiral centre of limonene, as depicted in Figure 1.7. They exhibit identical scalar physical properties, such as the melting points, but their chiral properties, like the optical rotation due to optical activity, have equal modulus, but opposite sign.

In liquid crystals, the introduction of chirality either through chiral elements directly within the mesogens (in most cases chiral centres), or through chiral dopants, leads to a multitude of novel structures, phases and

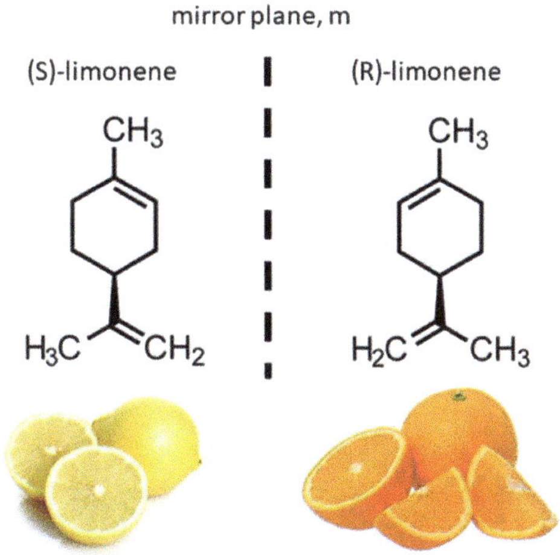

Figure 1.7 Chemical structures of the two chiral enantiomers of limonene. The two compounds have the same scalar properties, but opposite chiral properties, which makes R-limonene smell of orange, while S-limonene smells of turpentine with a lemon note.

effects.[28] In the chiral nematic, N*, and the chiral SmC* phase, one observes for example the spontaneous formation of helical superstructures, which can be left- or right-handed, depending on the handedness, and thus configuration, of the chiral centre or dopant. This is schematically illustrated in Figure 1.8. Nevertheless, it should be mentioned that at this time no prediction of the helical handedness in relation to that of the chiral centre or dopant can be made.

Novel phases can also be formed, which are only observable in chiral liquid crystal materials. These are specifically the Blue Phases, BP, which are located between the isotropic liquid and the chiral nematic or cholesteric phase.[29] They consist of a structure of mutually perpendicular double twist cylinders, which in three dimensional space necessarily has to host defects. These defects are arranged on a cubic lattice. Three different types of such BPs have so far been identified, BPIII, also called the Blue Fog, or the fog phase, which is macroscopically amorphous, but most likely locally of cubic symmetry. BPII has a simple cubic unit cell of lattice defects, while BPI is body centred cubic. The lattice parameters are of the order of a few hundred nanometres. A schematic presentation, together with an indication how BPs appear between crossed polarizers, is given in Figure 1.9(a). Since the Blue Phases are optically isotropic due to their cubic symmetry, the observed colours are due to different orientations of the system of double twist cylinders. The Blue Phases will be discussed in more detail below in chapters 12 and 13 in relation to polymer stabilization, as these materials potentially offer significant advances over displays of the common nematic type.

Another set of frustrated phases which can only be observed for highly chiral materials are the Twist Grain Boundary (TGB) phases.[30] These are

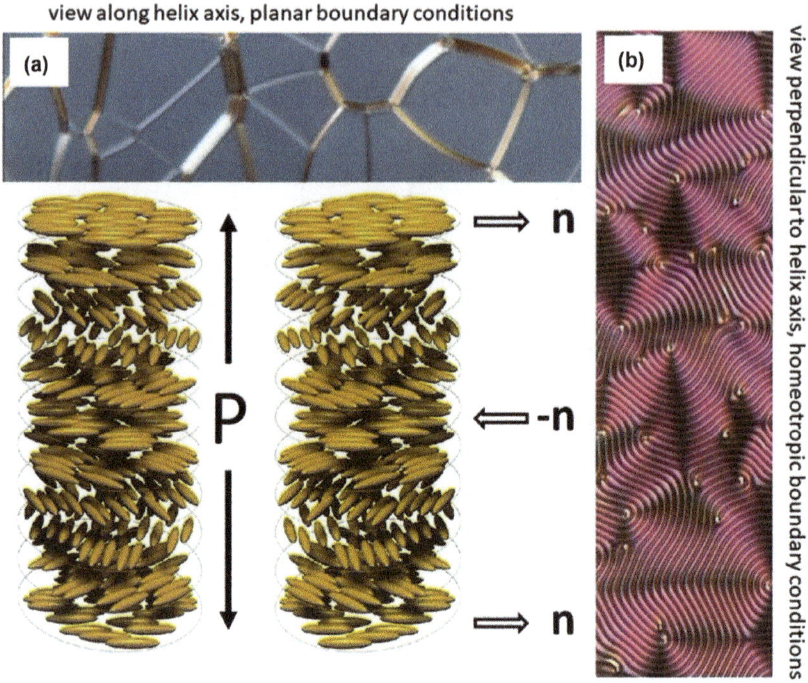

Figure 1.8 The chiral nematic or cholesteric phase can exhibit a right- or a left-handed helical superstructure, depending on the configuration of the chiral centre(s) or that of the chiral dopant. (a) When viewed along the helical axis, which is generally achieved by planar boundary conditions, an oily-streak texture is often observed. (b) For homeotropic boundary conditions, the helix axis orients perpendicular to the direction of light propagation, and a typical fingerprint texture is shown. Note that the periodicity of the cholesteric phase is P/2, due to the head-tail symmetry of the director $\mathbf{n} = -\mathbf{n}$.

located over small temperature intervals between the cholesteric and the respective smectic phases, namely SmA* and SmC*. TGB phases have attracted interest not so much for their applicational potential, but more for their fundamental insights into the description of condensed matter and soft matter systems. The TGB phase is the liquid crystal analogue of the Abrikosov flux lattice phase of a type II superconductor in an external magnetic field. A formal analogy between the SmA phase and superconductors, as introduced by de Gennes[31] was extended to chiral SmA* by Renn and Lubensky,[32] who predicted the TGB phase in 1988. The first experimental observation was reported one year later by Goodby and co-workers.[33] The TGB phase exhibits a helical superstructure, which can be commensurate or incommensurate. Blocks of layered SmA* (SmC*) are discontinuously twisted with respect to each other, mediated by grain boundaries of regular arrays of screw dislocations. Figure 1.9(b) illustrates

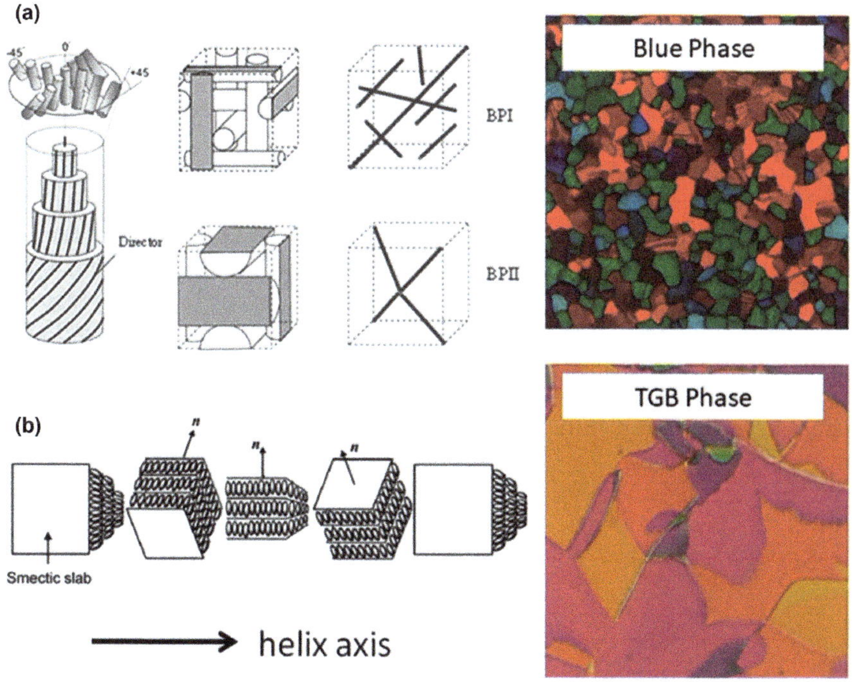

Figure 1.9 Novel, frustrated liquid crystal phases observed for chiral systems. (a) The structure and appearance of Blue Phases, which are observed in a narrow temperature interval between the isotropic liquid and the cholesteric liquid crystal. The double twist cylinders at the left arrange on a cubic lattice, shown in the middle, which gives rise to defects as depicted in the right of the figure. (b) The Twist Grain Boundary (TGB) phases appear between the cholesteric and the fluid smectic phases as a result of a competition between smectic layer formation and helical twist. Smectic blocks are mediated by grain boundaries of regular arrays of screw dislocations.

the schematic structure of the TGB phase, together with its appearance viewed using polarized microscopy.

Probably the two most prominent manifestations of novel effects of chirality in liquid crystalline systems are the observation of selective reflection from phases with a helical superstructure, and the occurrence of ferroelectricity in tilted chiral smectic phases. Selective reflection is due to the helical nature of the respective phase and can easily be observed with the naked eye when the pitch of the helix lies in the visible range of the spectrum. If one considers for simplicity incoming linear polarized light, this can be thought of as a superposition of a left-handed and a right-handed circular polarized wave. De Vries[34] solved Maxwell's equations for propagation of light along the helical axis of a continuously twisted birefringent medium and found that at a wavelength $\lambda = \lambda_0$, with $\lambda_0 = \Delta nP$, where Δn is the average refractive index and P the pitch of the helical structure, circular polarized light that matches the twist of the structure is reflected without phase change, while the other handedness is transmitted. A right-handed cholesteric structure thus reflects right-handed circular polarized light, when following the convention of Born that the electric field vector of right-handed circular polarized light rotates clockwise when viewed against the direction of light propagation. Figure 1.10 summarizes the optical behaviour of helical, cholesteric liquid crystals (Figure 1.10(a)), demonstrates the selective reflection of a short pitch cholesteric (Figure 1.10(b)), and provides an example where selective reflection is used in nature (Figure 1.10(c)).[35] Since the pitch of the helix is temperature dependent, so is the colour of the reflected light. This is employed in temperature imaging devices based on cholesteric liquid crystals. A rigorous treatment of the optics of helical birefringent phases can be found in ref. 36.

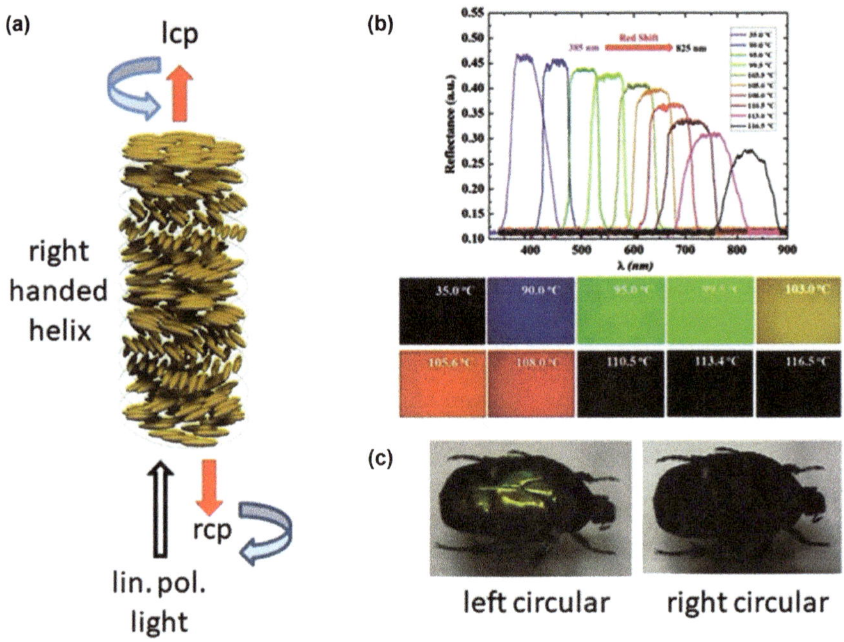

Figure 1.10 (a) A right-handed cholesteric structure reflects right-handed circular polarized (rcp) light without a phase jump, as this describes a left-handed helix in space when propagating. Left-handed circular polarized (lcp) light is transmitted. (b) Selective reflection is demonstrated on well oriented planar samples where the direction of light propagation is parallel to the helical axis. Different colours are observed, because the pitch of the helix is in general temperature dependent. Reproduced from ref. 37 with permission from the Royal Society of Chemistry. (c) Some beetles employ selective reflection, as can be demonstrated by viewing them through left- and right-handed circular polarizers.

A second very prominent effect of chirality may be observed in tilted chiral smectic phases, especially in SmC*, SmI*, SmF* and the anticlinic versions of these phases. It is the occurrence of a spontaneous polarization and thus ferroelectricity and antiferroelectricity.[38,39] This goes back to the symmetry arguments of Meyer *et al.*,[40] who showed that the tilted SmC phase has its symmetry reduced from C_{2h} for the non-chiral case to C_{2y} in the case of chiral molecules, SmC*, which allows for the formation of a local spontaneous polarization, P_S. All tilted chiral smectic phases are thus pyroelectric and can exhibit a macroscopic polarization in the absence of an applied electric field. In the case of the SmC* phase, the spontaneous polarization is generally macroscopically compensated through the formation of a helical structure; it is thus better called helielectric. It has been shown by Clark and Lagerwall[41] that true ferroelectricity may be observed when the helical superstructure is unwound through suitable boundary conditions, then called surface stabilized ferroelectric liquid crystals (SSFLC). For this to happen the pitch *P* is generally larger than the cell gap. The described behaviour is schematically illustrated in Figure 1.11 for the SmC* and the SmC_A* phases.

In terms of applications, ferroelectric liquid crystals (FLC), and especially their SSFLC variant, have many advantages over common nematic systems. The switching process is bistable, and a device only needs to be addressed

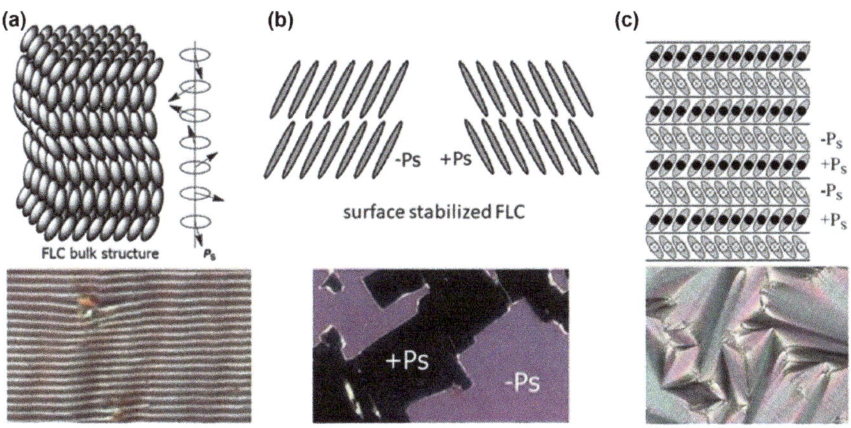

Figure 1.11 (a) In the bulk state the chiral SmC* phase exhibits a helical superstructure of its director, which is accompanied by a macroscopic compensation of the spontaneous polarization, P_S. Note that the periodicity here is the pitch *P*, in contrast to the cholesteric phase where it was P/2. (b) Surface stabilization suppresses the helical superstructure and the formation of ferroelectric domains may be observed, which is illustrated in the texture photomicrograph. (c) In the antiferroelectric SmC_A* phase the tilt of the molecules alternates from smectic layer to layer for a non-helical configuration. The spontaneous polarization in the field-free state is thus compensated locally on a molecular scale. The AFLC phase can be identified by the non-periodic striation of the smectic fans.

when changes are desired. This allows for very low power devices, which are optically binary. The switching dynamics is substantially faster than that of nematics, with response times about three orders of magnitude faster, in the region of tens of microseconds. Also, the switching process is active in both directions as it is achieved *via* a reversal of the electric field direction (switching $\sim$E), in contrast to the switching of nematics ($\sim E^2$), which cannot be influenced externally for the switching-off process, because that is elastically driven. Yet, disadvantages like cost and smectic layer orientation and instability problems have so far prevented large scale production of FLC displays, and these are currently confined to niches of specialized applications and micro-displays. Polymer stabilized FLCs and their properties will be discussed further in chapter 9, while chapter 11 will provide an overview of polymer stabilized antiferroelectric liquid crystals, AFLCs.

1.4 Polymer-modified Liquid Crystals

In order to shortly introduce the combination of polymers with liquid crystals to form composite systems, we have to look at both ends of the phase diagram. At large polymer concentrations, larger than about 70%, one speaks of polymer dispersed liquid crystals, PDLC.[42] In these systems a continuous polymer matrix incorporates droplets of liquid crystal, which so far have mostly been of the nematic, cholesteric or ferroelectric SmC* type. A typical example is shown in the scanning electron micrograph (SEM) of Figure 1.12(a), with the liquid crystal removed from the droplets by a suitable solvent. PDLCs are generally formed either by thermal polymerization, or by UV illumination, causing photo-initiated polymerisation of bifunctional monomers. During the polymerization process, phase separation between polymer and liquid crystal occurs. The operation principle of PDLCs is as follows:[44] in the field-free state the director adopts a certain configuration within the droplet, depending on boundary conditions. These are generally bipolar or radial for planar or homeotropic conditions, respectively. The optic axis of the liquid crystal droplets is randomly oriented and at the boundaries a strongly changing refractive index occurs, which leads to light scattering. The field-free state is thus the opaque, scattering configuration. Application of an electric field reorients the liquid crystal director and thus the optic axis, along the field. The refractive index $n_{\parallel}$ is usually matched with the isotropic refractive index n of the polymer, so that no sudden refractive index changes are observed, and scattering is kept to a minimum. The field-on state is thus the clear, transmitting state. This is schematically summarized in Figure 1.12(b). Commercial applications of such devices have been well known for about two decades for example privacy windows, which are often employed in office spaces or private houses, as exemplary shown in Figure 1.12(c). A detailed review of the state-of-the-art research and technology concerning PDLCs is given in chapter 5.

At the other end of the polymer-liquid crystal dispersion phase diagram, one finds the polymer stabilized liquid crystals, PSLC,[45] sometimes also

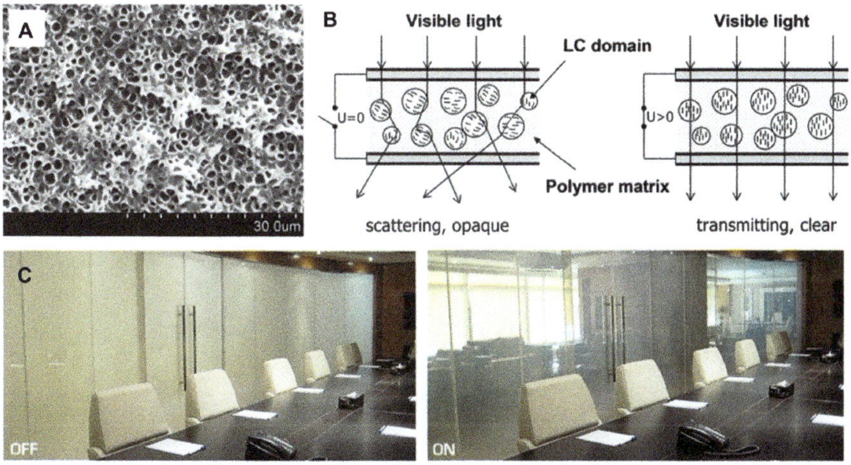

Figure 1.12 Example of a polymer dispersed liquid crystal, PDLC. (a) Scanning electron micrograph[43] of a continuous polymer matrix surrounding liquid crystalline droplets (here removed by a solvent for imaging). Reproduced from ref. 43 with permission from JCPT. (b) Working principle of a PDLC changing from light scattering at zero applied voltage to clear for an applied voltage which leads to refractive index matching. © 2012 Mouquinho A. I., Petrova K., Barros M. T., Sotomayor J. Published in ref. 4 under a CC BY 3.0 license. Available from: http://dx.doi.org/10.5772/48203. (c) A commercial example of a switchable PDLC privacy window from Magic Film (reproduced with permission from Magic Film, www.magic-film.com).

called polymer stabilized cholesteric textures, PSCT, if a cholesteric phase is involved as the liquid crystal.[46] Such systems are usually formed at a concentration of less than 10% polymer and more than 90% liquid crystal. A bifunctional photoreactive, often mesogenic, monomer is dispersed into the liquid crystal, together with a small amount of photoinitiator. The monomer molecules align with the director field of the liquid crystal and are then polymerized by UV illumination. The formed crosslinked polymer-network acts as a template of the phase it was formed in (see Figure 1.13(a)). We can illustrate the working principle of such a device on a long pitch PSCT with planar boundary conditions in the reflective reverse-mode. At zero applied electric field, the cholesteric phase will exhibit a Grandjean texture with the helix axis perpendicular to the substrates. The selective reflection wavelength will be far out of the visible spectrum, because the chosen pitch is typically in the order of $P\sim5$ μm. In this configuration the polymer network is formed, which will also form a helical structure, following the director field of the liquid crystal. The device is transparent, with zero reflectivity in the visible and very little scattering, thus representing the "black-state". Application of an electric field above a threshold will break up the helical superstructure, leading to many small domains of varying refractive index, which will be largely light scattering and thus represent the

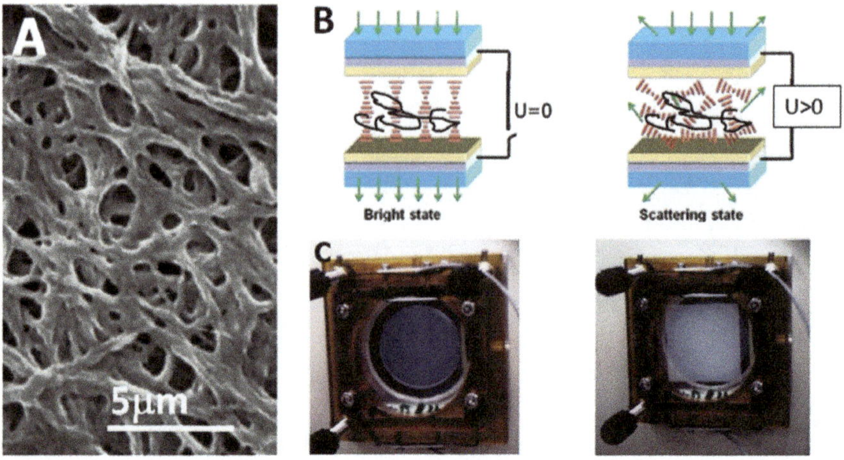

Figure 1.13 Example of a polymer stabilized cholesteric texture (PSCT) in reverse mode. (a) Scanning electron micrograph of the stabilizing cross-linked polymer network structure, formed by photo-polymerization within the liquid crystal phase. The LC is here removed from the bicontinuous system, for reasons of imaging. (b) Schematic illustration of the working principle of a reverse-mode PSCT, switching from a clear state at zero voltage to a scattering state at applied electric field. After field removal the polymer network drives the liquid crystal back to its original, stabilized configuration. Reproduced from ref. 47 with permission from The Optical Society. (c) Example of a PSCT prototype from 1996. Reproduced from ref. 48 with permission from John Wiley and Sons, © 2000 WILEY-VCH Verlag GmbH, Weinheim, Fed. Rep. of Germany.

"white-state". So far, the same behaviour would also be observed simply for a cholesteric liquid crystal. But on turning the field off, a non-stabilized cholesteric would stay in the scattering mode. In the case of the PSCT though, the elastic interaction of the liquid crystal with the helical polymer network will rapidly drive the system back to the non-scattering state, allowing for switching between the field-off and the field-on states. This is schematically illustrated in Figure 1.13(b), with an experimental prototype example from 1996 shown in part (c). The main application of PSLCs at that time was seen in reflective, paper-like displays.

It is worthwhile to note that most liquid crystal display devices rely on a set of two crossed polarizers. This is not the case for the polymer-modified systems introduced here. Both PDLCs and PSLCs are based on scattering effects, not birefringence modulation, and thus do not use polarizers, but rather ambient light. A detailed overview of different aspects of polymer stabilized liquid crystals will be given in the chapters to follow.

References

1. I. W. Hamley, *Introduction to Soft Matter*, (rev. ed.), Wiley, Chichester, 2007.

2. M. Kleman and O. D. Lavrentovich, *Soft Matter Physics*, Springer, New York, 2003.
3. *The Oxford Handbook of Soft Condensed Matter*, ed. E. M. Terentjev and D. A. Weitz, Oxford University Press, Oxford, 2015.
4. P. J. Collings and M. Hird, *Introduction to Liquid Crystals: Chemistry and Physics*, Taylor & Francis, London, 1997.
5. P. G. de Gennes and J. Prost, *The Physics of Liquid Crystals*, Clarendon Press, Oxford, 1993.
6. S. Chandrasekhar, *Liquid Crystals*, Cambridge University Press, Cambridge, 2nd edn, 1992.
7. *Liquid Crystals: Experimental Study of Physical Properties and Phase Transitions*, ed. S. Kumar, Cambridge University Press, Cambridge, 2011.
8. P. Oswald and P. Pieranski, *Nematic and Cholesteric Liquid Crystals*, Taylor & Francis, Boca Raton, 2005.
9. P. Oswald and P. Pieranski, *Smectic and Columnar Liquid Crystals*, Taylor & Francis, Boca Raton, 2006.
10. I. Dierking, *Textures of Liquid Crystals*, Wiley-VCH, Weinheim, 2003.
11. A. G. Petrov, *The Lyotropic State of Matter: Molecular Physics and Living Matter Physics*, Gordon and Breach Science Publishers, Amsterdam, 1999.
12. A. M. Figueiredo Neto and S. R. A. Salinas, *The Physics of Lyotropic Liquid Crystals*, Oxford University Press, Oxford, 2005.
13. *Liquid Crystalline and Mesomorphic Polymers*, ed. V. P. Shibaev and L. Lam, Springer, New York, 1994.
14. M. Warner and E. M. Terentjev, *Liquid Crystal Elastomers*, Oxford University Press, Oxford, 2003.
15. *Liquid Crystal Elastomers: Materials and Applications*, ed. W. H. de Jeu, Springer, Berlin, 2012.
16. *Cross-Linked Liquid Crystalline Systems*, ed. D. J. Broer, G. P. Crawford and S. Zumer, CRC Press, Boca Raton, 2011.
17. P. S. Drzaic, *Liquid Crystal Dispersions*, World Scientific, Singapore, 1994.
18. G. P. Crawford and S. Zumer, *Liquid Crystals in Complex Geometries*, Taylor & Francis, London, 1996.
19. S. Kumar, *Chemistry of Discotic Liquid Crystals: From Monomers to Polymers*, CRC Press, Boca Raton, 2011.
20. L. Lin, *Mol. Cryst. Liq. Cryst.*, 1987, **146**, 41.
21. *Liquid Crystals: Materials Design and Self-Assembly*, ed. C. Tschierske, Springer, Berlin, 2012.
22. H. Takezoe and A. Eremin, *Bent-shaped Liquid Crystals: Structures and Physical Properties*, CRC Press, Boca Raton, 2017.
23. *Micelles, Membranes, Microemulsions, and Monolayers*, ed. W. M. Gelbart, A. Ben-Shaul and D. Roux, Springer, New York, 1994.
24. I. Dierking and S. Al-Zangana, *Nanomaterials*, 2017, 7, 305.
25. *Chirality: From Weak Bosons to the Alpha-helix*, ed. R. Janoschek, Springer, Berlin, 1991.
26. W. J. Lough and I. W. Wainer, *Chirality in Natural and Applied Science*, Blackwell Science, Oxford, 2002.

27. G. H. Wagniere, *On Chirality and the Universal Asymmetry*, Verlag Helvetica Chimica Acta, Wiley-VCH, Zürich, 2007.
28. *Chirality in Liquid Crystals*, ed. H.-S. Kitzerow and C. Bahr, Springer, New York, 2001.
29. P. P. Crooker, in *Chirality in Liquid Crystals*, ed. H.-S. Kitzerow and C. Bahr, Springer, New York, 2001, ch. 7.
30. R. Dhar, *Phase Transitions*, 2006, **79**, 175.
31. P. G. de Gennes, *Solid State Commun.*, 1972, **10**, 753.
32. S. R. Renn and T. C. Lubensky, *Phys. Rev. A*, 1988, **38**, 132.
33. J. W. Goodby, M. A. Waugh, S. M. Stein, E. Chin, R. Pindak and J. S. Patel, *Nature*, 1989, **337**, 449.
34. H. de Vries, *Acta Crystallogr.*, 1951, **4**, 219.
35. I. Dierking, *Symmetry*, 2014, **6**, 444.
36. V. A. Belyakov, *Diffraction Optics of Complex-Structured Periodic Media*, Springer, Berlin, 1992.
37. Y. Wang, Z.-G. Zheng, H. Krishna Bisoyi, K. G. Gutierrez-Cuevas, L. Wang, R. S. Zola and Q. Li, *Mater. Horiz.*, 2016, **3**, 442.
38. S. T. Lagerwall, *Ferroelectric and Antiferroelectric Liquid Crystals*, Wiley-VCH, Weinheim, 1999.
39. I. Musevic, R. Blinc and B. Zeks, *The Physics of Ferroelectric and Antiferroelectric Liquid Crystals*, World Scientific, Singapore, 2000.
40. R. B. Meyer, L. Liebert, L. Strzelecki and P. Keller, *J. Phys. Lett.*, 1975, **36**, L69.
41. N. A. Clark and S. T. Lagerwall, *Appl. Phys. Lett.*, 1980, **36**, 899.
42. J. W. Doane, N. A. Vaz, B. G. Wu and S. Zumer, *Appl. Phys. Lett.*, 1986, **48**, 269.
43. M. Rao Darla, S. Hedge and S. Varghese, *J. Cryst. Process Technol.*, 2014, **4**, 60.
44. A. I. Mouquinho, K. Petrova, M. T. Barros and J. Sotomayor, in *New Polymer Networks for PDLC Film Applications*, ed. A. De Souza Gomes, New Polymers for Special Applications, InTech, London, 2012, ch 5.
45. R. A. M. Hikmet, *J. Appl. Phys.*, 1990, **68**, 4406.
46. D.-K. Yang, L.-C. Chien and J. W. Doane, *Appl. Phys. Lett.*, 1992, **60**, 3102.
47. Y.-C. Hsiao, C.-Y. Tang and W. Lee, *Opt. Express*, 2011, **19**, 9744.
48. I. Dierking, *Adv. Mater.*, 2000, **12**, 167.

CHAPTER 2

Phase Diagrams, Phase Separation Mechanisms and Morphologies in Liquid Crystalline Materials: Principles and Theoretical Foundations

EZEQUIEL R. SOULE[a] AND ALEJANDRO D. REY*[b]

[a] Institute of Materials Science and Technology (INTEMA), University of Mar del Plata and National Research Council (CONICET), J. B. Justo 4302, 7600 Mar del Plata, Argentina; [b] Department of Chemical Engineering, McGill University, Montreal, Quebec H3A 0C5, Canada
*Email: alejandro.rey@mcgill.ca

2.1 Introduction

Polymer–liquid crystal composites are heterogeneous materials that display a rich phase behavior and a variety of morphologies, and are typically used in electro-optical technological applications. This chapter presents a comprehensive general introduction to phase behavior and morphologies of these materials. The fundamental physics underlying the formation of different morphologies is described, with a strong emphasis on how the trajectory of the phase diagram determines the phase separation mechanism and emerging morphology. The effect of processing variables, chemical reaction and molecular structures on phase diagrams is rationalized through

Soft Matter Series No. 8
Polymer-modified Liquid Crystals
Edited by Ingo Dierking

Published by the Royal Society of Chemistry, www.rsc.org

its effect on phase diagrams and diffusion. The theories, models and computations presented in this introductory chapter are a critical contribution to achieving desired functionalities by identifying the optimal processing conditions for these soft matter materials.

Although a polymer–liquid crystal composite material (PLCM) can be homogeneous, the mixture usually phase-separates, leading to a heterogeneous material with complex morphologies and structures that determine its properties. The formation of multi-phasic materials can be driven by three phase separation methods: polymerization-induced- (PIPS), thermally-induced- (TIPS), and solvent-induced- (SIPS) phase separation. PIPS is the standard approach to produce most PLCMs as well as many other polymeric materials.[1,2] The analysis of a PIPS process includes the extra complexity that the chemical reaction has to be considered, simultaneously with phase transitions, but has the advantage that the final morphologies of the material can be controlled with relatively high precision, by manipulating the chemical reaction. In addition, it involves an initially low viscosity, easy-to-handle, monomer–LC mixture, and no solvent is required for processing. In the case of cross-linked polymers, the matrix behaves like a solid (it does not flow), so PIPS becomes the only reasonable processing method.

At the core, the structural richness in these mixtures emerges from the fact that PLCMs can undergo a regular liquid–liquid, or isotropic–isotropic (I–I), phase separation, as well as order–disorder or order–order phase transitions[3–5] The phase diagram may include multiple phases, and phase transition dynamics involves the evolution of conserved and non-conserved order parameters. Each order parameter can undergo a phase transition following different dynamic laws and with different kinetic coefficients, and they may evolve through different mechanisms; but on the other hand they are coupled and they can't be analyzed independently. This complex phase behavior leads to complex morphologies, and understanding it is crucial to controlling the final structure of the material.

The main objective of this chapter is to present a general introduction to phase behavior, phase transitions, and structure formation in PLCMs. Generic and simple "rules" and strategies for controlling the final morphologies of the material are defined, as universally as possible. The phase diagram and the trajectory of the system in the temperature–composition plane is emphasized as the main conceptual tool to achieve this crucial goal. The three processing methods TIPS, SIPS and PIPS are considered, highlighting their conceptual similarities, but special attention is paid to specific aspects of PIPS due to its technological importance. Mathematical details are given in the relevant references. To simplify the presentation the mesogenic component is denoted as the LC component.

2.2 Phase Transition Mechanisms

There are two well-known mechanisms of phase transitions: nucleation and growth, and unstable growth (when the system demixes, unstable growth is called spinodal decomposition). Generally speaking, when the curvature (second derivative) of the free energy function is negative, the system is unstable against infinitesimal fluctuations, so microscopic thermal fluctuations in composition or order parameter grow in time (unstable growth); when the curvature is positive, small fluctuations shrink while fluctuations larger than a threshold (nucleus) grow, which is the mechanism of nucleation and growth. In a system like a PLCM, where the free energy is a function of several order parameters (composition and one or more mesophase order parameter), the free energy can present saddle points so that the second derivative with respect to different variables has different signs.[4,6–9] This means that the system can be stable against fluctuations in one variable and unstable to fluctuations in another variable, or unstable in all the variables, or absolutely stable, which can lead to complex combinations of different mechanisms of phase separation.

Nucleation and growth typically gives rise to a morphology of disperse domains, usually with a regular geometric shape, in a continuous matrix, while spinodal decomposition may give rise to more complex morphologies like co-continuous structures, inter-connected or worm-like domains.[6,10] In addition, the equilibrium shape of dispersed domains depends on the type of phase considered; for isotropic domains it is a sphere, for nematic domains it can be an elongated sphere or a tactoid (although not very far from a sphere),[11,12] while for more ordered phases like smectics the domains can be highly elongated (bâtonnets).[13–15] In terms of the Gaussian curvature K_G, the equilibrium dispersed spheroidal domains have $K_G > 0$, while spinodal shapes are associated with $K_G < 0$ (saddle-like local curvature), a fact that can be exploited experimentally to discern mechanisms.

The mechanism of phase separation and consequently the final morphology depends strongly on the trajectory of the mixture in the temperature–composition plane, relative to the phase diagram. Despite the crucial importance of phase diagrams, it is not uncommon that PLCMs are synthesized and their morphologies are optimized by trial-and-error, without analyzing the impact of chemical or processing variables on phase behavior. In the following sections, the emergence of different types of morphologies in PLCMs will be analyzed in terms of phase diagrams and phase separation mechanisms, complemented with some basic knowledge about diffusion and polymerization reactions.

2.3 Integration of Phase Diagrams, Phase Separation Mechanisms and Morphology

Figure 2.1a shows a schematic of a generic phase diagram of a polymer–nematic (N) liquid crystal mixture with I–I and I–N phase coexistence regions. In addition to the binodal lines (which represent the phase coexistence in equilibrium), the I–I spinodal and the I-to-N stability limit are shown; the first one represents the limit of stability (*i.e.* second derivative

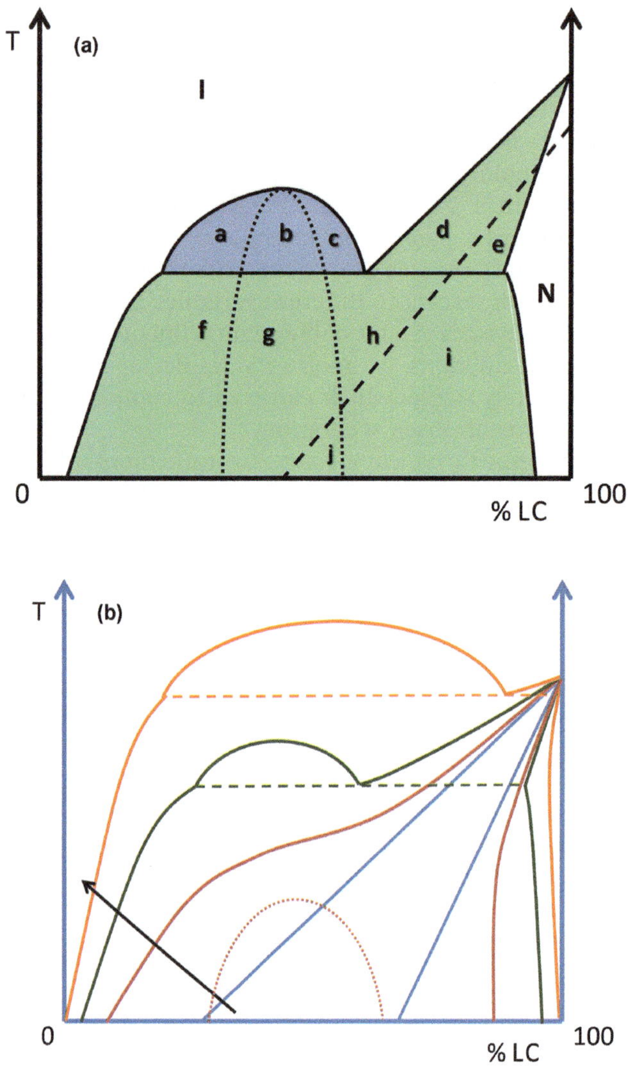

Figure 2.1 (a) Schematics of a generic phase diagram of a polymer–LC mixture. Areas of phase coexistence are shaded blue (I–I phase coexistence) and green (I–N phase coexistence), which are enclosed by the corresponding binodal lines. Non-shaded areas correspond to existence of a single phase as indicated. Letters a–j indicate regions where different phase separation mechanisms are expected, as indicated in the text. The dashed line is the I–I spinodal and the dotted line the I-to-N stability limit. (b) Effect of increasing immiscibility, as indicated by the direction of the arrow. In a reactive system, immiscibility increases with conversion. Matching colors correspond to the binodals at the same level of immiscibility. The blue curve corresponds to a case without I–I phase coexistence. The red curve shows the case with a metastable I–I equilibrium (dotted line), buried below the I–N equilibrium (the shape of the I–N binodal is affected). Green and orange curves correspond to the case with I–I phase coexistence.

changes sign) of an isotropic phase against composition fluctuations and the second one against nematic order parameter fluctuations. Different regions of the phase diagram (a–j), which correspond to different phase separation mechanisms, are identified. The I-to-N stability limit in PLCMs has not been considered in the bibliography until recently[9] (although this was very well known for pure LCs), while usually another spinodal, called the nematic spinodal, is specified.[6,7,16,17] The nematic spinodal (which is not shown in Figure 2.1) represents the limit of stability of a *nematic* phase against fluctuations in composition and order parameter. As TIPS or PIPS involve phase separation from an isotropic phase, the I-to-N stability limit, and not the nematic spinodal, is the relevant curve. Figure 2.1b shows how the phase diagram is modified by decreasing miscibility, indicated by the direction of the arrow. The origin of the arc on the top curve is the growth of the I–I dome, clearly seen in the dotted inverted parabola. In Figure 2.2, generic morphologies corresponding to different phase separation processes are illustrated, as will be discussed below.

As previously mentioned, phase separation can be induced by varying temperature (TIPS), conversion (PIPS) or evaporating solvent (SIPS). The general principle is the same for the three processes: at the beginning the mixture is in an isotropic state, then by modifying the corresponding variable (temperature, conversion or solvent content), the mixture moves in the temperature–composition plane such that it goes into a biphasic region of the phase diagram. In the case of a TIPS process, the system is (usually) a

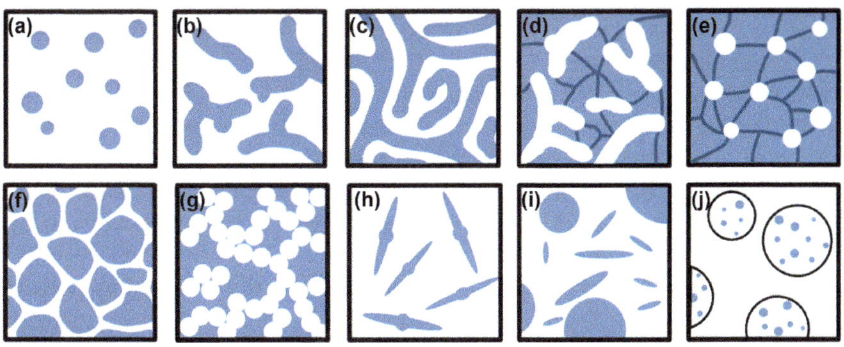

Figure 2.2 Schematics of different generic morphologies in a phase-separated polymer–LC mixture. LC-rich phase is represented with blue color. (a) Dispersed LC droplets (Swiss cheese morphology). (b) Worm-like/branched LC domains. (c) Co-continuous structure. (d) Worm-like/branched polymer domains (reverse morphology). (e) Dispersed polymer domains (reverse morphology). (f) Extreme case of LC dispersed domains, for large, deformed, non-coalescing domains (g) Network of polymer domains (reverse morphology-polymer ball morphology). (h) LC bâtonnets. (i) Mixed morphology of spherical and elongated domains. (j) Salami structure. Note that in general the mesophase will have an internal structure (texture, defects, domain walls). Only domain walls in (d) and (e) (where the mesophase is the matrix) are shown.

binary mixture, with a fixed phase diagram (like Figure 2.1a), and the coordinates of the mixture move down as the temperature is decreased. In the case of SIPS and PIPS, the mixture is a ternary or multi-component system; the process is usually isothermal, and the composition is varied until a biphasic region is reached. The case of SIPS can be represented in a solvent concentration *vs.* composition (in solvent-free basis) diagram (analogous to Figure 2.1a, but temperature is replaced by solvent concentration), where the phase diagram is fixed and the coordinates of the mixture move in the plane as solvent is evaporated, or in a temperature-solvent free composition diagram, where now the coordinates of the mixture are fixed and the phase diagram evolves as solvent is evaporated (Figure 2.1b). In the case of PIPS, the situation depends on the mechanism of polymerization. If the mechanism of polymerization is a chain-reaction (like free-radical polymerization), the polymer usually has a high molecular weight from the beginning of the reaction and it might remain more or less constant, so the system can be viewed as a ternary mixture of mesogen–monomer–polymer, and the reaction mixture moves in the ternary phase diagram as the reaction proceeds.[18] In the case of step-wise polymerization, the polymerization index increases continuously from 1 (monomer) to infinite (complete conversion) and the polymer is highly polydisperse, so the system has to be viewed as a polymer–LC mixture where the molecular weight distribution varies with conversion. Detailed analysis of phase diagrams with this consideration can be found in the literature for polymer mixtures,[18–23] and specifically for PLCM.[24,25] In any case, the system can be analyzed again as a pseudo-binary system in a conversion *vs.* mesogen content diagram (analogous to Figure 2.1a, except that higher conversion is equivalent to low temperature), or a temperature–mesogen concentration diagram where the phase diagram shifts to higher temperatures with conversion,[24–27] as in Figure 2.1b. It must be taken into account that, when the system is considered a pseudo-binary, the phase diagram not only "shifts" to higher temperature as the reaction evolves or the solvent is evaporated, but the shape can also change. Specifically, an I–I phase equilibrium is more strongly affected by conversion than an I–N equilibrium (this is specifically shown in Figure 2.1b),[3] which is explained by the lower entropy of mixing of systems containing polymers compared to those of low molecular weight species. It has to be noted that for ternary or polydisperse systems the full coexistence cannot be represented in a simple diagram like those shown here, but there is a distinction between cloud point (onset of phase separation), and coexistence curves, and there might also be multi-phase coexistence zones (a full discussion goes beyond the scope of this review and can be found in ref. 18–25). Despite this, the simple binary diagram is enough for qualitative analysis. So, in what follows, the analysis will be based on these binary diagrams, and "quenches" (that can be achieved by varying temperature, conversion or solvent content) to different regions are considered.

It has to be noted that, as polymer concentration is increased or temperature is decreased, molecular mobility decreases and the concentration and relative amounts of each phase can significantly depart from thermodynamic equilibrium.[1,28] If the system vitrifies, the phase separation process (and the chemical reaction in case of PIPS), might become, for practical purposes, completely arrested and the morphologies are frozen. If the polymer cross-links (physically or chemically), when a percolated gel (*i.e.* an infinite network) is formed, there is also a strong decrease in mobility. Usually, the effects of polymer composition on mobility are stronger than the effects of temperature, so they are more important for PIPS and SIPS.

If the mixture in quenched to regions *a, b* or *c* of the phase diagram shown in Figure 2.1a, an I–I phase-separated material is produced, and a further quench to the I–N region is needed to produce an N phase. This can be the case when the PLCM is prepared by PIPS using thermal polymerization at temperatures higher than the I–N transition temperature; then the nematic phase is generated when the material is cooled to room temperature[10,29,30–32]). In *a* and *c*, phase separation proceeds through nucleation and growth leading to a morphology of dispersed droplets (Figure 2.2a and 2.2e) and in region *b* through spinodal decomposition. Note that in *b* the mechanism of nucleation and growth can still take place and compete with spinodal decomposition, but as nucleation requires a finite fluctuation, while spinodal decomposition is initiated with infinitesimal fluctuation, usually the latter prevails. As discussed before, spinodal decomposition might produce different types of morphologies and this depends on the exact position within this region. In general, if the composition is close to the I–I critical composition, a co-continuous structure is formed (illustrated in Figure 2.2c);[10] as the composition departs from the critical concentration, the morphology moves from co-continuous to disperse domains, showing intermediate morphologies of interconnected or worm-like domains (Figure 2.2d and b).[6,7] The dispersed-domains morphology produced by spinodal decomposition is similar and often indistinguishable to that produced by nucleation and growth (Figure 2.2a and e).[7,33,34] It has to be noted that when the LC concentration is lower than the critical point (region *a* and left half of region *b*), a polymer-dispersed liquid crystal (PDLC), that is, the nematic phase dispersed in an isotropic matrix (Figure 2.2a), is produced. This is often referred to as "Swiss cheese morphology" and is probably the most widely observed.[26,27,29,35,36,37,38–41] But if the concentration is higher than the critical, the continuous phase is rich in liquid crystal, so further quenching induces the transformation of part of the matrix to a nematic phase. This might lead to a reverse morphology, that is, an isotropic phase dispersed in a nematic matrix. This morphology is shown in Figure 2.2d and e. In general, the mesophase phase will have some texture, defects, *etc.*; specifically, it shows domain walls between regions where the director has different orientations. Usually the isotropic domains are located along these domain walls. In ref. 15 and 34 morphology Figure 2.2e was studied in a TIPS process for nematic and smectic systems, while in ref. 7 Figure 2.2d and e were predicted by simulations.

If the mixture is quenched to *d, f* or *h*, in the N–I coexistence regions but above the spinodals, phase separation proceeds through nucleation and growth of a nematic phase in an isotropic matrix, giving rise to a morphology like Figure 2.2a. The difference between the different regions is the equilibrium concentration of each phase and, through the lever rule, the amount of each phase: the highest content of nematic phase is found in region *h* and deep in region *f*, close to region of homogeneous nematic phase (although it has to be noted that, as mentioned before, equilibrium conditions might not be reached because of kinetic limitations).

In region *g*, which is within the I–N coexistence region and below the I–I spinodal, phase separation may be initiated by spinodal decomposition, or by nucleation and growth of the nematic phase, depending on the relative kinetics of each process, and the final morphology will depend on the competition of both processes. In regions *e* and *i*, below the I-to-N stability limit, phase separation can proceed by nucleation and growth of the equilibrium nematic phase (this means with equilibrium concentration of LC), leading to Figure 2.2a, or by unstable growth of the order parameter in a homogeneous phase followed by I–N phase separation through nucleation and growth. Note that in this latter case, the isotropic phase will be nucleated from a nematic continuous phase leading to an inverted morphology.[6,7,42] The distinction between regions *e* and *i* is again the relative amounts and compositions of each phase. Finally, in region *j* (below both spinodals), the transition may evolve through nematic nucleation and growth, I–I spinodal decomposition or unstable growth of the nematic order parameter.

Some differences between TIPS and PIPS have to be pointed out. When a PDLC is formed (*i.e.* the nematic phase is dispersed), for high concentrations of LC, as the reaction proceeds, more and more LC is phase-separated from the matrix and at some point, the LC-rich phase can become the majority phase. The nematic domains can grow up to the point that they almost touch each other, but if the polymer matrix behaves like an elastic material, coalescence of the droplets will be prevented, and the droplets will be deformed. This structure will look like a lot of irregular geometric shaped LC domains, separated by thin walls of the polymer matrix which form a network-like structure. This is illustrated in Figure 2.2f and is rather commonly observed in PIPS-generated cross-linked systems,[35–37,26,27,39–41] where the matrix has a permanent elasticity. In non-cross-linked systems, this structure can be observed at an intermediate stage of phase separation due to temporary elastic deformations of the matrix, but as these elastic forces are relaxed, the interconnected structure of the polymer-rich phase breaks down into droplets and a reverse morphology like Figure 2.2d or 2.2e is formed (except if the morphology is kinetically arrested). This last mechanism is known as visco-elastic phase separation.[43]

Another possible situation in PIPS is when the polymer is highly immiscible with the monomer–LC mixture and phases separate at low conversion. The situation is initially similar to a quench to regions *a*, *b* or *c*, and phase separation evolves like described previously in the early stages, except that the disperse phase is polymer-rich. The binary phase diagram might be deceiving in this case, as the monomer/polymer partition cannot be seen. Figure 2.3 shows this situation in a ternary phase diagram (which represents the case of a chain-reaction polymerization), where the evolution of the reacting system is represented by an arrow (only the I–I region of the phase diagram in shown). Right after phase separation, as can be seen by applying the lever rule in the tie lines, the polymer-rich phase represents a small fraction of the total volume of the system, but as the reaction proceeds more and more polymer is produced and at some point, it might become the majority phase. Again, if the polymer cross-links (physically or chemically) or approaches vitrification, full coalescence of the droplets is prevented so a network of inter-connected polymer spheres is formed, that looks like Figure 2.2g. The LC fills the irregularly-shaped cavities, that can be highly interconnected (note that what seems to be dispersed domains in Figure 2.2g can be interconnected in 3 dimensions). This is a rather commonly observed structure, usually referred to as "polymer ball morphology".[44–48]

Other types of morphologies can be formed in more complex mixtures. For example, Hoppe *et al.* studied a mixture of 4-butyl-*N*-(4-ethoxybenzylidene)-aniline (EBBA), polystyrene (PS) and a polymerizable epoxy monomer.[10,30–32] They found that, due to a relatively low solubility of PS and EBBA in the epoxy-polymer, the system phase-separated at very low conversions forming domains with high concentrations of PS and EBBA. When the system was cooled, nematic EBBA phase-separated from the PS, but this secondary phase separation took place *within* the initially formed domains. This generated a "salami" structure, illustrated in Figure 2.2k, which is also very well known in other ternary polymer systems like high impact polystyrene.[50]

If the LC is smectic instead of nematic, the phase diagram is similar except that the nematic phase is replaced by the smectic one.[15,34,49] Phase

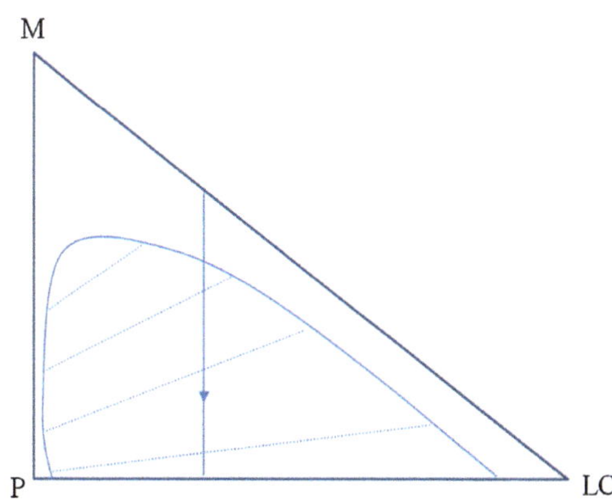

Figure 2.3 Evolution of a chain-reaction PIPS system, with very low LC-polymer miscibility, in a ternary phase diagram. The dotted lines represent tie lines between coexisting phases in equilibrium. The reacting system moves in the direction of the arrow as conversion increases. I–N phase equilibrium is not shown.

separation mechanisms are the same, but due to the higher anisotropy of the smectic phase, the equilibrium shape of the dispersed smectic domains might depart significantly from spherical. As the LC content is higher, the domains might become more ellipsoidal and even highly elongated bâtonnets (Figure 2.2h). Some curious cases might appear, for example a quench to region *a* followed by a quench to region *f* will produce spherical domains in the first stage, but anisotropic domains might be produced through a secondary-nucleation in the second stage, producing a mixed morphology (Figure 2.2i).[15] If the LC can form both nematic and smectic phases, then both regions are present in the phase diagram, and processes of nucleation and/or unstable growth of both nematic and smectic order parameters might compete.

It is worth mentioning that the process of phase separation might be actually a little more complex if metastable phases are considered. For example, in region *h*, right below the I–I coexistence region, I–I phase equilibria exist as a metastable state (also the full I–I coexistence region might be buried bellow the I–N curve as in the red curve of Figure 2.1b). So, if diffusion is faster than ordering, I–I phase separation might proceed first, followed by the I–N transition. In this case the situation would be similar to that of a quench to region *c* followed by a quench to region *h*. This situation has been analyzed by means of computer simulations and an extensive discussion can be found in ref. 5, 51–53. Also, in region *d* for low enough temperature (specifically, bellow the I–N transition line,[9] which is not shown in the diagram), a homogeneous nematic phase is metastable, so the system could evolve through nucleation and growth of a nematic phase, but with the same composition as the isotropic phase, and later through nucleation and growth of an isotropic phase within the nematic domains. Motoyama *et al.*[11] predicted this situation with computer simulations and they found that the isotropic domains were nucleated at the boundaries of the nematic domains and the morphology was similar to Figure 2.2e but with a very non-uniform spatial distribution of polymer domains. In smectic liquid crystals, within the conditions of existence of a smectic phase, the nematic phase is metastable so the formation of a smectic phase can be preceded by the formation of a metastable nematic phase which might lead to more symmetric domains.[54]

2.4 Competition Between Chemical Kinetics and Phase Separation

So far, we have discussed mainly the type of morphology produced in terms of shape and topology, but the size and number density of the domains has also a fundamental effect on the properties of the materials. This depends on thermodynamic factors (which determine the relative amount of each phase in equilibrium), but also very strongly on kinetic factors (the rate at which the domains can be formed and grow). In a polymerizing phase-separating system, these two effects are in competition as the polymer is generated; there is a decrease in both miscibility (driving force for phase separation), and mobility (so the process slows down). When the chemical reaction is slow compared to the phase transition, the dispersed domains have time to grow and coalesce, and the composition and relative amount of phases can approach equilibrium, while in the case of fast chemical reaction, the growth of phase domains can be severely restricted. Consequently, slow chemical reaction favors a smaller number of domains of larger size, a larger amount of dispersed phase, and compositions close to equilibrium. This effect of polymerization rate can be directly observed in a photo-polymerizing system by using different irradiation intensities; for example, in ref. 37, 55 it was observed that in PDLCs generated by photo-PIPS, increasing UV intensities (which lead to a faster polymerization rate) produced an increase in number and decrease in size of dispersed domains.

The effect of temperature and composition on morphology can be very complex, as the polymerization rate, miscibility, and mobility are all affected by those variables. The effect of LC content on morphologies is mostly dominated by solubility and is relatively straightforward, as thermodynamics prevails over kinetic factors; higher LC content leads to a larger amount of LC phase[37,55,56] (meaning larger size and usually larger number of domains). Of course, this trend only holds within a given region of the phase diagram, as crossing from one region to another can change the type of morphology. On the other hand, the effect of temperature is in general more difficult to predict. Li *et al.*[55] and Lu *et al.*[37] observed, for a nematic and a smectic PDLC respectively, that increasing temperature produced a non-monotonous effect on droplet size; below the Sm–I or the N–I transition temperature, droplet size increased with temperature, while above the transition temperature the opposite trend was observed. This is ascribed to the competing effects of increasing mobility and increasing polymerization rate with temperature, the first effect prevailing below the mesophase transition and the second one above. This is not, however, a universal trend. For example, West[57] analyzed two different PDLCs; in one of them he observed a similar trend, although the maximum domain size was not observed at the transition temperature but slightly above. In the other material he observed a monotonous decrease of domain size with temperature. Nwabunma *et al.*[58] also observed that, when the PDLC was cured above the N–I temperature, the dispersed LC domains have a more or less uniform size, while curing below the N–I temperature produces a collection of very large, and very small domains (bimodal distribution). This type of morphology is also observed in TIPS[31] when the mixture is continuously cooled into the coexistence region (remember that decreasing temperature is similar to increasing conversion), and it is explained as follows. At short times and shallow quenches, the composition of the isotropic phase is not very far from equilibrium, so the system evolves primarily by growth of the nucleated domains. But as the quench depth increases, the matrix becomes more and more supersaturated and diffusion to the existing domains is not fast enough to remove the excess of LC, so at some point a secondary phase separation is induced such that a new generation of domains is nucleated within the supersaturated matrix.

It has been observed that the droplet size of the cured material depends strongly on the distance between the temperature–concentration coordinates of the reactive system and the I–N binodal line of the un-reacted mixture, and not on the temperature or concentration independently.[26,27] This is because when the difference between the concentration of the mixture and the saturation concentration (*i.e.* the concentration of the I–N binodal) at the reaction temperature is the same for different mixtures, phase separation takes place at (approximately) the same conversion. As diffusivity is strongly dependent on polymer concentration and consequently on conversion, the droplet size also strongly depends on the phase separation conversion. This is valid within a relatively small range of compositions and temperatures, such that the effect of these variables on reaction rate and mobility is not relevant, and the region of the phase diagram under consideration is the same.

2.5 Effects of Chemical Structure on the Phase Diagram and Morphology

The phase diagram of a given mixture is determined by configurations and interactions between the components at a molecular level. I–I phase coexistence is in general more strongly affected than I–N equilibrium. This is because an I–I equilibrium is directly affected by binary interactions and polymer chain conformations, while an I–N equilibrium depends mostly on the interaction of the LC with itself. The I–N transition temperature of the pure LC depends on its chemical structure, but in the absence of specific directional interactions between the polymer and the LC and with high miscibility in the isotropic phase, the polymer only produces a "dilution effect", which decreases the I–N transition temperature proportionally to the polymer concentration. As the miscibility in the isotropic phase decreases, even when I–I phase equilibrium is not observed, the I–N phase equilibrium can be affected, shifting to higher temperatures and changing the shape of the I–N binodal. This can be seen as a metastable I–I phase coexistence, buried below the I–N coexistence that "pushes" it upwards, as illustrated in Figure 2.1b. With other mesophases the situation is the same.

There are some general rules regarding how chemical structures affect the phase behavior, besides "like dissolves like". First, for LCs in the same chemical family, higher molecular weight is observed to increase the NIT. This has been observed for example by Coles in the cyano-biphenyl family, although there is an "odd-even" effect and the increase is not monotonous.[59] The effect of this in mixtures is illustrated for example by the phase diagrams of polystyrene with 8-, 10- and 12-CB.[15,49]

As it is very well known in polymer systems, higher molecular weight means lower miscibility, due to the lower configurational entropy of long chains.[60] This shifts the I–I curve to higher temperature, as in Figure 2.1b.

For PDLC this was specifically investigated experimentally by Benmouna *et al.* for PS-8CB,[61] and theoretically by Benmouna *et al.*,[3] and Matsuyama and Kato.[62]

Cross-linked systems show less miscibility than non-cross-linked systems, although this is not just an effect of an "infinite molecular weight"; in addition, swelling of the polymer network by the LC introduces an elastic energy which decreases miscibility.[3,24,25] It has to be noted here that the existence of droplets dispersed in the matrix also implies a distortion of the network, so elasticity not only decreases miscibility, but also disfavors growth of the droplets. In addition, gelation of the system implies a very abrupt increase in viscosity and the appearance of viscoelasticity, so there is a strong influence on the morphology due to kinetic effects as well. In general, the evolution of phase separation becomes highly restricted and can be arrested after gelation.[1,28] All these effects are strongly dependent on cross-linking density. It was observed that by using cross-linkable monomers with similar chemical structure but different size, smaller monomers (which lead to a smaller distance between cross-linking points), produce smaller dispersed domains.[63,35] Meng *et al.*[29] analyzed the effect of varying the cross-linking density by using different amounts of epoxy monomer and hardener, observing a decrease in domain size by increasing cross-linking density. Zheng *et al.*[64] studied morphologies of a cross-linked PDLC using a mixture of fluorinated and non-fluorinated monomers, and they observed a maximum amount of phase separation with the largest droplets at some intermediate amount of fluorinated monomer. This was explained by the competition between the decrease in miscibility due to the chemical dissimilarity (prevalent at low amounts of fluorinated monomer), and the fact that the monomer they used induces "dead end polymerization" which decreases the cross-linking density thus increasing miscibility (as observed for high contents of fluorinated monomer).

Finally, chemical structure can affect not only phase transitions, but also the anchoring transition of the mesophase. For example, Amudson and Srinivasarao[65] found that by modifying the polymer side group, the anchoring transition temperature could be tuned over a wide range. Larger linear pendant groups favor homeotropic anchoring, while smaller or branched groups favor planar anchoring. This is understood by considering the pendant present at the interface between the matrix and the LC droplets; long linear groups would produce a "brush-like" interface which favors perpendicular orientation.

To conclude, PLCMs show a very rich phase behavior that leads to a variety of possible morphologies. The physics of phase separation is quite complex as it can involve not only phase diagrams with multiple phase coexistence regions, but also more than one order parameter that can evolve through different mechanisms. In the case of PIPS, this is further complicated as chemical reaction has to be accounted for as well.

The effect of quenching a mixture (by decreasing the temperature or increasing polymer concentration) on morphologies was discussed, considering the trajectory of the mixture in the phase diagram and the competition between thermodynamic, dynamic, and kinetic factors. Several regions in the phase diagram, associated with different phase transition mechanisms, were identified, and the expected morphologies for each case were described, accounting for how kinetic restrictions affect the size of the dispersed phase domains. The effect of chemical structure and processing conditions on morphologies can be explained by their effect on phase diagrams and mobility. The presented mechanisms of equilibrium and non-equilibrium phase separation taken together provide a firm foundation to target desired microstructures and hence functionalities of PDLCs.

Acknowledgements

A.D.R. is thankful to The National Research Council of Canada (NSERC) for financial support and to McGill University for support through the James McGill Professorship. E.R.S. thanks the National Research Council of Argentina (CONICET), and the Agency for Promotion of Science and Technology (ANPCyT) for financial support.

References

1. M. Mucha, Polymer as an important component of blends and composites with liquid crystals, *Prog. Polym. Sci.*, 2003, **28**, 837–873.
2. D. Coates, Polymer-dispersed Liquid Crystals, *J. Mater. Chem.*, 1995, **5**, 2063–2072.
3. F. Benmouna, L. Bedjaoui, U. Maschke, X. Coqueret and M. Benmouna, On the phase behavior of blends of polymers and nematic liquid crystals, *Macromol. Theory Simul.*, 1998, **611**, 599–611.
4. S. K. Das and A. Rey, Computational modelling of multi-phase equilibria of mesogenic mixtures, *Comput. Mater. Sci.*, 2004, **29**(2), 152–164.
5. E. R. Soulé and A. D. Rey, Modelling complex liquid crystal mixtures: From polymer dispersed mesophase to nematic nanocolloids, *Mol. Simul.*, 2012, **38**(8–9), 735–750.
6. S. K. Das and A. Rey, Texture formation under phase ordering and phase separation in polymer-liquid crystal mixtures, *J. Chem. Phys.*, 2004, **121**(19), 9733–9743.
7. S. K. Das and A. Rey, Kinetic structures in multi-phase liquid-crystalline composite materials, *Mol. Simul.*, 2005, **31**(2–3), 201–206.
8. E. R. Soulé, L. Reven and A. Rey, Thermodynamic Modelling of Phase Equilibrium in Nanoparticles – Nematic Liquid Crystals Composites, *Mol. Cryst. Liq. Cryst.*, 2012, **553**(1), 118–126.
9. E. R. Soulé and A. Rey, Modelling complex liquid crystal mixtures: from polymer dispersed mesophase to nematic nanocolloids, *Mol. Simul.*, 2012, **38**(8–9), 735–750.
10. C. E. Hoppe, M. J. Galante, P. A. Oyanguren and R. J. J. Williams, Polymer-Dispersed Liquid Crystals with Co-continuous Structures

Generated by Polymerization-Induced Phase Separation of EBBA-Epoxy Solutions, *Macromolecules*, 2002, **35**(16), 6324–6331.
11. M. Motoyama, *et al.*, Phase separation of liquid crystal-polymer mixtures, *Comput. Theor. Polym. Sci.*, 2000, **10**(3–4), 287–297.
12. A. Rey, Liquid crystal models of biological materials and processes, *Soft Matter*, 2010, **6**(15), 3402–3429.
13. I. Dierking and C. Russell, Universal scaling laws for the anisotropic growth of SmA liquid crystal bâtonnets, *Phys. B*, 2003, **325**, 281–286.
14. H. Chan and I. Dierking, Growth laws for the phase ordering dynamics of the B1 phase of a bent-core liquid crystal, *Phys. Rev. E*, 2004, **70**(2), 21703.
15. E. R. Soulé, N. M. Abukhdeir and A. Rey, Thermodynamics, Transition Dynamics, and Texturing in Polymer-Dispersed Liquid Crystals with Mesogens Exhibiting a Direct Isotropic/Smectic-A Transition, *Macromolecules*, 2009, **42**(24), 9486–9497.
16. C. C. Riccardi, J. Borrajo, R. J. J. Williams and I. Introduction, *Instability of a Nematic Phase Described by the Maier–Saupe Model*, 1998, vol. 108, no. 6, pp. 2571–2575.
17. P. Taylor and J. R. Dorgan, Spinodal decomposition in mixtures containing nematogens, *Liq. Cryst.*, 1991, **3**, 347–355.
18. E. R. Soulé, G. E. Elic, J. Borrajo, R. J. J. Williams and G. E. Eliçabe, Analysis of the Phase Separation Induced by a Free-Radical Polymerization in Solutions of Polyisobutylene in Isobornyl Methacrylate, *Ind. Eng. Chem. Res.*, 2007, **46**(23), 7535–7542.
19. E. R. Soulé, J. Borrajo and R. J. J. Williams, Thermodynamic Analysis of a Polymerization-Induced Phase Separation in Nanoparticle–Monomer–Polymer Blends, *Macromolecules*, 2007, **40**(22), 8082–8086.
20. E. R. Soulé, *et al.*, Thermodynamic Analysis of the Reaction-Induced Phase Separation of Solutions of Random Copolymers of Methyl methacrylate and N,N-dimethylacrylamide in the Precursors of a Polythiourethane Network, *J. Polym. Sci., Part B: Polym. Phys.*, 2006, **44**(19), 2821–2827.
21. S. Maiez-Tribut, J.-P. Pascault, E. R. Soulé, J. Borrajo and R. J. J. Williams, Nanostructured epoxies based on the self-assembly of block copolymers: a new miscible block that can be tailored to different epoxy formulations, *Macromolecules*, 2007, **40**(4), 1268–1273.
22. I. A. Zucchi, M. J. Galante, J. B. And and R. J. J. Williams, A Model System for the Thermodynamic Analysis of Reaction-Induced Phase Separation: Solutions of Polystyrene in Bifunctional Epoxy/Amine Monomers, *Macromol. Chem. Phys.*, 2004, **205**(5), 676–683.
23. M. Rico, J. Borrajo, M. J. Abad, L. Barral and J. Lopez, Thermodynamic analysis of phase separation in an epoxy/polystyrene mixture, *Polymer*, Jul. 2005, **46**(16), 6114–6121.
24. C. C. Riccardi, J. Borrajo, R. J. J. Williams, H. M. Siddiqi, M. Dumon and J. P. Pascault, Multiple Phase Equilibria in Polydisperse Polymer/Liquid Crystal Blends, *Macromolecules*, 1998, **31**(2), 1124–1132.
25. J. Borrajo, C. C. Riccardi, R. J. J. Williams, H. M. Siddiqi, M. Dumon and J. P. Pascault, Thermodynamic analysis of reaction-induced phase

separation in epoxy-based polymer dispersed liquid crystals (PDLC), *Polymer*, 1998, **39**(4), 845–853.
26. K. Amundson, A. Van Blaaderen and P. Wiltzius, Morphology and electro-optic properties of polymer-dispersed liquid-crystal films, *Phys. Rev. E*, 1997, **55**(2), 1646–1654.
27. S. Ohta, S. Inasawa and Y. Yamaguchi, Size control of phase-separated liquid crystal droplets in a polymer matrix based on the phase diagram, *J. Polym. Sci., Part B: Polym. Phys.*, 2012, **50**(12), 863–869.
28. R. J. J. Williams, J.-P. Pascault, J. Verdu and H. Sautereau, *Thermosetting Polymers*, Marcel Dekker, New York, 2002.
29. Q. Meng, H. Cao, M. Kashima, H. Liu and H. Yang, Effects of the structures of epoxy monomers on the electro-optical properties of heat-cured polymer-dispersed liquid crystal films, *Liq. Cryst.*, 2010, **37**(2), 189–193.
30. C. E. Hoppe, M. J. Galante, P. A. Oyanguren and R. J. J. Williams, Optical properties of novel thermally switched PDLC films composed of a liquid crystal distributed in a thermoplastic/thermoset polymer blend, *Mater. Sci. Eng., C*, 2004, **24**(5), 591–594.
31. C. E. Hoppe, M. J. Galante, P. A. Oyanguren and R. J. J. Williams, Polymer-Dispersed Liquid Crystals Based on Polystyrene and EBBA: Analysis of Phase Diagrams and Morphologies Generated, *Macromol. Chem. Phys.*, 2003, **204**(7), 928–935.
32. C. E. Hoppe, M. J. Galante, P. A. Oyanguren and R. J. J. Williams, Thermally Reversible Light Scattering Films Based on Droplets of a Liquid Crystal (N-4-Ethoxybenzylidene-4′-n-butylaniline)/Polystyrene Solution Dispersed in an Epoxy Matrix, *Macromolecules*, 2004, **37**(14), 5352–5357.
33. S. K. Das and A. Rey, Computational thermodynamics of multiphase polymer–liquid crystal materials, *Comput. Mater. Sci.*, 2006, **38**(2), 325–339.
34. M. Graca, S. A. Wieczorek and R. Hołyst, Growth of Polystyrene Domains in Isotropic, Nematic and Smectic Phase of 8CB Liquid Crystal, *Macromolecules*, 2003, **36**, 6903–6913.
35. M. Kashima, *et al.*, Effects of the chain length of crosslinking agents on the electro-optical properties of polymer-dispersed liquid crystal films, *Liq. Cryst.*, 2010, **37**(3), 339–343.
36. X. Ding, M. Cao, H. Liu, H. Cao, W. Li and H. Yang, A study of electro-optical properties of PDLC films prepared by dual UV and heat curing, *Liq. Cryst.*, 2008, **35**(5), 587–595.
37. Y. Lu, J. Wei, Y. Shi, O. Jin and J. Guo, Effects of fabrication condition on the network morphology and electro-optical characteristics of polymer-dispersed bistable smectic A liquid crystal device, *Liq. Cryst.*, 2013, **40**(5), 581–588.
38. G. W. Smith, Cure Parameters and Phase Behavior of An Ultraviolet-Cured Polymer- Dispersed Liquid Crystal, *Mol. Cryst. Liq. Cryst.*, 1991, **196**, 89–102.
39. G. B. Hadjichristov, Y. G. Marinov and A. G. Petrov, Single-Layered PDLC for Diffractive Optics, *Mol. Cryst. Liq. Cryst.*, 2010, **525**(1), 128–139.

40. F. Ahmad, M. Jamil, Y. J. Jeon, L. J. Woo, J. E. Jung and J. E. Jang, Investigation of nonionic diazo dye-doped polymer dispersed liquid crystal film, *Bull. Mater. Sci.*, 2012, **35**(2), 221–231.
41. C. Serbutoviez, J. G. Kloosterboer, H. M. J. Boots and F. J. Touwslager, Polymerization-Induced Phase Separation. 2. Morphology of Polymer-Dispersed Liquid Crystal Thin Films, *Macromolecules*, 1996, **29**(96), 7690–7698.
42. H. Nakazawa, *et al.*, Phase separation and gelation of polymer-dispersed liquid crystals, *Comput. Theor. Polym. Sci.*, 2001, **11**(6), 445–458.
43. H. Tanaka, *Viscoelastic Phase Sep.*, 2000, **12**, R207–R264.
44. G. Chidichimo, *et al.*, High Contrast Reverse Mode PDLC Films: A Morphologic and Electro-Optical Analysis, *Mol. Cryst. Liq. Cryst.*, 2009, **500**, 10–22.
45. R. A. Vaia, D. W. Tomlin, M. D. Schulte and T. J. Bunning, Two-phase nanoscale morphology of polymer/LC composites, *Polymer*, 2001, **42**(3), 1055–1065.
46. D. Cupelli, *et al.*, Self-adjusting smart windows based on polymer-dispersed liquid crystals, *Sol. Energy Mater. Sol. Cells*, 2009, **93**(11), 2008–2012.
47. S. Leclair, L. Mathew, M. Gigue, S. Motallebi and Y. Zhao, Photoinduced Alignment of Ferroelectric Liquid Crystals Using Azobenzene Polymer Networks of Chiral Polyacrylates and Polymethacrylates, *Macromolecules*, 2003, **36**, 9024–9032.
48. J.-H. Liu and F.-T. Wu, Synthesis of photoisomeric azobenzene monomers and model compound effect on electric-optical properties in PDLC films, *J. Appl. Polym. Sci.*, 2005, **97**(3), 721–732.
49. F. Benmouna, A. Daoudi, F. Roussel, J.-M. Buisine, X. Coqueret and U. Maschke, Equilibrium Phase Diagram of Polystyrene and 8CB, *J. Polym. Sci., Part B: Polym. Phys.*, 1999, **37**, 1841–1848.
50. G. R. Meira, C. V. Luciani and D. A. Estenoz, Continuous Bulk Process for the Production of High-Impact Polystyrene: Recent Developments in Modeling and Control, *Macromol. React. Eng.*, 2007, **1**(1), 25–39.
51. E. R. Soulé and A. Rey, Dynamics of transient metastable states in mixtures under coupled phase ordering and chemical demixing, *Eur. Phys. J. B*, 2011, **83**(3), 357–367.
52. E. R. Soulé, C. Lavigne, L. Reven and A. Rey, Multiple interfaces in diffusional phase transitions in binary mesogen-nonmesogen mixtures undergoing metastable phase separations, *Phys. Rev. E*, 2012, **86**(1), 11605.
53. E. R. Soulé and A. Rey, Formation and kinetics of transient metastable states in mixtures under coupled phase ordering and chemical demixing, *Europhys. Lett.*, 2009, **86**, 46006, p1–p5.
54. N. M. Abukhdeir and A. Rey, *Meta-stable Nematic Pre-ordering in Smectic Liquid Crystalline Phase Transitions*, 2009, pp. 3841–3844.
55. W. Li, *et al.*, Control of the Microstructure of Polymer Network and Effects of the Microstructures on Light Scattering Properties of

UV-Cured Polymer-Dispersed Liquid Crystal Films, *J. Polym. Sci., Part B: Polym. Phys.*, 2008, **46**, 2090–2099.
56. J. Han, Effects of Composition, Curing-Time, and Temperature on the Electro-Optical Characteristics of Polymer-Dispersed Liquid Crystal Films, *J. Korean Phys. Soc.*, 2000, **36**(3), 156–163.
57. J. L. West, Phase Separation of Liquid Crystals in Polymers, *Mol. Cryst. Liq. Cryst. Inc. Nonlinear Opt.*, 1988, **157**(1), 427–441.
58. D. Nwabunma, H. W. Chiu and T. Kyu, Morphology development and dynamics of photopolymerization-induced phase separation in mixtures of a nematic liquid crystal and photocuratives, *Macromolecules*, 2000, **33**(4), 1416–1424.
59. H. J. Coles and C. Strazielle, The Order-Disorder Phase Transition in Liquid Crystals as a Function of Molecular Structure I. The Alkyl Cyanobiphenyls, *Mol. Cryst. Liq. Cryst.*, 1979, **55**(1), 237–250.
60. R. T. DeHoff, *Thermodynamics in Materials Science*, McGraw-Hill, New York, 1993.
61. F. Benmouna, *et al.*, Effect of Molecular Weight on the Phase Diagram and Thermal Properties of Poly(styrene)/8CB Mixtures, *Macromolecules*, 2000, **33**, 960–967.
62. A. Matsuyama and T. Kato, Theory of binary mixtures of a flexible polymer and a liquid crystal, *J. Chem. Phys.*, 1996, **105**, 1654–1660.
63. W. Li, Y. Cao, H. U. I. Cao, M. Kashima, L. Kong and H. Yang, *Effects of the Structures of Polymerizable Monomers on the Electro-optical Properties of UV Cured Polymer Dispersed Liquid Crystal Films*, 2008, pp. 1369–1375.
64. Z. Zheng, J. Ma, W. Li, J. Song, Y. Liu and L. Xuan, Improvements in morphological and electro-optical properties of polymer-dispersed liquid crystal grating using a highly fluorine-substituted acrylate monomer, *Liq. Cryst.*, 2008, **35**(7), 885–893.
65. K. Amundson and M. Srinivasarao, Liquid-crystal-anchoring transitions at surfaces created by polymerization-induced phase separation, *Phys. Rev. E*, 1998, **58**(2), R1211–R1214.

CHAPTER 3

Photo-reactive Mesogens

INGO DIERKING

Department of Physics and Astronomy, University of Manchester, Oxford Road, Manchester M13 9PL, UK
Email: Ingo.Dierking@manchester.ac.uk

3.1 Introduction

The monomers used today in the process of UV-induced photopolymerization are generally bifunctional acrylates or methacrylates, that go back to the pioneering work of Broer *et al.*[1–5] in the 1980s–90s at Philips. In most cases the monomers also exhibit liquid crystalline behaviour, often a nematic phase at temperatures well above room temperature. Their elongated, cylinder-like shape, and their mesogenic behaviour ensure that the monomers are decently miscible with the liquid crystal host, and blend well into the director field of the matrix, which is a prerequisite for the polymer network to follow the liquid crystalline order. Despite the fact that the majority of reactive monomers are in fact mesogenic as well, examples of bifunctional acrylates which are non-mesogenic and can be used in polymer stabilization are also known. Nevertheless, their polymer network morphology is not as smooth as those observed for other bifunctional monomers, which affects the physical and electro-optic properties in a slightly adverse way.

A typical reactive mesogen thus combines a central, polarizable core, flexible side groups and at least two reactive groups at either end of the molecule in order to be able to form a crosslinked network, as schematically depicted in Figure 3.1.

Soft Matter Series No. 8
Polymer-modified Liquid Crystals
Edited by Ingo Dierking

Published by the Royal Society of Chemistry, www.rsc.org

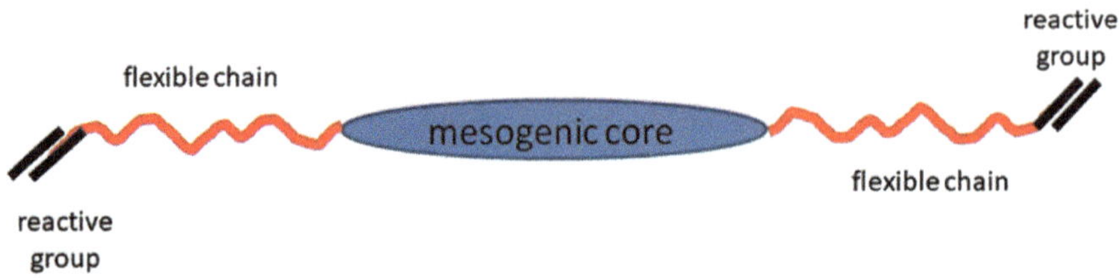

Figure 3.1 Schematic illustration of a typical bifunctional mesogenic photoreactive monomer.

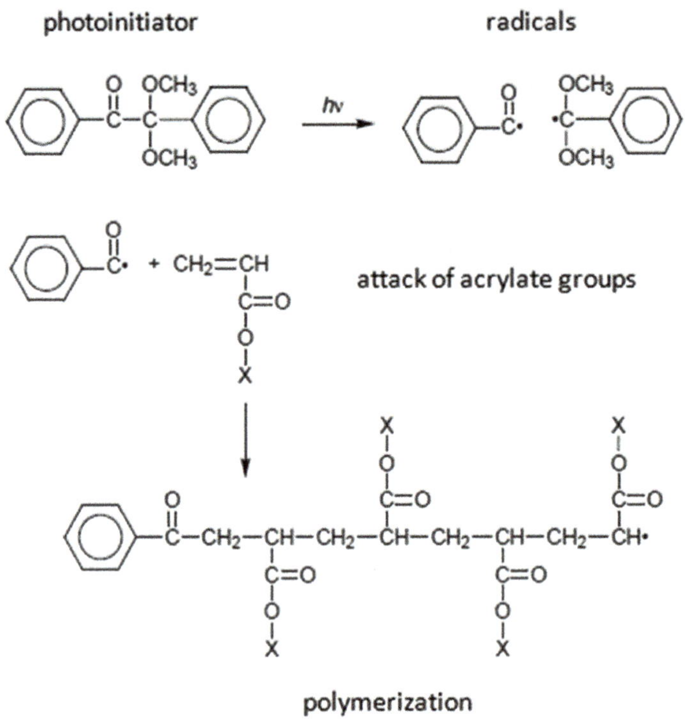

Figure 3.2 Schematic illustration of the polymerization process of an acrylate (after ref. 6). The photoinitiator (in this case 2,2-dimethoxy-2-phenylacetophenone, DMPA) is split into free radicals, which attack the double bonds at the ends of the reactive mesogen X and cause a polymerization process.

Molecules with a single reactive group terminate the polymerization and one can thus tune the crosslinking density by the choice of appropriate concentrations of bifunctional and monofunctional monomers. Nevertheless, in most cases exclusively bifunctional reactive mesogens are employed to stabilize the liquid crystal structure. For the induction of the UV polymerization, a small amount of photoinitiator is added. The action of UV light upon the photoinitiator creates radical species *via* photodissociation as shown in the top line of Figure 3.2. The free radicals attack the double bonds

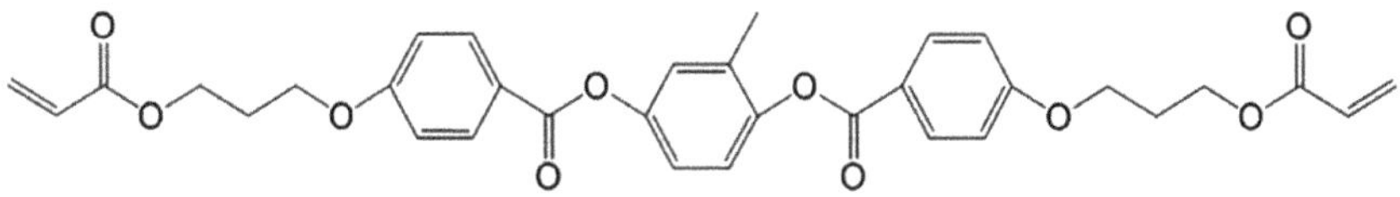

Figure 3.3 One of the most extensively employed bifunctional reactive mesogens is RM257, commercially available from Merck.

of the acrylate groups at the end of the reactive mesogens and cause a step-growth polymerization. The latter will terminate when two of the free radicals meet and recombine, the UV irradiation is turned off, or all of the monomer(s) are consumed. Under normal circumstances, the polymerization process takes approximately a few minutes, while it is safe to continue UV irradiation for about 15–30 minutes to ensure that all monomer material is converted.

As mentioned above, most of the normally-employed reactive mesogens are bifunctional acrylates or methacrylates, while other materials, such as diepoxides are used to a much lesser extent. A few materials are commercially available from Merck, and one monomer in particular, RM257, is extensively used by most research groups. It has a phase sequence of Cryst. 70 N 126 Iso. (Figure 3.3)

3.2 Bifunctional Photo-reactive Monomers

We will keep the list of discussed monomers relatively brief, firstly, because most research is done using RM257 or very similar monomers for nearly all the standard general applications of polymer stabilization, and secondly, because much of the chemical work has already been summarized in some quite detailed review articles by Hikmet and co-workers.[6,7]

As is common for calamitic mesogens and also for reactive mesogens, the core is typically a two-, three-, possibly even a four-ring system, which generally largely influences the clearing point into an upward direction. This is demonstrated by the two reactive mesogens in Figure 3.4 ($n = m = 6$), where the clearing temperature of the three-ring compound is about one hundred degrees higher than that of the two-ring material.

The introduction of side groups, R, on the middle phenyl ring decreases the clearing point, T_{iso}, and further, a larger group decreases T_{iso} more than a smaller group. An example is given for the reactive mesogen in Figure 3.5 with side groups R = H and R = CH_3. The example is given for $n = 6$, but does hold for all $n = 4$–11.[7] Naturally, other substituents can also be chosen, for example, the fluorinated compounds.

The bridging groups between the phenyl rings add a certain flexibility to the core and can, of course, also be varied considerably (Figure 3.6, groups X_1 and X_2). Variation between the isomers of the bottom compound of Figure 3.6 with $n = 6$ shows that the clearing temperature is hardly influenced and neither are the phase sequences in any significant way. In the

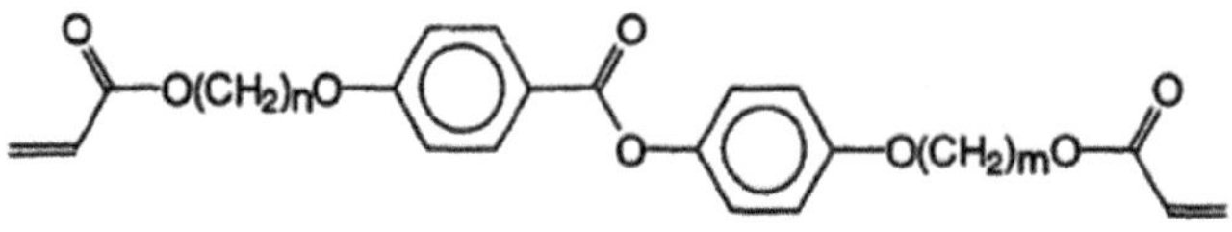

Cryst. 54 Iso. (Iso. 48 N 45 SmA)

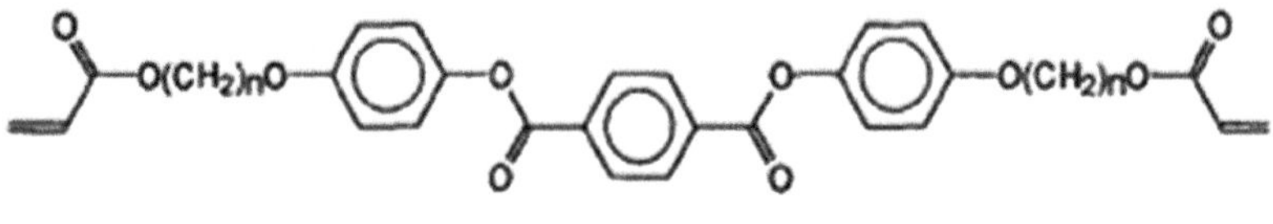

Cryst. 117 SmA 140 N 145 Iso.

Figure 3.4 Comparison of phase transition temperatures and phase sequences of two otherwise very similar photoreactive bifunctional mesogenic monomers. As expected from the behaviour of standard calamitic molecules and also for reactive monomers, the clearing temperature is much lower for two-ring systems as compared to three-ring cores.

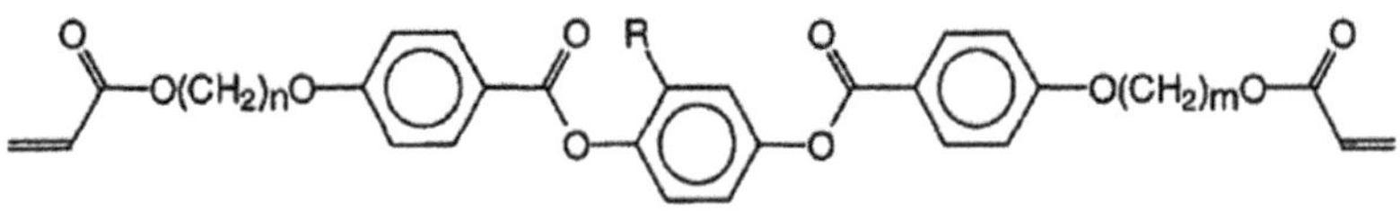

R=H: Cryst. 115 N 155 Iso.
R=CH_3: Cryst. 86 N 116 Iso.

Figure 3.5 Side group substituents on the phenyl rings of the core system reduce the clearing temperature considerably.

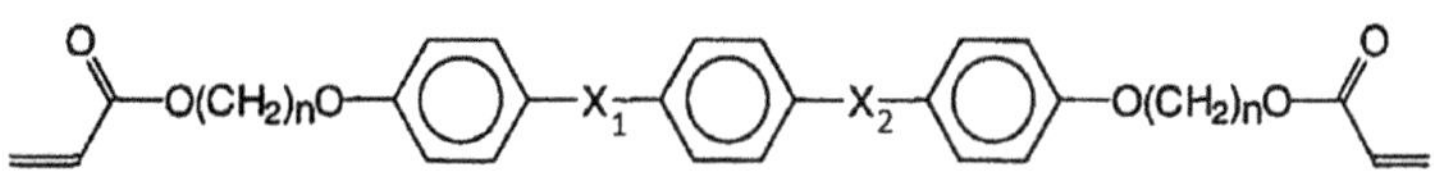

X_1=CO_2; X_2=OCO	Cryst. 115 N 155 Iso. (155 N 87 SmC)
X_1=OCO; X_2=OCO	Cryst. 76 SmA 138 N 150 Iso.
X_1=OCO; X_2=CO_2	Cryst. 117 SmA 140 N 145 Iso.

Figure 3.6 Isomeric diacrylates exhibit practically equivalent clearing temperatures and phase sequences, where $n = 6$.

example of Figure 3.6, it is $n = 6$, and one of the reactive monomers only shows a smectic phase on cooling, thus monotropic behavior. For the $n = 11$ homologue the phase sequence is practically unaltered between the three isomers.[7] On the other hand, if the aromatic rings are changed for cyclohexane rings, the transition temperatures to the isotropic phase decrease considerably due to increased flexibility and the phase sequences do change drastically, losing mesogenic behaviour altogether or retaining only a higher ordered smectic phase, as pointed out in[7] (Figure 3.7).

Despite the fact that these molecules are not mesogenic, they can be employed to form polymer networks in a liquid crystalline host. Nevertheless, the formed networks are generally not as efficient and reproducible in morphology as their liquid crystalline counterparts.

Lastly, one can of course change the length of the flexible side group of the reactive mesogens. This results in a very similar behaviour to that often observed for standard calamitic compounds. An example molecule that has been studied by the Philips group[7] is presented in Figure 3.8.

For short spacers, $n = m \leq 6$, no liquid crystalline behaviour was observed on heating, while the Cryst.–Iso. transition temperature was lowered with increasing spacer length. For $n = \mathrm{m} = 5{,}6$ monotropic nematic and smectic A phases were observed. For homologues with long spacers, $n = m = 11$, the nematic phase was lost, but the smectic phase was observed also on heating. This is in compliance with the general behaviour observed for non-reactive mesogens.

So far, we have only discussed diacrylates, which are the by far most commonly employed bifunctional photo-reactive monomers. Nevertheless, we have not presented their synthesis, which is discussed in detail in ref. 7. Without doubt, there have also been numerous other, slightly different molecules developed and synthesised, which for reasons of space we have not covered. The materials presented above were the first ones to be em-

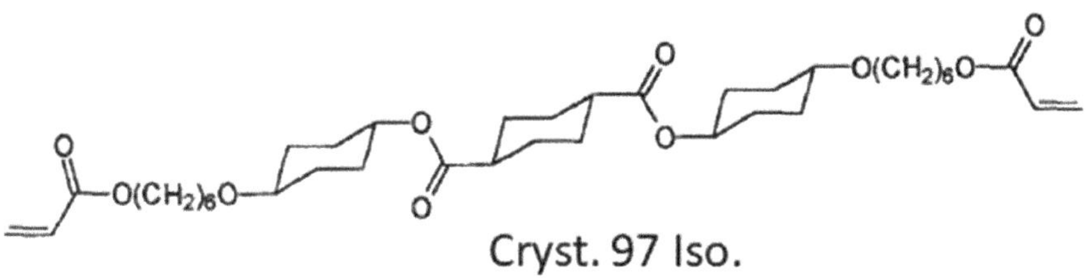

Figure 3.7 Loss of liquid crystalline behaviour is shown for reactive monomers where the aromatic rings were substituted by cyclohexane rings.

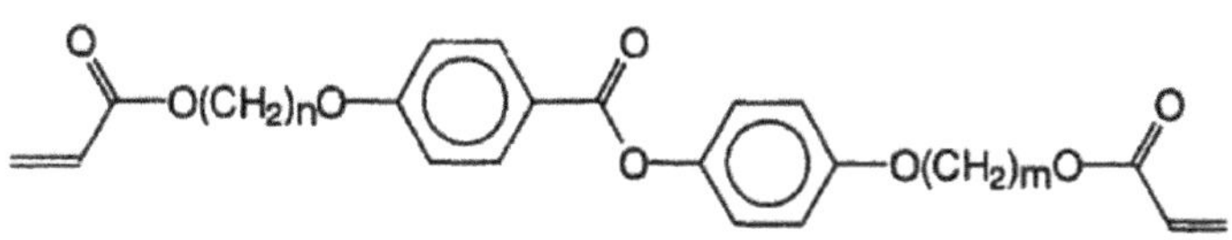

Figure 3.8 Sample molecule used to investigate the effects of a variation in length of the flexible side group, $n = m = 4$–11, as described in the text.

ployed in many of the fundamental investigations on polymer stabilized nematics,[8–14] cholesterics,[15–25] and ferroelectric liquid crystals[26–34] and are still used to date.

Besides diacrylates, there have also been attempts to use diepoxides as reactive monomers,[7] but it was found that these are not very stable towards elevated temperatures, and also the formed networks were quite unstable and not reproducible, due to poor crosslinking and monomer conversion. An exemplary molecule is depicted in Figure 3.9.

An interesting aspect of the formulation of reactive mesogens is the introduction of chirality, not only *via* the liquid crystalline host, but also through the bifunctional reactive monomer. An example[7,35] is given in Figure 3.10, with R^1, R^2 and $R^3 = H$ or CH_3.

The respective monomers themselves form a chiral nematic or cholesteric phase with a helical pitch in the range of optical wavelengths, due to the spacers containing a chiral methyl group. Selective reflection in the visible part of the spectrum was observed, and the handedness could be changed by changing the chirality of the molecule to its opposite configuration. Such chiral reactive mesogens can lead to interesting effects and applications when combined with an achiral or, indeed, a chiral liquid crystalline host.

Another reactive mesogen with intriguing properties has recently been reported,[36] a fluorescent monomer, as is shown in Figure 3.11.

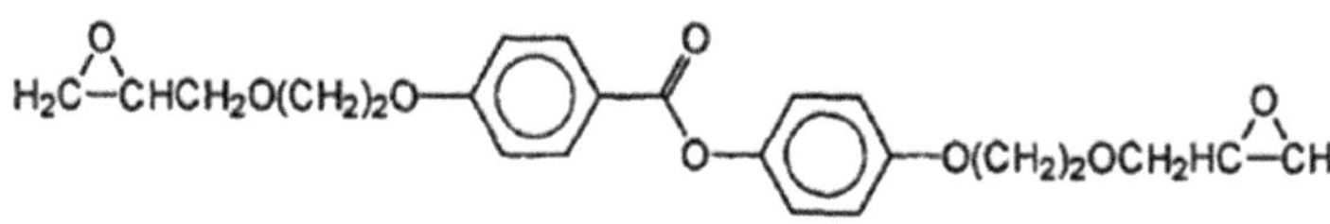

Figure 3.9 Mesogenic diepoxides appear not to be very suitable for the formation of polymer networks, which is attributed to poor crosslinking and thermal instability.

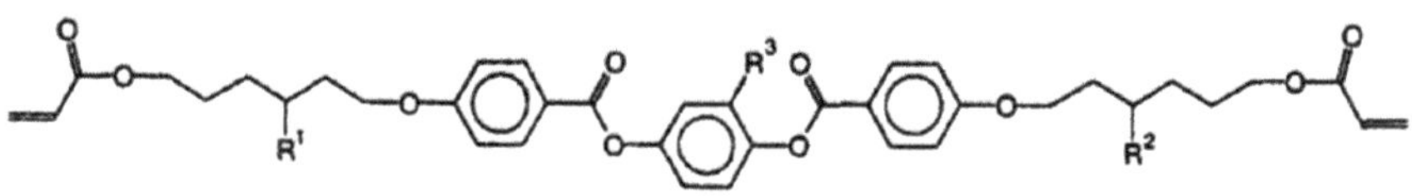

Figure 3.10 A chiral reactive diacrylate mesogenic monomer.

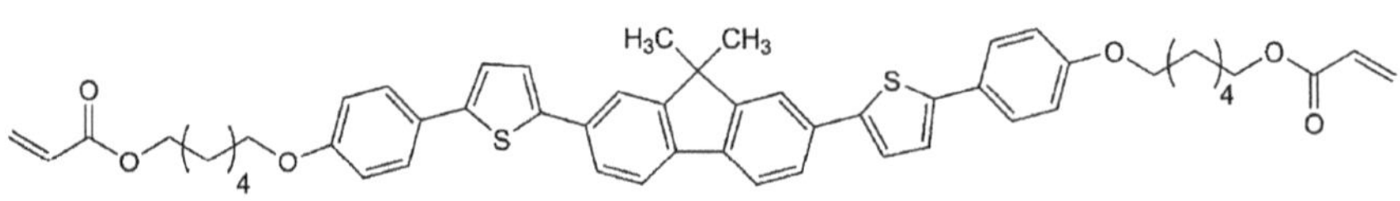

Figure 3.11 A fluorescent reactive mesogenic monomer can form an oriented network structure as a co-polymer in combination for example with RM257. Such polymer networks exhibit a very strong optical anisotropy.

3.3 Summary

In conclusion from all the examples given above, one can infer that bi-functional reactive monomers behave very similarly to standard calamitic molecules with respect to changes in the aromatic core, the flexible side groups and the length of the spacers. However, they exhibit the additional possibility of being polymerized into a continuous network, reflecting the molecular order of the liquid crystalline host phase and director configuration, when subjected to UV illumination in the presence of a photoinitiator. To obtain smooth and reproducible networks with significant beneficial influence on the liquid crystalline system, it is of importance to employ monomers of similar length to the liquid crystal mesogen, as well as of similar phase transition temperatures and phase sequences.

References

1. D. J. Broer, H. Finkelmann and K. Kondo, *Makromol. Chem.*, 1988, **189**, 185.
2. D. J. Broer, G. N. Mol and G. Challa, *Makromol. Chem.*, 1989, **190**, 19.
3. D. J. Broer, J. Boven, G. N. Mol and G. Challa, *Makromol. Chem.*, 1989, **190**, 2255.
4. D. J. Broer, R. A. M. Hikmet and G. Challa, *Makromol. Chem.*, 1989, **190**, 3201.
5. D. J. Broer, G. N. Mol and G. Challa, *Makromol. Chem.*, 1991, **192**, 59.
6. R. A. M. Hikmet, *J. Mater. Chem.*, 1999, **9**, 1921.
7. R. A. M. Hikmet and J. Lub, *Prog. Polym. Sci.*, 1996, **21**, 1165.
8. R. A. M. Hikmet, *J. Appl. Phys.*, 1990, **68**, 4406.
9. R. A. M. Hikmet, *Liq. Cryst.*, 1991, **9**, 405.
10. R. A. M. Hikmet, *Mol. Cryst. Liq. Cryst.*, 1992, **213**, 117.
11. R. A. M. Hikmet, *Adv. Mater.*, 1992, **4**, 679.
12. Y. K. Fung, D.-K. Yang, S. Ying, L.-C. Chien, S. Zumer and J. W. Doane, *Liq. Cryst.*, 1995, **19**, 797.
13. A. Y.-G. Fuh, M.-S. Tsai and C.-Y. Huang, *Jpn. J. Appl. Phys.*, 1996, **35**, 3960.
14. Y. K. Fung, A. Borstnik, S. Zumer, D.-K. Yang and J. W. Doane, *Phys. Rev. E*, 1997, **55**, 1637.
15. D.-K. Yang, L.-C. Chien and J. W. Doane, *Appl. Phys. Lett.*, 1992, **60**, 3102.
16. C. V. Rajaram, S. D. Hudson and L.-C. Chien, *Chem. Mater.*, 1995, **7**, 2300.
17. D. S. Muzic, C. V. Rajaram, L.-C. Chien and S. D. Hudson, *Polym. Adv. Technol.*, 1996, **7**, 737.
18. I. Dierking, L. L. Kosbar, A. Afzali-Ardakani, A. C. Lowe and G. A. Held, *Appl. Phys. Lett.*, 1997, **71**, 2454.
19. C. V. Rajaram, S. D. Hudson and L.-C. Chien, *Chem. Mater.*, 1996, **8**, 2451.
20. C. V. Rajaram, S. D. Hudson and L.-C. Chien, *Polymer*, 1998, **39**, 5315.

21. I. Dierking, L. L. Kosbar, A. Afzali-Ardakani, A. C. Lowe and G. A. Held, *J. Appl. Phys.*, 1997, **81**, 3007.
22. M. Mitov, A. Boudet, P. Sopena and P. Sixou, *Liq. Cryst.*, 1997, **23**, 903.
23. I. Dierking, L. L. Kosbar, A. C. Lowe and G. A. Held, *Liq. Cryst.*, 1998, **24**, 387.
24. I. Dierking, L. L. Kosbar, A. C. Lowe and G. A. Held, *Liq. Cryst.*, 1998, **24**, 397.
25. G. A. Held, L. L. Kosbar, I. Dierking, A. C. Lowe, G. Grinstein, V. Lee and R. D. Miller, *Phys. Rev. Lett.*, 1997, **79**, 3443.
26. J. Pirs, R. Blinc, B. Marin, S. Pirs and J. W. Doane, *Mol. Cryst. Liq. Cryst.*, 1995, **264**, 155.
27. R. A. M. Hikmet and J. Lub, *J. Appl. Phys.*, 1995, **77**, 6234.
28. R. A. M. Hikmet and M. Michielsen, *Adv. Mater.*, 1995, **7**, 300.
29. R. A. M. Hikmet, H. M. J. Boots and M. Michielsen, *Liq. Cryst.*, 1995, **19**, 65.
30. C. A. Guymon, E. N. Hoggan, D. M. Walba, N. A. Clark and C. N. Bowman, *Liq. Cryst.*, 1995, **19**, 719.
31. J. Li, Z. Wang, Y. Cai and X. Huang, *Ferroelectrics*, 1998, **213**, 91.
32. J. Nourry, A. Vigouroux, A. Magnaldo, P. Sixou, M. Mitov, A. Boudet, M. Glogarova and A. M. Bubnov, *Ferroelectrics*, 1998, **212**, 203.
33. I. Dierking, M. A. Osipov and S. T. Lagerwall, *Eur. Phys. J. E*, 2000, **2**, 303.
34. I. Dierking, L. Komitov, S. T. Lagerwall, T. Wittig and R. Zentel, *Liq. Cryst.*, 1999, **26**, 1511.
35. J. Lub, D. J. Broer, R. A. M. Hikmet and K. G. J. Nierop, *Liq. Cryst.*, 1995, **18**, 319.
36. V. Kumar Baliyan, V. Kumar, J. Kim and S.-W. Kang, *Opt. Mater. Express*, 2016, **6**, 2956.

CHAPTER 4

Electron Beam Curing of Monomer/Liquid Crystal Blends

MOHAMMED BOUCHAKOUR,[a,b,c] YAZID DEROUICHE,[a,c] ZOHRA BOUBERKA,[a,d] CHRISTOPHE BEYENS,[a] PHILIPPE SUPIOT,[e] FRÉDÉRIC DUBOIS,[f] FARID RIAHI[b] AND ULRICH MASCHKE*[a]

[a] Unité Matériaux et Transformations – UMET (UMR CNRS N°8207), Bâtiment C6, Université de Lille – Sciences et Technologies, 59655 Villeneuve d'Ascq Cedex, France; [b] Laboratoire de préparations, modifications et applications des matériaux polymériques multiphasiques, Faculté de Technologie, Université Sétif 1, 19000 Sétif, Algeria; [c] Faculté des Sciences Exactes et Informatique, Université Ziane Achour de Djelfa, 17000 Djelfa, Algeria; [d] Laboratoire Physico-Chimie des Matériaux-Catalyse et Environnement (LPCM-CE), Université des Sciences et de la Technologie d'Oran «USTO», BP 1505, El M'naouer, 31000 Oran, Algeria; [e] IEMN-P2M group, UMR 8520 (CNRS), Université de Lille – Sciences et Technologies, 59655 Villeneuve d'Ascq Cedex, France; [f] Unité de Dynamique et Structure des Matériaux Moléculaires (UDSMM), Université du Littoral - Côte d'Opale (ULCO), 62228 Calais, France
*Email: ulrich.maschke@univ-lille1.fr

4.1 Introduction

Polymer dispersed liquid crystals (PDLCs) have been studied extensively within the past decades.[1–3] They are generally made of micron-sized droplets dispersed in a solid polymer matrix. They have remarkable electro-optical

Soft Matter Series No. 8
Polymer-modified Liquid Crystals
Edited by Ingo Dierking

Published by the Royal Society of Chemistry, www.rsc.org

responses since they can be switched from one optical state to another simply by the application of magnetic or alternate electric fields. PDLC films are useful for various applications including optical shutters, intelligent windows, telecommunications and information displays. The realization of these films is often based on polymerization induced phase separation (PIPS) processes induced mainly by UV–visible light but less frequently with accelerated electron beams (EB). EB curing leads to high monomer conversions without a photoinitiator which acts as an impurity that might have a negative impact on the electro-optical performance of the obtained PDLC films. Some photosensitive monomers such as styrene, thiol-enes, mono-, di-, and multifunctional acrylates and methacrylates are used to prepare PDLCs. The ultimate control of the phase separation of the polymer/LC systems is necessary to obtain different morphologies, depending essentially on the polymerization conditions during which the LC becomes less miscible with the growing polymer, and finally the mixture will separate into two phases.[2]

Andrzejewska,[4] Decker *et al.*,[5,6] Scherzer *et al.*,[7] and Anseth *et al.*[8] studied extensively the photopolymerization of multifunctional monomers. They found that these show complicated reaction kinetics which involve phenomena not observed for linear polymerizations, particularly the immediate onset of autoacceleration and radical trapping at early stages, as well as the dominance of reaction diffusion by means of migration of radical reactive species. At later stages of polymerization, autodeceleration or incomplete conversion takes place because of the extremely low mobility of the propagating species in the crosslinked network. Generally, the polymerization is governed by the glass transition temperature of the growing polymer chains.

Defoort *et al.*,[9] Patacz *et al.*,[10,11] Maschke *et al.*,[12] and Knolle *et al.*[13] thoroughly studied the polymerization of tripropyleneglycoldiacrylate (TPGDA) using EB. They found that it follows the same reaction scheme as for UV-induced polymerization. The main differences between the two initiation processes lie in the energy deposition and the pathway for free radical generation. Polymer networks elaborated by EB are usually uniformly cured due to the full-depth penetration of the electrons. Meanwhile, their UV counterparts tend to be more or less heterogeneously cured due to the presence of photoinitiator molecules which absorb more of the UV irradiation at the surface rather than in the core, resulting in non-uniform polymer networks especially for thick films.

The electro-optical properties of PDLC films have been subject of many studies;[14–22] they were found to be controlled by several factors including the type of LC and monomers or pre-polymers, the method of preparation, film thickness, morphology, and temperature.

In addition to work that had been done at UMET and elsewhere,[14–16] a comparison between methods of rapid polymerization/crosslinking processes by UV-light and EB was necessary. This was realized in terms of investigation of polymerisation/crosslinking and phase separation kinetics, sample morphologies and electro-optical responses of systems involving three diacrylate monomers, namely TPGDA and polypropyleneglycoldiacrylate (PPGDA540) and (PPGDA800), possessing molecular weights of 300, 540 and 800 $g\,mol^{-1}$, respectively, with and without adding the nematic LC E7. The aim behind the use of a very rapid UV process (UV–Dr. Hönle system) was to verify whether the steep electro-optical responses previously seen in EB samples could be obtained or not.

4.2 Experimental

4.2.1 Materials

Tripropyleneglycoldiacrylate (TPGDA) with $M_n = 300$ $g\,mol^{-1}$ ($n = 3$, corresponding to three propyleneglycol repeating units) and polypropyleneglycoldiacrylate (PPGDA540) (PPGDA800) with $M_n = 540$ ($n = 7$) and $M_n = 800$ $g\,mol^{-1}$ ($n = 12$), respectively (Figure 4.1(a)), were purchased from Sigma Aldrich. E7 represents a eutectic mixture (Merck, Japan) containing four cyanoparaphenylene derivatives; namely: 51wt% 4-cyano-4′-pentylbiphenyl (5CB), 25wt% 4-cyano-4′-heptyl-biphenyl (7CB), 16wt% 4-cyano-4′-octyloxybiphenyl (8OCB), and 8wt% 4-cyano-4″-pentyl-*p*-terphenyl

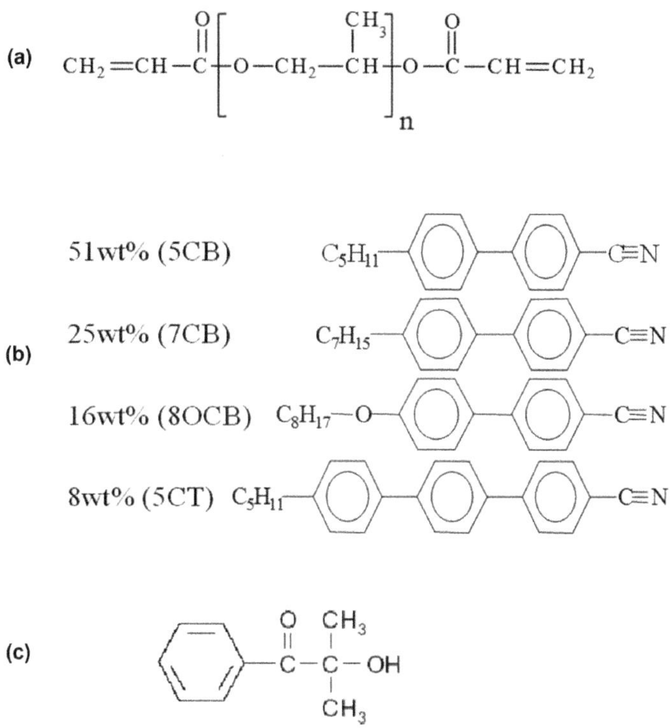

Figure 4.1 Chemical structures of (a) Polypropyleneglycoldiacrylate (PPGDA) monomers, (b) Nematic LC E7 mixture and (c) Photoinitiator 2-hydroxy-2-methyl-1-phenyl-propane-1-one (Darocur 1173).

(5CT). It exhibits a nematic-isotropic transition temperature at $T_{NI} = 61\ ^{\circ}C$ and a positive dielectric anisotropy $\Delta\varepsilon = 13.8$ ($\varepsilon_{\parallel} = 19.0$) at a frequency of 1 kHz, where ε stands for the parallel dielectric constant. The refractive indices of E7 at $T = 20\ ^{\circ}C$ are given as $n_o = 1.5183$; $n_e = 1.7378$ ($\lambda = 632.8$ nm), leading to a birefringence of $\Delta n = n_e - n_o = 0.2195$.[23–25]

The chemical structures of the E7 components are shown in Figure 4.1(b). Darocur 1173 (2-hydroxy-2-methyl-1-phenyl-propane-1-one) from Ciba (Italy) was used as a photoinitiator (Figure 4.1(c)). All chemicals were used as received without any purification. For UV curing, 2wt% of Darocur 1173, compared to that of the monomer, the LC E7 and the monomer were weighed and mixed overnight before use. For EB curing, no photoinitiator was needed.

Samples for all kinds of studies were prepared by sandwiching the initial reactive mixture between two supports, allowing a uniform penetration of the applied dose through the depth of the sample.

The ITO (indium tin oxide)-coated poly(ethyleneterephthalate) (PET) foils of thicknesses 50 and 100 μm were donated by Renker (Germany), and 13 μm thick PET foils were purchased from Goodfellow. ITO-coated glass plates used for electro-optic measurements was purchased from AWAT/Poland. The film thicknesses were measured by a micrometer caliper (Mitutoyo, with uncertainty of ±1 μm). No attempt has been made to control the temperature during the irradiation processes.

4.2.2 Sample Preparation

All UV–Dr. Hönle–cured samples were made by placing a small amount of the mixture between two ITO-coated glass plates, spaced by PET 13 μm thick film, and then cured under UV light. The EB samples were made by sandwiching a small droplet of the mixture between ITO-coated glass plates and ITO-coated 50 μm thick PET sheets, spaced by 13 μm thick PET film or a double-phase adhesive tape, and exposed to a dose of 120 kGy for about 2.45 s.

4.2.2.1 *Ultra-violet Curing*

A rapid light source was applied (Dr. Hönle AG, UV-Technologie, Germany), equipped with a UVH medium pressure mercury arc lamp rated at minimum 240 W cm^{-1}. The Dr. Hönle UV lamp provides fast curing processes with an output power of 10 kW which can be regulated from 20 to 100%. In this study, the power was set to 70% to obtain a dose rate of 100 mJs^{-1} cm^{-2} for better control of photopolymerization and phase separation kinetics. The maximum exposure time under the UV source did not exceed 3 s. This equipment was used to give very fast processes for both photopolymerization and phase separation, comparable to those under EB curing.

4.2.2.2 *Electron Beam Curing*

The generator used in these experiments was an Electrocurtain Model CB 150 (Energy Sciences Inc.) with an operating high voltage up to 175 kV used for electron acceleration to reach a penetration depth of 100 μm. The samples were covered by 50 μm thick PET films and put on a tray that moved on a conveyor belt. The received dose can be controlled by adjusting the intensity of the beam current and the conveyor belt speed. The maximum exposure time during one pass was measured to be around 2.45 s.

4.2.2.3 *Polarized Optical Microscopy*

The polarizing optical microscope (POM) used was an Olympus BX-41 model, equipped with a heating/cooling stage Linkam LTS 350, a Linkam TMS 94 temperature control unit, and a digital camera that can record images with high resolution, equipped with a computer. A small droplet of the monomer/E7 mixture was placed on a glass plate inside an oven set at a temperature around 50 °C, so that the mixture became isotropic; then the sample was immediately placed under the POM which was set at 60 °C. An upper glass plate was put on the drop and heated for an extra 5 min, after which the sample was cooled down at a rate of 1.5 $°C\,min^{-1}$ to −20 °C, kept for one minute at this level, then heated up to 60 °C by 1.5 $°C\,min^{-1}$. The presented transition temperatures were obtained by taking averages from cooling and heating cycles. The morphologies of the obtained PDLC films were observed under the POM at room temperature.

4.2.2.4 *Infrared Spectroscopy*

FTIR spectra of less than 10 μm thick films (about 5 mg) were recorded in the absorbance mode using a Perkin Elmer 2000 model with a spectral resolution of 4 cm^{-1} over 16 scans. A small drop of the mixture was put on a NaCl plate and covered by PET films with thicknesses of 100 μm and 50 μm for UV and EB curing, respectively. Small cumulated doses were applied for both methods of elaboration, the period of time between the end of the exposure and the infrared analysis was kept constant. The experiments were repeated three times for reproducibility reasons.

4.2.2.5 *Electro-optical Measurements*

The electro-optical properties were determined at 20 °C using the set-up described in ref. 26. In this set up, a linearly collimated beam from unpolarized He-Ne laser red light ($\lambda = 632.8$ nm) passes perpendicularly through a fixed sample, and the transmitted intensity is measured by a silicon photodiode. The electro-optical measurements were performed on PDLC samples sandwiched between two ITO-coated glass plates or an ITO-glass plate and a 50 μm thick ITO-PET film for UV- and EB-cured specimens, respectively. The samples were polymerized under optimum conditions found after studying the polymerization kinetics. The applied alternate sinusoidal voltage at a frequency of 1 kHz was increased linearly up to a desired maximum value V_{max}, then subsequently decreased in the same manner to 0 Volt. The entire up and down cycle lasted 120 s with an additional 60 s provided for sample relaxation. The same procedure was repeated for all other values of the applied voltages, which ranged from 20 to more than 300 Volts using increments of 20 Volts.

The light transmittance was calibrated by means of glycerol sandwiched between two ITO-glass plates or between one ITO-glass plate and an ITO-PET film for UV and EB samples, respectively.

4.3 Results and Discussion

4.3.1 Phase Diagrams by POM

Figure 4.2 illustrates the phase diagrams of the three binary monomeric mixtures, determined by POM. The full triangular, circular and square symbols relate to PPGDA800/E7, PPGDA540/E7 and TPGDA/E7, respectively. The (Nematic + Isotropic)/Isotropic (N + I)/(I) transition temperature ($T_{N+I/I}$) decreases upon adding the monomer to the LC E7. The depression of $T_{N+I/I}$ follows a similar trend for the three monomers; they differ from each other because of the difference in the molecular weights of TPGDA, PPGDA540, and PPGDA800. POM observations were performed by varying temperature from 60 °C to −20 °C, and concentration of the nematic LC E7, covering the range from 50wt% to 100wt% of E7, to detect the segregated LC domains in

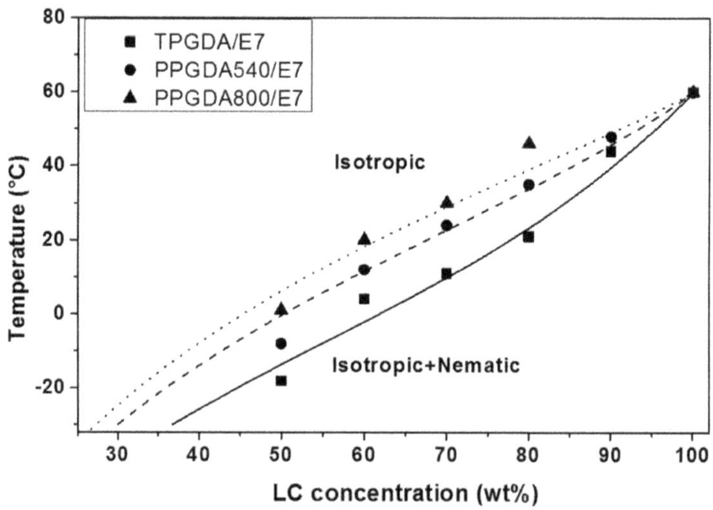

Figure 4.2 Phase diagrams of the monomeric mixtures: PPGDA800/E7 (triangular symbols), PPGDA540/E7 (circular symbols) and TPGDA/E7 (square symbols). Continuous, dashed, and dotted lines represent guides for the eye.

the (N + I) phase. It has been shown that at 20 °C, the solubility limits of E7 were detected to be around 60, 65 and 75wt% in PPGDA800, PPGDA540 and TPGDA, respectively.

Figure 4.3 shows the morphologies corresponding to the monomeric mixtures TPGDA800/60wt% E7, PPGDA540/60wt% E7, and PPGDA800/60wt% E7 at different temperatures. The micrographs reveal a clear variation of the transition temperatures from isotropic to isotropic + nematic states. These transition temperatures were observed around 0, 12, and 20 °C for TPGDA/60wt% E7, PPGDA540/60wt% E7, and PPGDA800/60wt% E7, respectively. The choice of the right monomeric mixture is important in terms of the phase separation induced by polymerization since good electro-optical results can generally be obtained if the initial reactive blend is in the homogeneous isotropic phase close to the solubility limit of the LC.[27] The composition of 60wt% E7 was chosen here deliberately for all mixtures in order to carry out the investigation at the same E7 concentration.

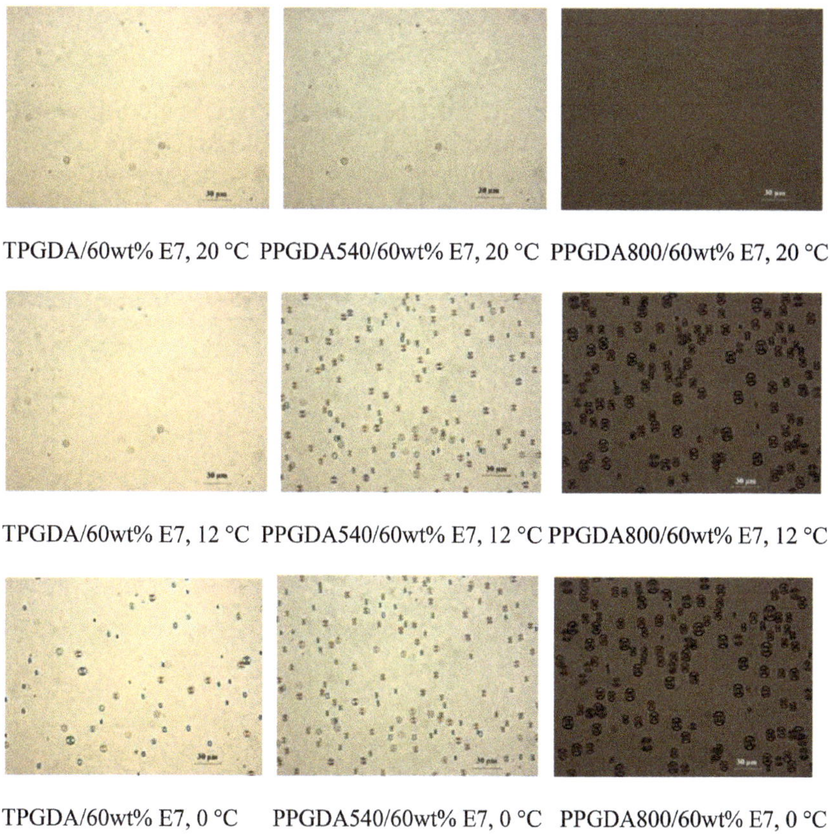

Figure 4.3 Morphologies of monomeric mixtures of TPGDA/60wt% E7, PPGDA540/60wt% E7, and PPGDA800/60wt% E7, observed by POM at temperatures 0, 12, and 20 °C.

4.3.2 Infrared Spectroscopy

It is well known that high acrylic double bond (monomeric) conversions should be reached to minimize undesired effects of unreacted or partially reacted monomer molecules. Polymerization/crosslinking reactions and phase separation kinetics of monomer/LC mixtures are likely to govern the morphologies of the obtained PDLC networks.

At least two absorption bands are available to measure the acrylic double bond conversions, one corresponding to the =C–H out-of-plane deformation at 810 cm^{-1}, and another C=C valence vibration about the bond axis at around 1638 cm^{-1}. However, since the aromatic groups of the LC E7 exhibit a strong absorption band near 810 cm^{-1}, originating from the vibration of two adjacent hydrogens in the phenyl groups,[28–30] the calculations of the acrylic double bond conversions were carried out by taking into account the difference of the peak heights of the absorption band at 1638 cm^{-1}. The conversion ratio C was calculated using the following equation:

$$C(\%) = \frac{(A_{1638})_{(D=0)} - (A_{1638})_{(D)}}{(A_{1638})_{(D=0)}} \tag{4.1}$$

where $(A_{1638})_{(D=0)}$ is the height of the peak at 1638 cm^{-1} before irradiation, and $(A_{1638})_{(D)}$ stands for the corresponding height for the mixture exposed to a dose D for a certain exposure time. Figures 4.4 and 4.5 represent the results of the polymerization/crosslinking kinetics experiments for each system

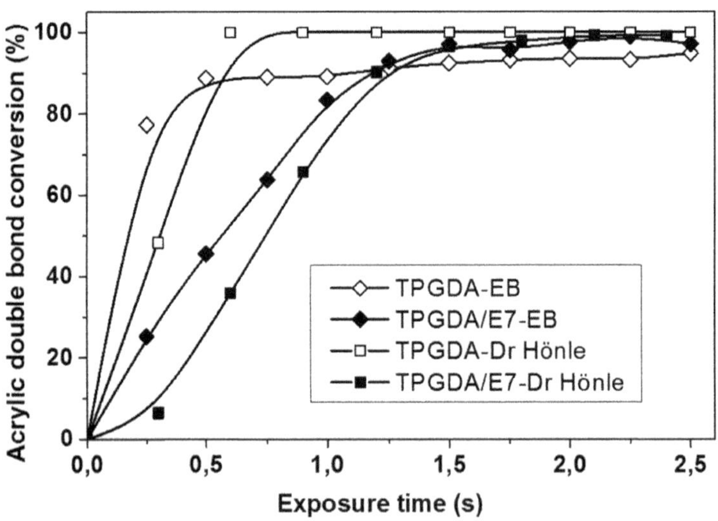

Figure 4.4 Acrylic double bond conversion *versus* exposure time: empty square symbols (TPGDA) and full square symbols (TPGDA/60wt% E7) for EB curing, empty rhombus for TPGDA and full rhombus for TPGDA/60wt% E7 for UV–Dr. Hönle curing.

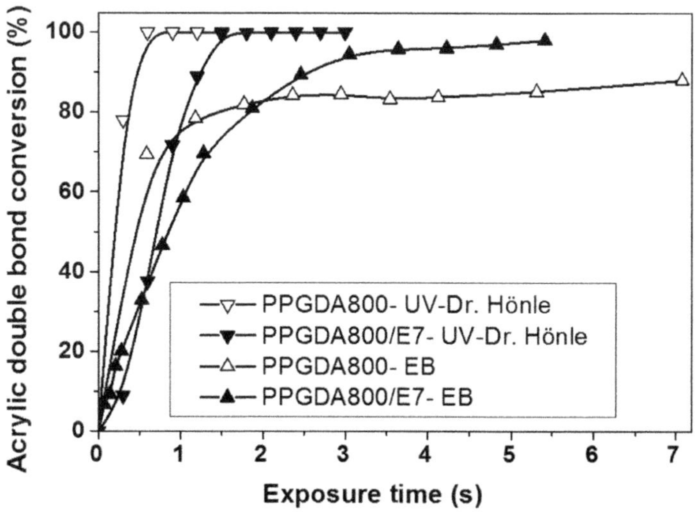

Figure 4.5 Acrylic double bond conversion *versus* exposure time: empty up triangular symbols (PPGDA800) and full up triangular symbols (PPGDA800/60wt% E7) for EB curing; empty down triangular symbols for PPGDA800 and full down triangular symbols for PPGDA800/60wt% E7 for UV-Dr. Hönle curing.

based on TPGDA/60wt% E7 and PPGDA800/60wt% E7 exposed to the UV-Dr. Hönle system and EB irradiation, respectively.

It was noticed that increasing irradiation doses (exposure times) lead to an increase in acrylic double bond conversion. The LC E7 acts as a solvent where the reaction is governed by the diffusion of the initiator radicals in this solvent towards the acrylic double bonds. When phase separation begins, E7 will confine in small domains and the reaction will then be enhanced towards higher conversion by pushing the reactive species into ultimate contact. In the later stages, an autodeceleration was noticed, and some reactions could reach 100% conversion, while still having a certain mobility of the reactive species.

For the UV-Dr. Hönle system, one observes very fast reactions, in such a way that no induction period was detected for the neat monomers. Similar trends were noticed for PPGDA800/60wt% E7 and TPGDA/60wt% E7, although these monomers have different initial acrylic double bond concentrations.

During the photopolymerization process, the reaction starts first around the photoinitiator radicals, forming tiny network spots which might be different from one spot to another depending on the solubility of the initiator in the monomeric mixture. As a consequence, this will induce some heterogeneity not only in the obtained polymer networks but also in the morphologies of the phase separated PDLCs. However, under EB exposure, the electrons directly hit the double bonds of the diacrylate monomers that are more or less homogenous in the mixture and as a result, homogenous networks and morphologies will be formed throughout the PDLC film thickness.

The limited acrylic double bond conversion of TPGDA cured by EB was mainly due to high crosslinking density of the resulting polymer network, whereas in the case of PPGDA800, this effect was attributed to higher initial viscosity of the monomer which restricted the mobility of the propagating polymer chains.

4.3.3 Morphologies

The samples for the morphology studies were prepared under the optimum conditions that were derived from the curves of acrylic double bond conversion *versus* exposure times as shown in Figures 4.4 and 4.5. For the PDLC samples prepared by the UV–Dr. Hönle system, the conveyor belt speed was fixed at 125 mm s^{-1} and an exposure time of 2.4 s was applied to reach a dose of about 240 mJ cm^{-2}. The EB samples were obtained by adjusting the intensity to 2 mA and the conveyor belt speed to 20.4 mm s^{-1} in order to get 2.54 s of exposure time and 105 kGy of irradiation dose.

The morphologies are illustrated in Figure 4.6 using POM at room temperature. When using a very rapid photopolymerization process (UV–Dr. Hönle system), smaller LC domains of PPGDA800/60wt% E7 were found,

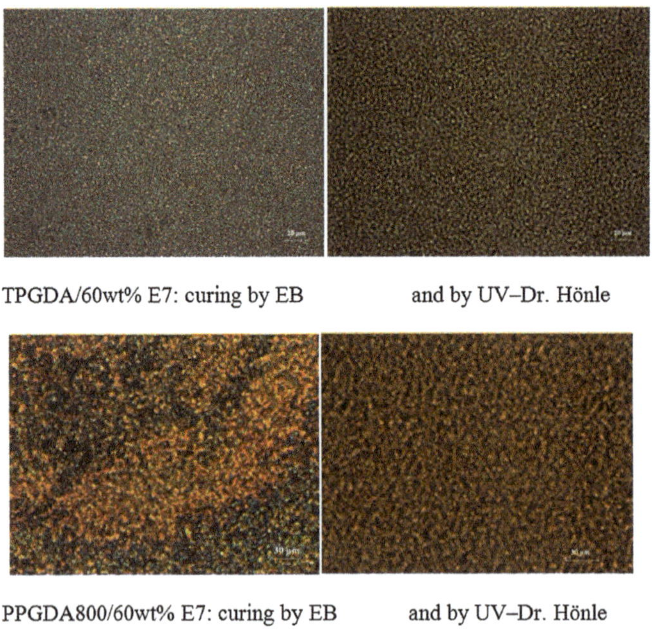

Figure 4.6 Morphologies of TPGDA/60wt% E7 and PPGDA800/60wt% E7 with EB and UV–Dr. Hönle curing.

compared to the same system under a slow photopolymerization process (Philips TL08 UV-light source); this might be attributed to a lack of sufficient time to form big domains due to the very rapid reaction time. In the case of EB samples, the lack of phase separation observed for the PPGDA800/60wt% E7 system was probably due to the slack network (high number of initiating chains and low initial double bond concentration).

For the EB-cured TPGDA/60wt% E7 system, small and regular phase separated LC domains were obtained, probably related to the fact that homogeneous polymerization took place under EB.

4.3.4 Electro-optical Responses

Figure 4.7 illustrates the electro-optical responses of 17 μm thick PDLC films of TPGDA/70wt% E7, PPGDA540/65wt% E7, and PPGDA800/60wt% E7, obtained by the UV–Dr. Hönle system, and EB processing. Unfortunately production of PPGDA540 was stopped while the study was in progress, so that the corresponding electro-optical properties of the UV–Dr. Hönle system could not be investigated.

The plateau of the ON-state transmission, T_{ON}, of the UV systems was reached at higher voltages than that of EB systems. The threshold V_{10} (10% transmission) and saturation voltages V_{90} (90% transmission) increased greatly for the UV systems compared to the EB systems; *i.e.* the UV-cured samples required higher voltages to be activated. This might be related to the different types of morphologies obtained by the two curing methods as shown in Figure 4.6; more regular morphologies were obtained in the case of EB.[31–33] The higher crosslinking density of TPGDA/70wt% E7 and lower spacing between the reactive double bonds yielded the smallest droplet sizes compared to PPGDA540/65wt% E7 and PPGDA800/60wt% E7 systems; for this reason, the OFF-state light transmission was reduced while the width of the hysteresis loop was narrowed. For the EB-cured systems, it was found that threshold V_{10} and saturation voltage V_{90} of TPGDA/70wt% E7 were much lower than those of the PPGDA540/65wt% E7 and PPGDA800/60wt% E7 systems; T_{ON} of 90% can be achieved at a voltage as low as 17 Volts and a very narrow hysteresis loop was obtained. In the case of the PPGDA800/60wt% E7 system, the allowed reaction conditions would not permit high enough conversion and the efficient phase separation required to obtain good electro-optical responses.

The results for the TPGDA/70wt% E7 blend confirmed the results obtained by Gyselinck *et al.*[34] and Benkhaled *et al.*:[35] EB-cured PDLC films have better electro-optical responses compared to UV-cured systems, even though a very fast UV curing method (UV–Dr. Hönle system) was employed. In the case of EB-cured PPGDA800/60wt% E7 films, inferior electro-optical responses were found compared to those containing TPGDA and PPGDA540; this might be attributed to the slack or weak network formed, which deformed easily upon application of an external electric field. It is thought that E7 bloomed out from the core to the surface of PPGDA800/60wt% E7 due to the large domain

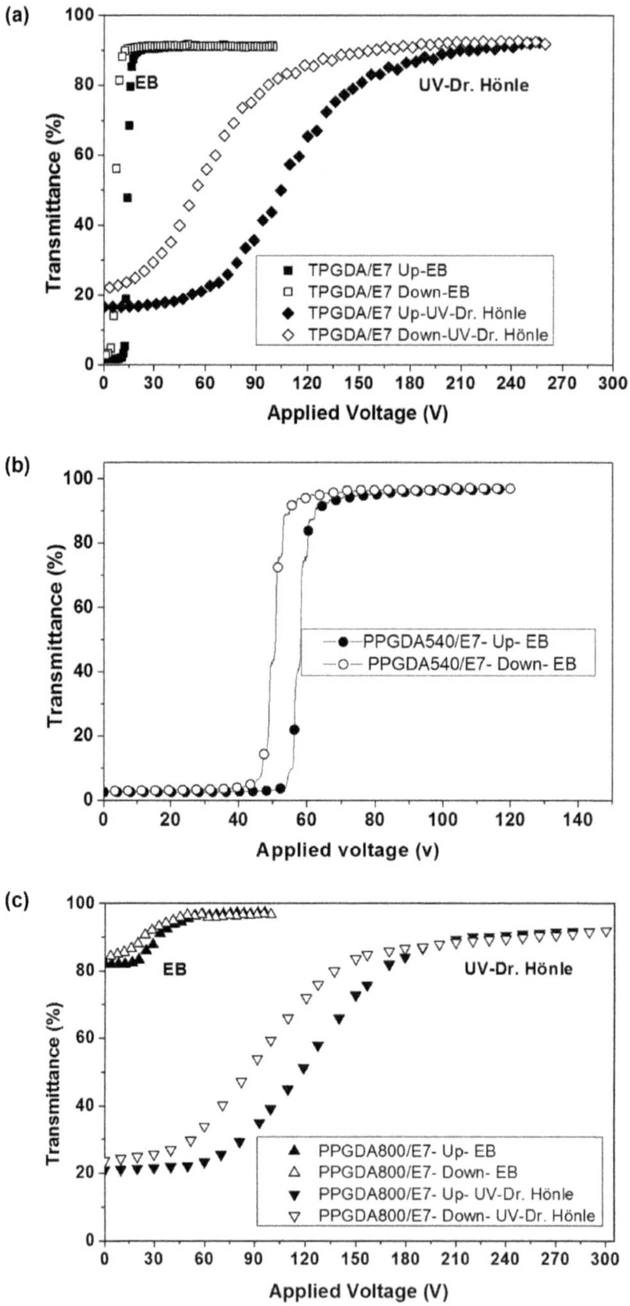

Figure 4.7 (a) Electro-optical responses of PDLC films made of TPGDA/70wt% E7, prepared by EB and UV–Dr. Hönle curing. (b) Electro-optical response of an EB-cured PPGDA540/65wt% E7 film. (c) Electro-optical responses of PDLC films composed of PPGDA800/60wt% E7, elaborated by EB and UV–Dr. Hönle curing.

sizes. This migration phenomenon is not wanted because it can lead to poor dimensional stability of the obtained PDLC films.

4.4 Conclusions

In this investigation, a comparison was made between two methods of elaboration: the UV–Dr. Hönle system (rapid photopolymerization), and EB (rapid curing) having comparable exposure times for TPGDA/E7, PPGDA540/E7, and PPGDA800/E7 systems. This comparison was performed in terms of some physical properties such as phase diagrams, kinetics of polymerization/crosslinking and phase separation, morphologies and electro-optical responses.

The phase diagrams of the monomeric systems indicated that 60wt% E7 was the limit of solubility for the mixture PPGDA800/E7. This concentration was deliberately chosen for TPGDA/E7 and PPGDA540/E7 to work with the same E7 content. In the case of the UV–Dr. Hönle system, all monomers followed the same trends of rapid photopolymerization. PPGDA800/60wt% E7 showed faster photopolymerization than TPGDA/60wt% E7 due to the higher mobility of the reacting species. The rapid photopolymerization always leads to very small LC droplet sizes.

In the EB-cured system, the shorter the spacing between the two acrylic double bonds in the monomer (the lower the viscosity), the higher the number of reactive sites which are in ultimate contact with each other and the higher the rate of polymerization/crosslinking reactions. More regular morphologies for TPGDA/60wt% E7 were mainly related to the more homogeneous reaction under EB leading to regular morphologies. The very slack or weak network formed under EB was the main reason for the clear phase separation seen in PPGDA800/60wt% E7.

The electro-optical responses of the systems TPGDA/70wt% E7 and PPGDA800/60wt% E7, elaborated by the UV–Dr. Hönle system, did not lead to the steep responses due to their more heterogeneous morphologies. The accelerated EB would be a promising technique to prepare PDLC films based on TPGDA/70wt% E7 and PPGDA540/65wt% E7. The steeper and better electro-optical responses revealed in the latter systems were the result of more regular morphologies obtained using EB curing.

References

1. J. P. F. Lagerwall and G. Scalia, A new era for liquid crystal research: Applications of liquid crystals in soft matter nano-, bio- and microtechnology, *Curr. Appl. Phys.*, 2012, **12**, 1387–1412.
2. M. Mucha, Polymer as an important component of blends and composites with liquid crystals, *Prog. Polym. Sci.*, 2003, **28**, 837–873.
3. Y. J. Jeon, Y. Bingzhu, T. June, D. L. Cheung and M. Jamil, Application and new developments in Polymer-Dispersed Liquid Crystal simulation studies, *Macromol. Theory Simul.*, 2007, **16**, 643–659.

4. E. Andrzejewska, Photopolymerization techniques of multifunctional monomers, *Prog. Polym. Sci.*, 2001, **26**, 605–665.
5. C. Decker, Photoinitiated crosslinking polymerisation, *Prog. Polym. Sci.*, 1996, **21**, 593–650.
6. C. Decker and B. Elzaouk, Photopolymérisation de monomères multifonctionnels—VII. Evaluation des constantes de vitesse de propagation et de terminaison, *Eur. Polym. J.*, 1995, **31**, 1155–1163.
7. T. Scherzer and U. Decker, The effect of temperature on the kinetics of diacrylate photopolymerizations studied by Real-time FTIR spectroscopy, *Polymer*, 2000, **41**, 7681–7690.
8. K. S. Anseth, C. M. Wang and C. M. Bowman, Kinetic Evidence of Reaction-Diffusion during the Polymerization of Multi(Meth)Acrylate Monomers, *Macromolecules*, 1994, **27**, 650–655.
9. B. Defoort, D. Defoort and X. Coqueret, Electron-beam initiated polymerization of acrylate compositions, 2. Simulation of thermal effects in thin films, *Macromol. Theory Simul.*, 2000, **9**, 725–734.
10. C. Patacz, B. Defoort and X. Coqueret, Electron-beam initiated polymerization of acrylate compositions 1: FTIR monitoring of incremental irradiation, *Radiat. Phys. Chem.*, 2000, **59**, 329–337.
11. C. Patacz, X. Coqueret and C. Decker, Electron-beam initiated polymerization of acrylate compositions 3: compared reactivity of hexanediol and tripropyleneglycol diacrylates under UV or EB initiation, *Radiat. Phys. Chem.*, 2001, **62**, 403–410.
12. U. Maschke, X. Coqueret and C. Loucheux, Electron beam processing for polymerization induced phase separation: preparation of polymer dispersed liquid crystal films, *Nucl. Instrum. Methods Phys. Res.*, 1995, **B105**, 262–266.
13. W. Knolle and R. Mehnert, Primary reactions in the electron-induced polymerization of acrylates, *Nucl. Instrum. Methods Phys. Res.*, 1995, **B105**, 154–158.
14. U. Maschke, N. Gogibus, A. Traisnel and X. Coqueret, Electron beam cured liquid crystal-polymer composite materials: electro-optical enhancement effect, *Liq. Cryst.*, 1997, **23**, 457–461.
15. U. Maschke, A. Traisnel, J. D. Turgis and X. Coqueret, Influence of liquid crystal concentration on the electro-optical behavior of polymer dispersed liquid crystal films prepared by electron beam processing, *Mol. Cryst. Liq. Cryst.*, 1997, **299**, 371–378.
16. F. Z. Abdoune, L. Benkhaled, L. Méchernène and U. Maschke, Investigation of the electro-optical behavior of UV-cured polymer/liquid crystal systems, *Phys. Procedia*, 2009, **2**, 643–648.
17. M. Kashima, H. Cao, Q. Meng, H. Liu, D. Wang, F. Li and H. Yang, The influence of crosslinking agents on the morphology and electro-optical performances of PDLC films, *J. Appl. Polym. Sci.*, 2010, **117**, 3434–3440.
18. Z. Yang, D. Li, C. Pan and Y. Wang, Effects of hyperbranched prepolymers prepared from butyl acrylate and butyl methacrylate on the

electro-optical properties of polymer dispersed liquid crystal, *Polym. Adv. Technol.*, 2012, **23**, 1321–1327.
19. J. He, B. Yan, B. Yu, S. Wang, Y. Zeng and Y. Wang, The effect of molecular weight of polymer matrix on properties of polymer-dispersed liquid crystals, *Eur. Polym. J.*, 2007, **43**, 2745–2749.
20. J. H. Ryu, S. G. Lee, J. B. Nam and K. D. Suh, Influence of SMA content on the electro-optical properties of polymer-dispersed liquid crystal prepared by monodisperse poly(MMA-co-SMA)/LC microcapsules, *Eur. Polym. J.*, 2007, **43**, 2127–2134.
21. M. M. Dzhons, S. A. Bulgakova, I. A. Pantyukhina and I. A. Kazantzeva, Effects of chemical structure and composition of the polymer matrix on the morphology and electro-optical performance of polymer-dispersed liquid crystal films, *Liq. Cryst.*, 2011, **38**, 1263–1268.
22. P. Malik and K. K. Raina, Droplet orientation and optical properties of polymer dispersed liquid crystal composite films, *Opt. Mater.*, 2004, **27**, 613–617.
23. Merck Liquid Crystals, Licrilite Brochure. 1994.
24. J. Li, C. H. Wen, S. Gauza, R. Lu and S. T. Wu, Refractive Indices of Liquid Crystals for Display Applications, *IEEE/OSA J. Disp. Technol.*, 2005, **1**, 51–61.
25. H. A. Tarry, *The Refractive Indices of Cyanobiphenyl Liquid Crystals*, Merck Ltd, Merck House, Poole, Great Britain, 1967.
26. L. Méchernène, L. Benkhaled, D. Benaissa and U. Maschke, Studies of optical transmission properties of electron beam cured polymer/liquid crystal systems, *Opt. Mater.*, 2009, **31**, 632–639.
27. K. Amundson, A. van Blaaderen and P. Wiltzius, Morphology and electro-optic properties of polymer-dispersed liquid-crystal films, *Phys. Rev. E*, 1997, **55**, 1646–1654.
28. G. W. Bentley and J. L. Koenig, Studying the Curing Kinetics of a Diacrylate by Using Infrared Spectroscopy, *Appl. Spectrosc.*, 1997, **51**, 1453–1459.
29. Y. Derouiche, F. Dubois, R. Douali, C. Legrand and U. Maschke, Some Properties of Nematic Liquid Crystal E7/Acrylic Polymer Networks, *Mol. Cryst. Liq. Cryst.*, 2011, **541**, 439–448.
30. Y. Derouiche, S. A. K. Ghefir, F. Dubois, R. Douali, C. Legrand, M. Azzaz and U. Maschke, Preparation and Characterization of selected Polypropyleneglycoldiacrylate/nematic Liquid Crystal Materials, *IREPHY*, 2010, **4**, 188–193.
31. A. M. Lackner, J. D. Margerum, E. Ramos and K. C. Lim, Droplet size control in polymer dispersed liquid crystal films, *SPIE*, 1989, **1080**, 53–61.
32. A. M. Lackner, J. D. Margerum, L. J. Miller, F. G. Yamagishi, E. Ramos, K. C. Lim, W. H. Smith Jr. and C. I. van Ast, Methods to obtain lower-voltage-activated polymer-dispersed LCDs, *Proc. SID.*, 1991, **32**, 173–176.
33. S. A. Carter, J. D. LeGrange, W. White, J. Boo and P. Wiltzius, Dependence of the morphology of polymer dispersed liquid crystals on the UV polymerization process, *J. Appl. Phys.*, 1997, **81**, 5992–5999.

34. F. Gyselinck, U. Maschke, A. Traisnel and X. Coqueret, PDLC films prepared by electron beam and ultraviolet curing: influence of curing conditions on the electro-optical properties, *Liq. Cryst.*, 2000, **27**, 421–428.
35. L. Benkhaled, L. Méchernène, A. Traisnel, M. Benmouna, J. M. Gloaguen, X. Coqueret and U. Maschke, Electro-optical properties of electron-beam cured polymer dispersed liquid crystal systems, *Mol. Cryst. Liq. Cryst.*, 2002, **375**, 651–658.

CHAPTER 5

Polymer Dispersed Liquid Crystals

MARIACRISTINA RUMI,[a,b] TIMOTHY J. BUNNING*[a] AND LUCIANO DE SIO[c,d]

[a] Materials and Manufacturing Directorate, Air Force Research Laboratory, Wright Patterson Air Force Base, OH 45433, USA; [b] Azimuth Corporation, 4027 Colonel Glenn Highway, Suite 230, Beavercreek, OH 43531, USA; [c] Department of Medico-Surgical Sciences and Biotechnologies, Sapienza University of Rome, Corso della Repubblica 79, 04100, Latina, Italy; [d] CNR-Lab. Licryl, Institute NANOTEC, 87036 Arcavacata di Rende, Italy
*Email: timothy.bunning@us.af.mil

5.1 Introduction

Low molar mass liquid crystals (LCs) are typically not soluble in polymer systems to any great degree. When the two different materials are mixed, a two-phase system is obtained whose morphology depends on a variety of factors including, primarily, the concentration. The resulting two-phase structures can have inclusions with nanometer through macroscopic dimensions. Although there are a large number of variants, these structures are generically called 'polymer dispersed liquid crystals' (PDLCs) when the resulting morphologies lead to light-scattering systems. In truth, sometimes the structures are solid polymer fibrils or balls dispersed in a continuous liquid crystal fluid and sometimes they are a liquid crystal fluid dispersed in discrete domains surrounded by continuous solid polymer. For this review, all these systems will be referred to as PDLC structures. Although lots of work has been done across the concentration region from 0–100%, for

Soft Matter Series No. 8
Polymer-modified Liquid Crystals
Edited by Ingo Dierking

Published by the Royal Society of Chemistry, www.rsc.org

systems where there is 30–70% LC, morphologies with large mesoscale inclusions are typically formed, which leads to substantial light scattering. This is in general where the term PDLC is most often used. Although technically when the amount of polymer is small (10%), the liquid crystal (90%) is still dispersed throughout, the community has termed these systems 'polymer stabilized liquid crystals' (PSLCs). The delineation between PSLCs and PDLCs is arbitrary and again, in general, PDLCs historically refer to systems where there is appreciable scatter. If the refractive index matching is done correctly, upon application of an electric field, the scattering can be turned off, leading to dynamic transparency. This field-controlled scattering is the root application that caused the field to explode some 20+ years ago. This is a review of past literature with a focus on the type of morphologies that can be exhibited. Basic electro-optic properties are discussed as are the large variety of morphologies that can be induced. The first part (Section 5.2) describes standard structure-property relationships of 'isotropic' PDLC systems, discussing nuances about their formation, morphology, optical, and electro-optical properties. The section ends with some articulation of related areas including nanosized PDLCs, nanoparticle containing PDLC systems, and dye-doped PDLC systems. The second part (Section 5.3) describes 'periodic' PDLC systems, wherein the phase separation process is induced spatially. This leads to anisotropic systems where an electric field can control diffraction, instead of scattering. These systems have attracted recent attention due to the large variety of dynamic diffractive and photonic effects that can be observed.

5.2 Non-patterned Polymer Dispersed Liquid Crystals

PDLC systems are two-phase composites of a small-molecule liquid crystal and a polymeric binder. Depending on the relative content of the two components and their miscibility, the liquid crystal component can exist as isolated droplets embedded in a continuous network of the polymer phase, the polymer can form ball-like structures either isolated or coalesced into a network surrounded by a continuous liquid crystal phase, or an interpenetrating polymer network can be present with the liquid crystal filling the empty spaces in the network. The term PDLC is most often used for composites in which the polymer or polymer precursor possesses no molecular order, but mesogenic monomer systems have also been used. PDLCs with non-mesogenic polymer components will be the main focus of this chapter. This section will discuss PDLC systems wherein no periodic or regular arrangement is imposed on the droplets or polymer network during the fabrication process.

5.2.1 Fabrication Methods and Working Principles

A large number of studies have been devoted to the characterization of the electro-optic (EO) properties and morphologies of PDLC devices and to establishing correlations with the system composition and preparation conditions. Several comprehensive reviews on this topic already exist.[1–5] We will discuss here only the devices' general characteristics and the origin of their EO behavior. A PDLC device with the droplet morphology is shown schematically in Figure 5.1a. These devices are highly scattering if the dimensions of the droplets are on the order of the wavelength of light. This is a result of the fact that the alignment of the liquid crystal molecules is not uniform across the droplets and therefore there is a mismatch between the average refractive index of the polymer matrix and the liquid crystal droplets. The droplets typically assume a spherical or ellipsoidal shape driven by liquid-liquid phase separation and surface tension during the growth process. The liquid crystal droplets are confined by the polymer matrix and the curved polymer surface acts locally as an alignment layer.

The alignment assumed by nematic liquid crystals in PDLC droplets is often tangential to the droplet walls and the droplets commonly exhibit a bipolar configuration, in which the liquid crystal molecules align along the droplet walls and form two diametrically opposed point defects.[6] For molecules located near the symmetry axis of the droplet (the line connecting the two point defects), the LC director is aligned along the axis. Elsewhere in the droplets, the LC molecules arrange themselves in such a way as to minimize the elastic free energy of the system due to splay and bend deformations within the confined volume. Because the droplets are isolated from one another, there is no correlation between their symmetry axes. For spherical droplets the symmetry axes are oriented randomly in the PDLC device.

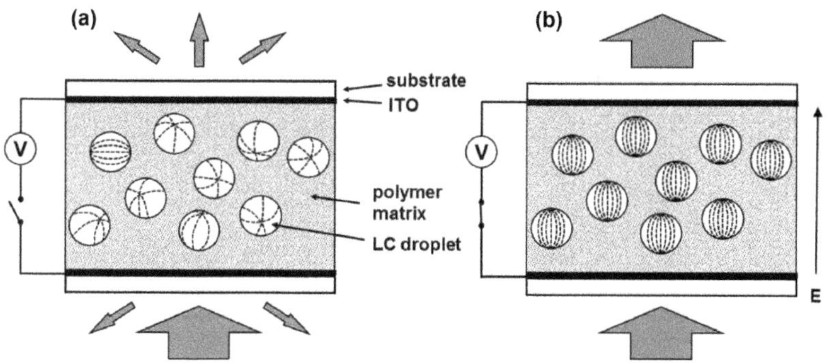

Figure 5.1 Schematic of a PDLC with droplet morphology and containing a positive dielectric nematic LC. (a) Device in the *off* state (no electric field); the LC molecules are aligned in each droplet, but the symmetry axes of the bipolar droplets are randomly oriented. The film is strongly scattering (small grey arrows in the forward and backward directions) for light incident along the film normal (large grey arrow at the bottom of the diagram). (b) Device in the *on* state (electric field directed perpendicular to the film surface); the alignment is the same in every droplet and is along the field direction. The film is transparent for light incident along the normal direction.

For ellipsoidal droplets, the point defects are located in the regions with largest curvature, but the axes can assume any orientation in the two remaining directions. Figure 5.2 shows a micrograph of bipolar droplets of this type in a polymer matrix. It can be seen that the droplet axes can assume any orientation in the plane of the image. As a result of the random orientation of the droplets, light propagating through a PDLC device experiences large variations of refractive index over short distances, resulting in efficient scattering for all incident directions and the device appears opaque (Figure 5.1a).

When an electric field of sufficient strength is applied across the device, the LC molecules in the droplets align along the field direction if the LC has a positive dielectric anisotropy ($\Delta\varepsilon > 0$). In the case of strong anchoring, the overall configuration within the droplet does not change significantly, but the point defects migrate so that the droplet symmetry axis coincides with the field direction (Figure 5.1b); this minimizes the elastic energy in the presence of the field (the situation where the point defects remain in the same position as in the field-off state and the LC molecules align along the electric field in the center of the droplet has higher energy). Thus, in the *on*-state, all droplets have the same symmetry axis and light propagating along the field direction experiences the same effective refractive index for each droplet. If the polymer matrix is chosen to have refractive index, n_p, close to the ordinary refractive index of the liquid crystal, n_o, the scattering is minimized and the device becomes transparent. If the anchoring is weak, reorientation of the LC towards the electric field direction can also involve the molecules adjacent to the polymer walls, resulting in a more uniform vertical alignment within each droplet.

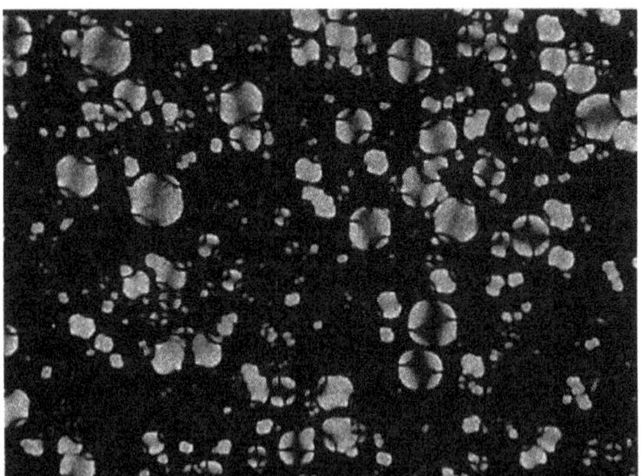

Figure 5.2 Micrograph of bipolar LC droplets in a polyvinyl alcohol film, under crossed polarizers. The diameter of the largest droplets in this view is in the order of a few microns.
Reproduced from *Liquid Crystal Dispersions*, P. S. Drzaic, Copyright © 1995 World Scientific Publishing.[2]

PDLC devices can be prepared by various methods including the emulsion method, the phase separation method, and the infiltration of an LC into porous structures.[2] Historically, the first working PDLCs were prepared by the emulsion method, in which a liquid crystal is mixed in an aqueous solution or dispersion of a polymer.[7–9] The dispersion is then stirred vigorously to form an emulsion and used to coat a conductive substrate. When the system is allowed to dry, the liquid crystal droplets of the emulsion remain trapped in the solid polymer film. A second conductive substrate is then laminated on top of the film. The substrate can be rigid or flexible. Examples of polymers used with this method are polyvinyl alcohol (which is water soluble) and latex (which forms a colloidal suspension in water). This type of composite has been called 'nematic curvilinear aligned phase' (NCAP).[2]

A variety of phase separation methods can be utilized to form two-phase composites as well. The phase separation can be triggered by the evaporation of a solvent in which the LC and polymer are dispersed (*s*olvent-*i*nduced *p*hase *s*eparation, SIPS), by changes in temperature for mixtures of a LC and a thermoplastic polymer (*t*hermally-*i*nduced *p*hase *s*eparation, TIPS), or by initiating the polymerization of a polymer precursor (monomer or oligomer) in which the LC is dissolved (*p*olymerization-*i*nduced *p*hase *s*eparation, PIPS). In all three cases, as time progresses, an LC-rich phase and a polymer-rich phase are formed due to the growing incompatibility between the pure LC and the polymer.[2,10] The classic morphology typically expressed depends on the approaches used to initiate the phase separation and the properties of the starting materials. Simplistically, it can be one phase coalescing into isolated domains surrounded by a continuum of the other phase (Figure 5.3a; image obtained by scanning electron microscopy, SEM), or the two phases forming a bicontinuous interconnected network (Figure 5.3b). The PIPS process is arguably the approach most often used in the literature as it allows the greatest control of the morphology and thus the

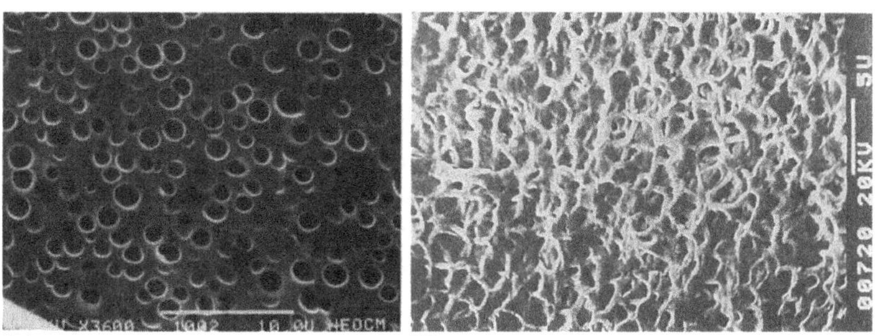

(a) (b)

Figure 5.3 Example of morphologies of PDLC films (SEM images of samples after removal of the LC component). (a) Droplet morphology. Reprinted with permission from J. L. West, *ACS Symposium Series*, 1995, **435**, 475. Copyright 1995 American Chemical Society. (b) Film with interpenetrating polymer and LC phases. Reproduced from *Liquid Crystal Dispersions*, P. S. Drzaic, Copyright © 1995 World Scientific Publishing.[2]

characteristics of the final film by an appropriate selection of monomer and polymerization conditions. The polymerization can be initiated thermally or by irradiation.

In the TIPS and PIPS approaches, the initial one-phase solution is typically used to fill the space between two conductive substrates held at a desired distance by spacers, after which the temperature is reduced or the polymerization initiated to obtain the final two-phase product. In systems which possess the isolated droplet morphology, the material in the droplets is close to pure LC, as evidenced by the fact that the clearing temperature is similar to that in the bulk. However, the LC component does not phase separate completely from the polymeric phase and a fraction of it typically remains trapped in the polymer matrix.[2] This can account for a final phase-separated volume fraction of droplets smaller than the LC content in the starting composition. The droplet volume fraction for any system is governed by the fundamental thermodynamic compatibility between the LC and the final polymer material and the amount of kinetic trapping that occurs driven by the polymerization.

The PIPS method affords large flexibility in polymer precursor materials and polymerization conditions, allowing good control of the morphology and properties of the resulting PDLC films. The polymerization is usually irreversible and this can provide stability to the device. Various materials and polymerization chemistries have been employed over the years and they have been discussed in several reviews.[1,2,5] We only briefly mention the most commonly used materials and methods. PDLCs have been prepared extensively by photo-induced PIPS using thiol-ene-based photochemistry. Typical polymer precursors are allyl-functionalized monomers and thiol-based cross-linking compounds, mixed with an appropriate photoinitiator. The commercial optical adhesive NOA-65 (Norland) is one material system of this class, photocurable by exposure to UV light, which has been widely used to prepare PDLCs. Photopolymerization of acrylate and methacrylate monomers has also been used to fabricate PDLCs by PIPS, often based on combinations of monofunctional and multifunctional monomers in order to achieve a desired degree of crosslinking and control of the dynamics of the phase separation process. The photopolymerization of acrylate monomers proceeds *via* a free radical chain mechanism in the presence of radical initiators. Examples of acrylate monomers used include 2-ethylhexyl acrylate, butanediol diacrylate, hexanediol diacrylate and dipentaerythrol hydroxy pentaacrylate. Specific prepolymer mixtures have been developed commercially for use with certain classes of nematic LCs including PN393, SAM114, and MXM035 (Merck). PIPS has also been demonstrated in epoxy-based systems (usually with thermal initiation), where an epoxy resin (such as Epon 828) is mixed with a polythiol hardener (for example Capcure 3800) and a soluble LC.

The characteristic behavior of PDLCs derives from the fact that the polymer and liquid crystal components phase separate at some point during the preparation of the device. The details of how this process starts and proceeds determine the final morphology and domain sizes of the devices. In the simple case of a binary system consisting of a nematic liquid crystal and an isotropic polymer, the TIPS process can be described using the simplified phase diagram in Figure 5.4.[2,4,11]

At low fractions of LC in the mixture, ϕ, the LC is soluble in the polymer at all temperatures and the system is in a single homogeneous phase. At high ϕ, the polymer is soluble in the LC and the system is in a single phase, either ordered (nematic) at low temperatures or disordered (isotropic) at high temperature (a small nematic-isotropic two-phase region is often present between the two). At intermediate ϕ, the LC is only soluble at high temperatures. When the temperature is decreased from the initial temperature, T_{start} (dotted arrow in the figure), the solubility of the LC in the polymer decreases. The system remains in a single phase until the temperature reaches the upper critical solution temperature or binodal curve (solid line in the figure). Under the binodal curve, a single phase is no longer stable and the system becomes metastable. Here, phase separation can occur in the presence of heterogeneities, as the process requires overcoming an activation barrier. Thus, in the presence of nucleation sites, droplets of the less soluble component can form and then grow over time (nucleation and growth process). If the temperature is further reduced below the spinodal curve (dashed line in the figure), the system becomes unstable and phase separation proceeds spontaneously without activation, resulting in domains of the two phases of irregular shapes and interpenetrating one another

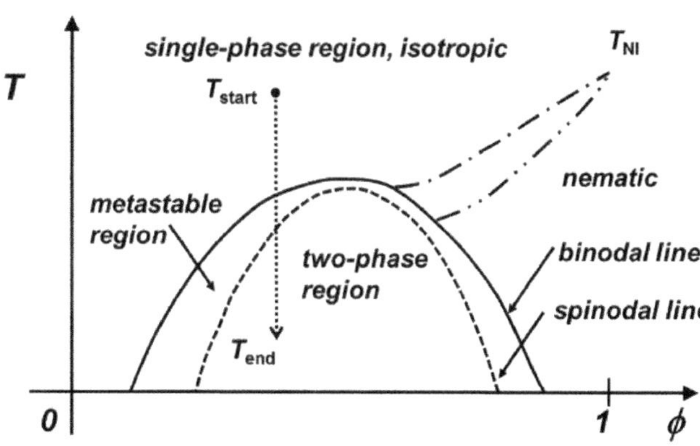

Figure 5.4 Phase diagram for a binary mixture of a nematic liquid crystal and an isotropic monomer. The abscissa is the fraction of LC in the mixture, ϕ, the ordinate the temperature, T. The solid line is the binodal curve and the dashed line is the spinodal curve. The area under the binodal and spinodal curves is biphasic with different morphology (see text). Above the binodal curve, the system is in a single phase and isotropic above the dash-dot line, or single-phase and nematic below the dash-dot-dot line (a mix of nematic and isotropic phase is present between these two lines). T_{NI} is the temperature for the nematic-isotropic phase transition at $\phi = 1$.

(spinodal decomposition). These domains can grow in size and become more sharply defined (larger concentration gradient at the interface) as the temperature is further reduced or as time elapses. In the binodal and spinodal biphasic regions, the domains of each type typically do not consist of pure LC and polymer components, but some partial mixture of the two, as dictated by the tie lines across the phase diagram. The LC-rich phase will be nematic if the temperature is below the clearing temperature of the LC in the bulk (due to the presence of dissolved residual monomer in the LC-rich phase, the actual nematic-isotropic transition temperature is usually depressed relative to the pure LC case). The polymer-rich phase is instead isotropic. The details of the morphology of the phase-separated system depends on where the final temperature, T_{end}, is located relative to the binodal and spinodal lines and the rate at which the temperature was changed from T_{start} to T_{end}. A morphology characterized by isolated droplets (such as those in Figure 5.3a) will generally be obtained if the system underwent binodal, but not spinodal, decomposition. An interpenetrating network can be obtained if the temperature change is rapid so that the system spends limited time in the nucleation and growth phase before the onset of spinodal decomposition (Figure 5.3b).

In the case of PIPS, although the outcomes can be the same, the phase separation process is much more complicated as one is essentially dealing with a ternary phase diagram (monomer, LC, growing polymer). Initially the LC is miscible in the monomer. After the polymerization is initiated and as the monomer is being consumed, the solubility of the LC molecules decreases until the equivalent of the binodal curve is reached (the binodal line moves to higher temperature with conversion). At this point, nucleation and growth of LC-rich domains is observed. Depending on the rate of polymerization and the initial LC content, the LC-rich domains can grow and remain isolated or merge with adjacent ones. If the spinodal regime is reached, spontaneous phase separation throughout the system occurs, typically leading to an interconnected network of LC-rich and polymer-rich phases. The LC domains, either isolated or interconnected, grow until the system reaches the gel point and the viscosity becomes too large to allow further migration of molecules. It is also possible that polymer chains become insoluble in the LC/prepolymer mixture at some point during their growth and phase separate from the initial mixture. Although typically not discussed much in the conventional literature, this is what occurs most of the time for free-radical polymerizations and the ‘assumption’ that LC always phase separates from the initial mixture is false. Polymer particles that have crashed out of solution can aggregate and merge into structures of irregular shape and large interconnectivity, often referred to as the ‘polymer ball’ morphology. In PIPS, the phase separation process is typically irreversible, as it is driven by the polymerization. However, due to incomplete conversion or side reactions caused by remaining active species, some variation of PDLC device properties over time can be observed.

In PIPS, the morphology exhibited by PDLC devices is related to the type of polymerization process through which the polymer was formed[12] and some details of how this emerges are discussed here. In systems with polymerization by a step mechanism, the monomer is consumed quickly, yet the system at early times contains mainly low-molecular weight oligomers, which are soluble in the initial mixture. As time progresses, the LC component is typically the one that phase separates from the bulk phase as its solubility decreases in the growing oligomer mixture. A droplet morphology, revealed *via* several techniques in Figure 5.5, is usually obtained under these conditions and the droplet shapes are nearly spherical, driven mostly by surface tension effects from the two immiscible fluids. In Figure 5.5a, TEM images clearly show spherical-like domains of LC (dark phase) separated from each other and dispersed in a polymer matrix (light phase). Depending on the nature of the phase separation conditions and specific materials used, a variety of droplet sizes, dispersity, and shapes can be formed, even with isotropic polymerization. A corresponding 'classic' SEM micrograph is shown in Figure 5.5b where in this case a large density of droplets has been generated in the polymer matrix. In classic SEM techniques, the sample is freeze-fractured, the LC removed, and the corresponding remaining surface is 'painted' with a thin layer of conductive coating. In Figure 5.5c, cryo-SEM was used to image a fracture surface where the LC was not removed. The white domains in the holes are frozen LC. This type of imaging has never been published before. The sample was fractured under vacuum at cryo temperatures in the microscope and then immediately imaged.

In systems where polymerization proceeds by a chain (free-radical) mechanism instead, high-molecular weight polymer chains appear early on in the process. These high molecular weight particles typically phase

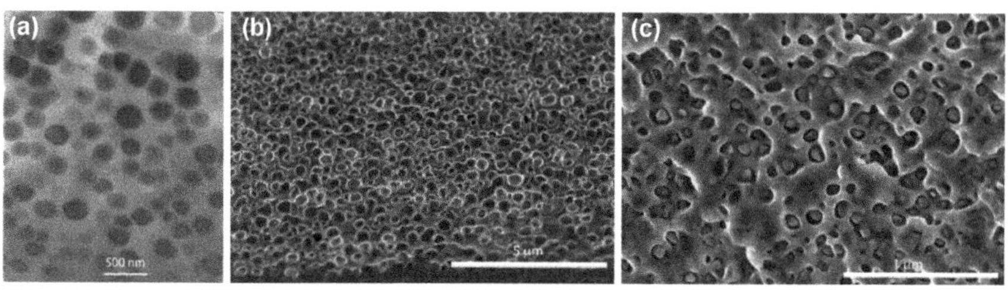

Figure 5.5 Morphologies of isotropic PDLCs from typical step-growth polymerization chemistries with approximately 50/50 concentration of monomer/LC in the starting reactive mixture. (a) TEM image showing small LC domains (dark) dispersed in a continuous polymer matrix. (b) SEM of a system where the dispersed LC domains are considerably larger and have a much larger overall density. Differences between morphologies in (a) and (b) are dictated by the nuances of how the reaction was conducted. (c) Cryo-SEM image, where the fracture and imaging was done in a cryo-stage. The bulk of the small domains possess frozen LC globules consisting of the pure LC below its freezing point. In all three images, considerable non-spherical shaping of the LC domains occurs.

separate out of the original fluid mixture early in the reaction. The monomer is consumed slowly as it adds to the growing chains. The monomer and LC fluid co-exist as a fluid until late in the polymerization process. A polymer-ball morphology is typically obtained in this case and a connected network of irregular polymer particles usually develops, due to reactive species on the phase-separated balls. Some detailed imaging of such morphology is shown in Figure 5.6. Figure 5.6a is the TEM image of a microtomed sample and clearly shows an irregular and non-spherical phase-separated structure. The light areas in this case represent the regions where the LC resides, dispersed between the many small polymer inclusions. This is consistent with the SEM morphology in Figure 5.6b, which clearly shows a variety of polymer nodules interconnected with irregularly shaped spaces in between, where the LC resides. In this case, the polymer phase separates from the remaining reactive liquid, which eventually becomes rich in LC and changes from an isotropic liquid to a nematic liquid crystal phase. The simple electro-optic models that are present in the literature are inadequate to accurately predict structure/property relationships in such complex systems.

Differences in morphology and feature sizes can be obtained by varying the relative content of LC and prepolymer in the starting mixture, or the polymerization conditions. For example, from a prepolymer containing a penta-acrylate monomer, a photoinitiator, *N*-vinyl-2-pyrrolidone (acting as a homogenizer), and E7, a range of morphologies was observed depending on the content of E7 in the mixture.[13] For less than 20% E7, no discrete LC domains were formed, due to the rapid generation of a crosslinked polymer network that impeded the growth of droplets and froze the density fluctuations brought about by spinodal decomposition. For 20–35% E7, small LC droplets were present in the polymerized films and scattering data were

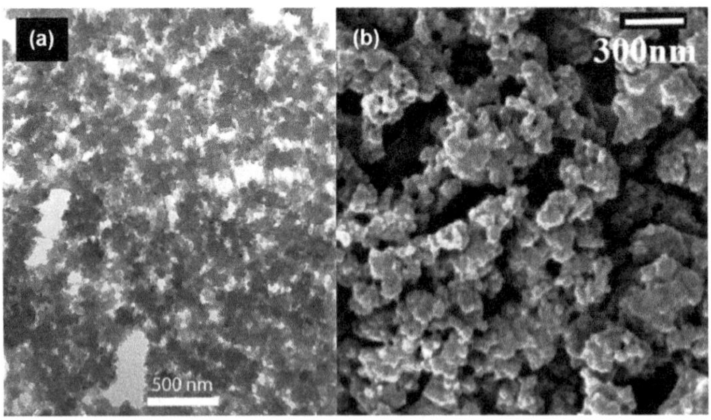

Figure 5.6 Typical morphologies of PDLC systems from free-radical polymerization. (a) TEM image showing the irregular polymer nodules (dark), which are surrounded by irregularly shaped LC domains. (b) SEM showing the polymer 'nodules' which phase separate out of the growing reactive mixture.

consistent with a hierarchical morphology with a two-phase structure at short length scales and aggregation at longer range. For E7 content above 35%, the films exhibited bicontinuous morphologies and the polymer component was characterized by progressively larger and smoother features with increasing LC content. These features were the result of the lower crosslinking density in samples with lower monomer content, which delayed the onset of gelation, thus providing more time for the LC component to phase separate and for the polymer beads to grow and aggregate.

Characterization techniques in addition to SEM are very useful for the analysis and interpretation of PDLC morphologies, because the removal of the LC component from the polymerized film can actually affect the morphology. For example, it has been shown using small-angle X-ray scattering (SAXS) that the polymer/LC interfaces appear to be rougher for a filled PDLC relative to one after leaching the LC fraction, which is the case usually accessible *via* SEM.[14] SAXS and light scattering are non-invasive techniques that can be used *in situ* on pristine samples and can provide detailed information on complementary length scales about the two-phase morphology of PDLCs and, if used in real time, on the kinetics of the phase separation process.[14,15]

The droplet morphology is often observed following polymerization of epoxy resins and thiol-ene systems, which proceeds by a step mechanism.[2,12] One of the material systems that has been used widely for PDLCs is NOA65, a photopolymerizable adhesive by Norland Products, often with 5CB, other cyanobiphenyl mesogens, or eutectic mixtures of cyanobiphenyls and a cyanoterphenyl, such as E7. The exact composition of NOA65 is not reported in the open literature, but is has been described as containing diallyl ethers and multifunctional thiols.[2] In an extensive study by Lovinger *et al.*, mixtures of E7 and NOA65 were found to exhibit a droplet morphology only at intermediate compositions (about 20 to 65% E7 content) and for a low curing temperature (Figure 5.7a).[16] For content of E7 above 70%, phase separation occurred before irradiation. For E7 content below 20% or at high temperature, E7 remained soluble in the polymer, but ordered structures that resemble spherulites formed over time. It was observed that droplet formation started very early in the polymerization process at low doses of UV and, after the initial formation, the droplet sizes grew slowly.[16] In the composition range where phase separation occurs, the thiol conversion at the onset of phase separation was found to increase with increasing temperature and decreasing fraction of E7 in the mixture, as determined from real-time FTIR spectroscopy (Figure 5.7b).[17] The droplets initially contained liquid crystal in the isotropic phase and the transition to the nematic phase was observed later in the course of the reaction (Figure 5.7b).[17]

For the 7CB/NOA65 system, it was found that the droplet size increased with time during photopolymerization (following a power law) until it reached a saturation value at the point where cross-linking of the polymer impeded further coalescing of droplets.[18] This saturation point occurred at shorter times in mixtures containing larger fractions of NOA65, resulting in

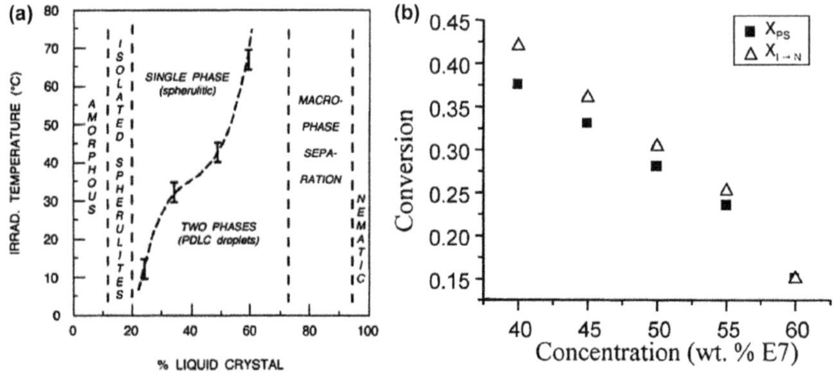

Figure 5.7 (a) Phase diagram for the system E7/NOA65 as a function of E7 fraction and polymerization temperature. Reprinted with permission from A. J. Lovinger, K. R. Amundson and D. D. Davis, *Chem. Mater.*, 1994, **6**, 1726. Copyright 1994 American Chemical Society. (b) Conversion of the thiol-ene reaction as a function of E7 content in E7/NOA65 at the onset of phase separation (squares) and the ordering transition in the LC-rich droplets (triangles). Reprinted with permission from R. Bhargava, S.-Q. Wang and J. L. Koenig, *Macromolecules*, 1999, **32**, 8989. Copyright 1999 American Chemical Society.

a finer morphology. At high 7CB content, the droplets can lose their spherical shape and become polygonal, with a thin polymer wall separating them. In other systems a larger variation of droplet sizes was observed as a function of polymerization conditions. For example, with thermal polymerization of a mixture of Capcure 3-800, MK107, and Epon 828 with E7, the droplet size was found to decrease with increasing of the polymerization temperature (following approximately an exponential dependence on $1/T$).[19] In the photopolymerization of acrylates in E7, the intensity of the curing UV light is also important, as this controls the initiation rate for the polymerization. Indeed, for the same initial composition, high curing intensities can lead to small polymer domains, because the growing polymer chains become insoluble and sufficiently cross-linked at short times. At lower intensities, polymer domains continue to grow for longer times, leading to larger features.[15] Some control of the phase separation process can also be achieved by varying the functionality of the reactive species. In PDLCs based on thiol-ene chemistry, increasing the functionality of the thiol or ene species was shown to lead to a decrease in droplet size and in the fraction of phase-separated LC, as a result of changes in the polymerization rate and the gel point.[20] In the case of free-radical polymerization of acrylates, greater monomer conversion and extent of phase separation of the LC component can be achieved using a diacrylate, relative to a triacrylate or higher multifunctional acrylates.[21] This stems from the fact that the crosslinking density is lower with diacrylates, allowing for a longer time before the diffusion of LC molecules and unreacted monomer is hindered by the gelled network.

As mentioned above, a PDLC device is strongly scattering in the unpowered state, due to the random orientation of the droplets and the mismatch between n_p and the effective refractive index of the LC (a value between n_o and n_e, depending on the orientation of the droplet axis, where n_e is the extraordinary refractive index of the LC). For droplet of sizes between about 0.5 and 3 µm, the scattering efficiency does not have a strong wavelength dependence in the visible and near infrared ranges and devices appear milky white.[2] When voltage is applied between the top and bottom substrate of the device, the droplet axes reorient along the field, light propagating parallel to the electric field direction only experiences the refractive index n_o, and the device becomes transparent (Figure 5.8a). Figure 5.8b shows the dependence of the haze for a set of NCAP films of different thicknesses as a function of applied voltage. It can be seen that the haze value, which is high at zero and low voltage, starts to decrease sharply above a certain threshold voltage and then levels off around 20% at 40–60 V. Because of the effect of surface alignment and of the curvature of the droplet surface, the reorientation voltage increases with a decrease in droplet size. The voltage is often found to scale approximately with the inverse of the droplet diameter, as in the case shown in Figure 5.9.[8] When the electric field is turned off, the orientation of the LC molecules in the droplets returns to the initial configuration, with random orientation of the droplet axes, and the device becomes opaque again.

In PDLCs, the local orientation of the LC director is determined by the interface between the phase-separated LC domains and the local polymer surface. As such, the device does not require alignment surfaces during the preparation stage. Another property of PDLC devices is that the transmission characteristics are polarization independent in both the *off*-state (because

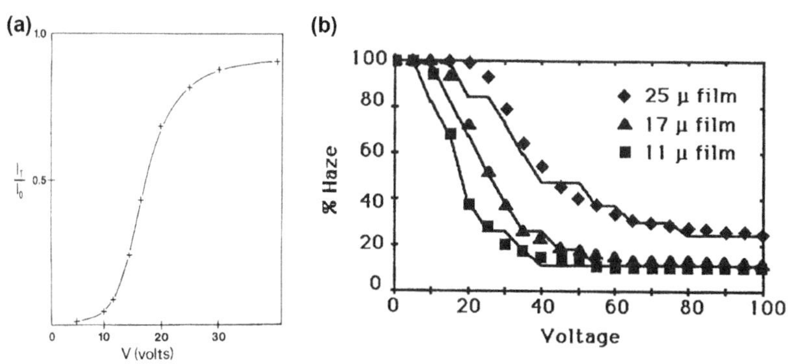

Figure 5.8 Electro-optic properties of PDLC devices. (a) Transmittance through a 24 µm device prepared by the PIPS method (starting materials: 5CB, epichlorohydrin, bisphenol A, and polyaniline). Reprinted from J. W. Doane, N. A. Vaz, B.-G. Wu and S. Žumer, *Appl. Phys. Lett.*, 1986, **48**, 269, with the permission of AIP Publishing. (b) Haze values for three films prepared by the emulsion method (components: ZLI2061 and PVA) as a function of applied voltage. Reprinted from P. S. Drzaic, *J. Appl. Phys.* 1986, **60**, 2142, with the permission of AIP Publishing.

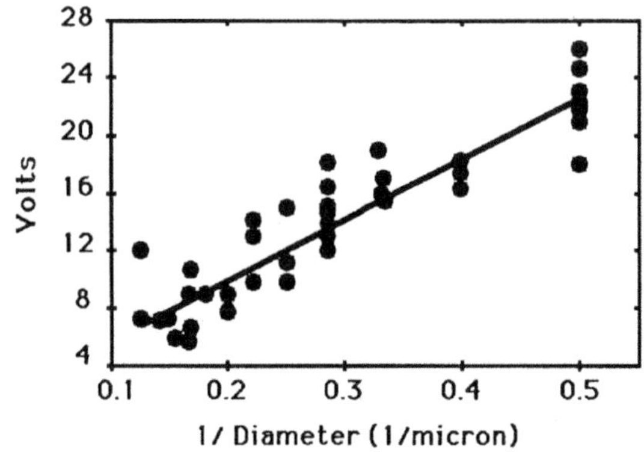

Figure 5.9 Voltage required to reorient LC droplets in a PDLC film prepared by the emulsion method (components: ZLI2061 and PVA) as a function of droplet diameter.
Reprinted from P. S. Drzaic, *J. Appl. Phys.* 1986, **60**, 2142, with the permission of AIP Publishing.

of the random orientation of the droplet optical axes) and in the *on*-state (when the droplet optical axes are parallel to the electric field). Thus, the operation of these devices does not require the use of polarizers. The main applications for these types of devices are displays and privacy windows.[9]

The confinement of the LC molecules in small volumes is also responsible for the fast dynamics of the electro-optic response in PDLCs. Multiple response times for both the rise and decay side have often been observed, in part reflecting a distribution of droplet sizes. Drzaic has reported that in ZLI 1840/PVA films (prepared with the emulsion method), a large response is recorded within a millisecond of applying a voltage. A saturation of the response is however seen only after 0.1–1 s from the voltage turn-on.[22] On the decay side, a small response is observed in the timescale of 1–10 ms and it is followed by a larger change with a characteristic time of 0.1–1 s. Fast and slow components of the decay dynamics have also been reported in PDLCs prepared by the PIPS[10,23] and SIPS[24] methods. It is believed that the fast response time corresponds to the reorientation of the LC molecules in the inner parts of the droplets (rotation toward the electric field direction on the rise side and back to the initial alignment on the decay side) and the slow response time corresponds to the reorientation of the droplet optical axis.[2]

The rise time in PDLCs typically decreases with increasing driving voltage V and increasing temperature, the latter as a result of changes in viscosity. In certain cases, a dependence of t_{rise} on V^{-2} has been observed,[23] but often the trends are more complex because of the influence of droplet size and shape. Models for the reorientation dynamics of confined LCs have been developed to help explain the electro-optic behavior of PDLC devices. One of the expressions derived for the rise time in such models is the following:

$$t_{\text{rise}} = \frac{\eta}{\varepsilon_0 \Delta\varepsilon E^2 + K(l^2 - 1)/a^2} \tag{5.1}$$

where η is the rotational viscosity, E is the applied field, K the elastic constant, and $l = a/b$ is the ratio of the half-lengths of the long (a) and short (b) axes of the droplets.[2] The second term in the denominator goes to zero for spherical droplets. This term can also be thought of as a threshold for reorienting the LC droplets. The anchoring strength to the polymer surface is also known to affect the response time but it is not accounted for in eqn (5.1). Based on the same assumptions used to derive t_{rise} in eqn (5.1), the decay time t_{decay} is given by:[2,3]

$$t_{\text{decay}} = \frac{\eta}{K(l^2 - 1)/a^2} \tag{5.2}$$

The rise and decay times can vary with the polymer/LC content and the chemical nature of the PDLC components, as these affect not only the morphology of the composite, but also the physical properties of the interfaces. It has been reported that in PDLCs prepared by PIPS using a cyanoethyl acrylate and a biphenyl-based LC, rise and decay times shorter than 1 ms could be achieved.[25] Understanding the dynamics of PDLCs also requires taking into account the voltage drop that can occur through the polymer matrix, which is responsible for the local electric field at the LC droplet not necessarily being given by V/d, where V is the potential difference between the top and bottom electrodes and d is the device thickness. Because of the different conductivities of the polymer and LC phases, space-charge build up could occur at the interfaces, which could lead to a frequency dependence of the device rise time[26] (a frequency dependence of the rise time also results from a dispersion in $\Delta\varepsilon$).

In general, the decay time in PDLCs should not depend on the driving voltage (see eqn (5.2)), as the relaxation is driven by the restoring forces provided by the droplet surfaces. A voltage dependence of t_{decay} has however been observed in several instances, for example, in devices that exhibit different operating regimes at high and low voltages,[24] possibly because of droplet asymmetry or broad size distribution. Faster decay times are seen in devices with large threshold voltages.[2] Other approaches used to affect PDLC response times include the addition of dopants or surfactants to the system.[2] For example, Lu and Yang have shown that small amounts of a chiral dopant added to the LC decreased t_{decay} from 150 ms to *ca.* 10 ms in an epoxy-based PDLC.[27] At the chiral dopant concentrations used in that study, the helical pitch was larger than the droplet diameter, but the presence of the dopant affected the equilibrium orientation of the LC molecules in the droplets and the restoring energy. The addition of surfactants, for example, octanoic acid, to PDLC mixtures has been shown to lead to lower threshold voltages and quicker turn-on times.[28] However, the presence of octanoic acid can also lead to a lower fraction of phase-separated LC in the sample[21] or a depression of the LC phase transition temperature.[29] Thus various competing effects need to be considered when optimizing specific performance parameters. *N*-vinylpyrrolidone had been used as an additive in acrylate-based

PDLC formulations, to reduce the viscosity of the starting mixture and drive the acrylate polymerization to higher conversions.[21]

For all the devices described so far, the *off*-state is characterized by large scattering and random orientation of the LC droplets axes. In the *on*-state, the LC molecules are aligned along the electric field direction and the film becomes transparent. This type of operation is often called "normal mode". Devices that exhibit the opposite behavior, that is, transparent in the *off*-state and becoming opaque when powered on, could also be useful in certain applications (for example, to reduce power consumption in cases when transparency is needed for extended periods of time or is desirable in case of a power failure).[3,30] Various strategies have been explored to realize such devices, which are commonly referred to as "reverse-mode" PDLCs. An approach based on LCs with negative dielectric anisotropy ($\Delta\varepsilon < 0$) was described by Ma *et al.* and involved the use of a surfactant to change the type of anchoring between the LC and the polymer surface to achieve droplets with an axial alignment and axes oriented normal to the film substrate.[31] A vertical alignment of the droplet axis has also been achieved by placing a mixture in a magnetic field during the photopolymerization step.[32] If $n_o \sim n_p$, the resulting film is transparent in this aligned state. Application of an electric field leads to reorientation of the LC molecules perpendicular to the field direction, which renders the film opaque, due to refractive index mismatch between polymer matrix and droplets. Reverse mode PDLCs have also been realized using dual frequency LCs,[30,33] *i.e.* LCs in which the dielectric anisotropy changes sign with frequency ($\Delta\varepsilon > 0$ at low frequency and $\Delta\varepsilon < 0$ at high frequency; the frequency at which $\Delta\varepsilon$ changes sign is called cross-over frequency, f_c). During the preparation of the device, a field with frequency $f < f_c$ is applied in order to align the LC molecule along the field and normal to the film surface. The molecules at the LC–polymer interface are locked into this orientation during the photopolymerization step. After fabrication, these molecules act as an alignment layer for the rest of the LC in the droplet. As such, the homeotropic orientation is retained at zero field and, if $n_o \sim n_p$, the film has high transmittance. Scattering is activated by applying a field with $f > f_c$, so that the LC molecules reorient perpendicular to the field and the droplets exhibit an effective refractive index larger than that of the polymer matrix. Transparency is recovered after the field is turned off. If dual-frequency LCs are used in normal mode PDLC devices, they can provide a means to control the decay time;[2] the scattering state can be recovered by switching the frequency of the driving voltage to a value $f > f_c$, where $\Delta\varepsilon < 0$ (a low frequency voltage is used instead to turn the device on). In this manner, t_{decay} is determined by eqn (5.1) and not eqn (5.2) and it can be decreased by increasing the amplitude of the high frequency signal.

5.2.2 Nano-PDLCs

The main characteristic of the first generation of PDLC devices was the ability to switch them from an opaque to a transmitting state. As discussed in the previous section, this is a result of the presence of meso-scaled LC-rich domains in the device, which are randomly oriented in the *off*-state and become aligned along a common direction in the *on*-state. The scattering efficiency in the *off*-state depends on the contrast in refractive index between the LC-rich phase and polymer, and the dimension of the droplet. Films characterized by droplets of sizes significantly smaller than the wavelength of light, often referred to as "nano-PDLCs" exhibit a different electro-optic response from the typical PDLC with micron-sized droplets.

As early as the 1990s, it was shown that LC–polymer composites with droplet sizes less than 100 nm could be used in devices that exploit the field-induced reorientation of LC molecules in a different way, such as through the optical Kerr effect. For example, Sansone *et al.*[34,35] showed that if the polymerization rate was sufficiently large, a mixture of Norland NOA60 and the nematic liquid crystal E7 could be cured to yield a PDLC with a droplet size distribution centered around 40 nm. For 15% LC loading, the sample had transmission above 80% in the visible and near-IR ranges without any applied field. The LC in the droplets could be reoriented by an electric field directed along the film thickness, leading to a change in birefringence (Figure 5.10a). This change was found to be dependent on the square of the electric field magnitude, E, at low fields, as expected for an optical Kerr effect where $\Delta n = \lambda B E^2$ (Δn is the birefringence, λ is the probe wavelength, and B the Kerr constant). The effective Kerr constant, B, was several orders of magnitude larger in the nano-PDLCs than in the reference (isotropic) material CS_2, and increased with temperature up to the clearing point of the LC in the nano-droplets. The response of the sample was reported to be fast, with rise and decay times of less than 0.1 ms, but it slowed with temperature

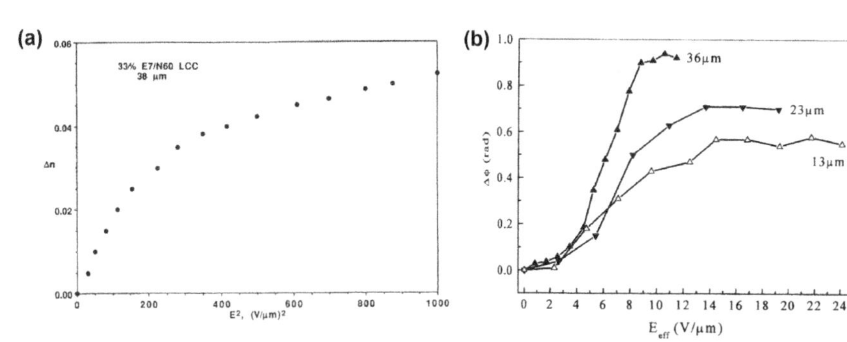

Figure 5.10 (a) Birefringence Δn as a function of E^2, where E is the magnitude of the applied electric field for a nano-PDLC based on NOA60 and E7. Reprinted from M. J. Sansone, G. Khanarian, T. M. Leslie, M. Stiller, J. Altman and P. Elizondo, *J. Appl. Phys.*, 1990, **67**, 4253, with the permission of AIP Publishing. (b) Induced phase shift as a function of applied field for a device with initial composition: 35% E7, 49% dipentaerythrol hydroxy pentaacrylate, 0.5% benzophenone, 15.5% *N*-vinylpyrrolidone. Reprinted from D. E. Lucchetta, R. Karapinar, A. Manni and F. Simoni, *J. Appl. Phys.*, 2002, **91**, 6060, with the permission of AIP Publishing.

and LC loading. The fast response time relative to bulk LC and traditional PDLCs is due to the LC confinement in very small volumes.

Devices with droplets smaller than 100 nm were also obtained by PIPS starting from a mixture of NOA81 (Norland) and the liquid crystal BL24.[36,37] Electric field-induced birefringence was recorded in those samples, but the magnitude of the changes was relatively small, because the volume fraction of the droplets was small, suggesting that the phase separation was not complete. Large phase shifts (in the order of $\lambda/2$ in a 10 μm cell) were instead achieved in a device containing 28% E7 in NOA81, even if very large driving voltages were needed.[38] Lucchetta *et al.* showed that, for droplet sizes between 100 and 250 nm, devices can have sufficiently large transmittance in the *off*-state to be used for phase-only modulation.[39] In their implementation, the prepolymer mixture contained dipentaerythrol hydroxy pentaacrylate (44 to 54%) as the monomer, benzophenone as the photoinitiator, *N*-vinylpyrrolidone as a stabilizer, and the liquid crystal E7. Relatively large field-induced phase shifts were observed in these samples, with some dependence on cell thickness and mixture composition (Figure 5.10b). The response times were about 0.5 ms for the rise and 0.7 ms for the decay. The rise time decreased slightly with increasing field amplitude. The phase modulation resulting from the field-induced birefringence change in nano-PDLCs is independent of light polarization, due to the random orientation of the droplet axes in the *off*-state, which is not the case with devices based on bulk LCs in aligned cells.

For a nano-PDLC formed from curing NOA81 (Norland) in the presence of 27 to 37% of E7, which exhibited droplets of an average size of 170 nm and Kerr coefficients in the range $0.1–1.0\times10^{-11}$ V m^{-2}, light scattering was used to confirm that the LC assumes nematic order in a wide temperature range below 50 °C.[40] The correlation length obtained from the scattering data followed a different trend relative to bulk E7 and approached the saturation value of 140 nm deep into the nematic phase (called "IsoN" phase by the authors, as the sample is optically isotropic at the macroscopic level). The fact that the correlation length was of the same order as the droplet size was taken as an indication of the short-range order experienced by LC molecules confined in the nanodroplets, which in turn is responsible for the fast response time in the EO behavior.

A different use of refractive index changes in PDLCs with nano-sized droplets has been demonstrated by Ren *et al.* with the generation of tunable microlenses.[41] In this investigation, an array of microlenses was generated by filling an appropriately shaped NOA65 mold with a mixture of NOA65 and E7, and then curing the system. The device was transparent because the droplet size was sufficiently small. Because of the shape of the mold, the phase shift induced by the refractive index contrast between polymer and randomly oriented LC droplets in each microlens is not uniform in space (it is largest at the center of the microlens and it decreases moving toward the lens edge). This leads to a bending of the wavefront propagating through the device and a focusing of a light beam by each microlens. When an electric field was applied, the refractive index contrast decreased and the microlens focal length increased, providing tunability to the device.

5.2.3 PDLCs Doped with Nanoparticles

Over the years, formulations for PDLCs containing various polymer precursors and LCs have been investigated to study the correlation between composition, morphology, and EO performance. In order to provide greater flexibility in device characteristics and address perceived drawbacks such as high operating voltages or slow response times, the use of additives to the basic components was also explored. One class of versatile materials that has been extensively investigated as a component in PDLC devices is that of nanoparticles or, in general, nano-sized objects. These have been considered, for example, as means to affect the dielectric constant of the medium (and thus the operating voltage), the refractive index of the LC or the polymer matrix (so as to change the scattering efficiency or the on/off contrast ratio), or the anchoring strength to the polymer interface (which can influence the operating voltage and response time of a device). A few recent examples of the properties exhibited by nanoparticle-doped PDLC are discussed below.

Hinojosa *et al.* have investigated the effect of inclusion of gold nanoparticles (Au NPs) in PDLCs.[42] The composite contained E44 as the liquid crystal and an acrylate-based prepolymer. The size of the Au NPs was about 14 nm. It was found that the threshold voltage was lower in devices containing nanoparticles relative to those without nanoparticles, and it decreased with increasing content of Au NPs (Figure 5.11a). This behavior was attributed to local field effects due to the metal nanoparticles.

A decrease in threshold voltage has also been observed when non-metallic ZnO nanoparticles were used.[43] In this investigation, small amounts of 10 nm ZnO particles were added to a 1 : 1 mixture of E7 and NOA65, which was then cured with UV light. It was observed that the threshold and operating voltages (voltages to achieve 10% and 90% of the maximum transmittance, respectively) decreased significantly in PDLCs containing ZnO (Figure 5.11b) relative to devices without nanoparticles. The presence of ZnO also affected the sample transmittance in the *on*-state and the contrast ratio was best at intermediate content of ZnO. The EO behavior was explained by the observation that PDLCs containing larger amounts of ZnO exhibited larger droplets, for which lower voltages are required to induce reorientation of the LC molecules, as discussed earlier.

In a different study, it was reported that in the presence of silica nanoparticles (7 nm in size), acrylate-based PDLCs exhibited an increase in threshold and driving voltages, as well as a decrease in the transmittance at saturation.[44] This was attributed to the nanoparticles being preferentially dispersed in the LC phase, where they form aggregates, and lead to an increase in the anchoring strength of the LC. Because the presence of nanoparticles has the potential to influence the morphology and properties of the

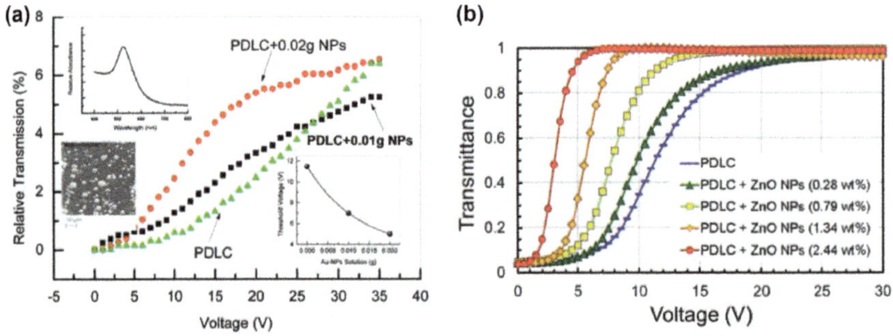

Figure 5.11 Transmittance as a function of applied voltage in nanoparticle-doped PDLCs: (a) Samples with two different amounts of Au NPs and a sample with no Au NPs. The inset on the bottom-right shows the threshold voltage of these devices as a function of Au NPs content; the inset on the top-left is the absorption spectrum of the NPs in water. Also shown is an SEM image of a representative PDLC. Reprinted from A. Hinojosa and S. C. Sharma, *Appl. Phys. Lett.*, 2010, **97**, 081114, with the permission of AIP Publishing.[42] (b) Samples with and without ZnO NPs. Reproduced from ref. 43 with permission from The Optical Society of America.

PDLCs in multiple ways, it is usually difficult to predict how doped PDLCs will perform relative to undoped systems. As seen in the cases above, for example, the trend in threshold voltages can be traced back to different nanoparticle effects. It is also difficult to perform comparisons between nanoparticles in independent investigations and extrapolate results to more general cases, as the set of parameters monitored are not necessarily the same or the PDLC composition varies.

Nanoparticles can also be used to modify the response time and frequency behavior of PDLCs. An inverse relationship between rise time and dielectric anisotropy is often seen in PDLCs (the response time depends also on droplet size and aspect ratio, on the applied voltage, and on the LC viscosity, see eqn (5.1)).[3] A similar behavior has been seen in nanoparticle-doped PDLCs, with shorter times observed in systems with larger dielectric anisotropy.[45] For example, PDLCs based on polyvinyl acetate and 5CB exhibited the largest dielectric anisotropy ($\Delta\varepsilon = 13-14$) and the shortest rise time ($t_{rise} = 1.0-1.2$ ms) when doped with Ag/Si or $TaSi_2$/Si nanoparticles. For the undoped sample, the corresponding time was 1.8 ms. Doping with SiO_2 instead decreased $\Delta\varepsilon$ to 2.1 and slowed the response ($t_{rise} = 2.6$ ms) relative to the undoped case.

Shim *et al.* have shown that barium titanate ($BaTiO_3$) nanoparticles can have a large effect on the frequency response.[46] For this investigation, PDLCs were prepared by PIPS from mixtures of TL205 and NOA65 with up to 1wt% $BaTiO_3$ nanoparticles (size ~600 nm). It was shown that in the presence of 0.5wt% $BaTiO_3$ the maximum frequency at which the device could be switched on increased by about an order of magnitude. The threshold voltage decreased when $BaTiO_3$ nanoparticles were present. No significant morphological changes were caused by the nanoparticles. The observed behavior was correlated with an increase in the film capacitance for the devices doped with the high permittivity, ferroelectric nanoparticles.

In some cases, the presence of nanoparticles can affect directly the phase separation process and thus the morphology of the final PDLC. For example, when 5wt% of SiO_2 nanoparticles grafted with LC molecules were added to a mixture of nematic LC BL24 and prepolymer NOA81 and processed by PIPS, smaller droplets were formed but they accounted for a larger volume of the film relative to the sample without nanoparticles, indicating that a smaller fraction of LC molecules remains mixed with the polymer if nanoparticles are present.[47] This was thought to be a result of the LC-grafted nanoparticles acting as nucleation centers and leading to more complete phase separation.

5.2.4 Dye-doped PDLCs

Early in the development of PDLCs, researchers considered the effect of inclusion of dyes in the devices. When the dyes are preferentially located within the LC droplets, the dye molecule orientation can be switched together with that of the LC, as in the case of bulk guest-host LC devices. The systems are referred to as dye-doped, dyed, or dichroic PDLCs and have been considered for use in display applications.[1–3] For positive dichroic dyes, or pleochroic dyes, the transition dipole moment for the lowest electronic transition is directed along the long molecular axis. The probability that a photon of appropriate wavelength is absorbed is at a maximum when the polarization is along this direction and at a minimum when it is perpendicular to it (the dependence is on $\cos^2 \beta$, where β is the angle between the transition dipole moment and the polarization direction). If the dyes align with the LC in the droplets, then in the device *on*-state the dyes have the long molecular axis along the direction of propagation of light and thus they attenuate the light only weakly in the region of their absorption band (and at other wavelengths the dye-doped PDLC is transparent, if $n_o \cong n_p$). In the *off*-state, the dyes are directed along the droplet axis, which is randomly oriented in the cell, and thus the contribution to absorption will be larger (in the case of spherical shapes, the droplet ensemble should behave as an isotropic medium). The light propagating through the film in the *off*-state will typically undergo multiple scattering events, thus increasing the path length inside the medium and leading to additional attenuation of light (or, equivalently, smaller amounts of dyes are needed to provide sufficient light attenuation). The film performance does not depend on the polarization direction when probed at normal incidence.

The preparation of dye-doped PDLCs by the emulsion method is relatively straightforward. If the dye is selected to be soluble in the LC component but not in the polymer, it will reside in the LC phase after the evaporation of the solvent.[48,49] In the phase separation method, a fraction of the dye may remain trapped in the polymer-rich phase, decreasing the amount of dye that can be switched, and thus the contrast between the *on*- and *off*-states of the device. Despite this problem, with proper selection of dye, PDLC composition, and preparation conditions, devices with relatively large contrast were demonstrated early on.[50,51] Dye-doped PDLCs prepared by the emulsion method also have another advantage relative to other preparation methods in that the droplets have been shown to have an elliptical rather than spherical shape, with the long axis in the plane of the film. For a given order parameter of the dye in the LC phase, a larger on/off contrast can be obtained if the dye molecules are oriented in the plane of the film instead of randomly in three dimensions (as in the case of spherical droplets).[48]

Reflective displays can be obtained by combining a PDLC film with a colored reflector placed behind the film.[2,48] For dye-doped PDLCs, both the absorption and scattering process contribute to the light attenuation in the *off*-state and the display appears dark. Better color contrast is achievable than in purely absorbing displays, though, due to the backscattering of the incoming white light. In the *on*-state, scattering and absorbance are low and the back reflector is visible.

Starting from these general ideas, a wide range of dye-doped PDLCs have been investigated over the years in order to optimize contrast, to achieve reverse mode operation, and to enable the ability to control the device response by means other than with an electric field. Various types of dyes have been tested in a range of PDLC recipes. In several instances, the dye was found to affect the morphology of the final PDLC film yielding droplets of different size, size distribution, or symmetry.[52–55] The presence of dye molecules can also alter the electro-optic response through indirect effects on the anchoring strength of the droplet walls. These effects are usually difficult to generalize among classes of dyes, LCs, and monomers. A certain degree of composition tailoring is often needed for optimization of the device characteristic of interest.

If the dyes added to the PDLC formulation are photoactive, the PDLC device may acquire additional functionality. If the dye is photochromic, the device color can be changed by exposure of the system to light of appropriate wavelength, while the device can still be switched between a scattering and transparent state by an electric field.

For example, a spiro-indolino-oxazine derivative has been used to fabricate a reverse-mode PDLC with switchable color.[56–58] Using a nematic LC with negative dielectric anisotropy (ZLI4788-000) and the urethane diacrylate prepolymer CN965, a reverse mode PDLC was obtained if the liquid crystal was aligned during the curing step (in this work by a magnetic field).[56] The device had a clear *off*-state and it became scattering when an electric field was applied. It was shown that if the spiro-indolino-oxazine was added to the LC and prepolymer mixture, the dye retained its photochromic behavior (for the case under investigation, the absorption band of the material was in the UV range in the dark and around 550 nm with exposure to UV light). The film became absorbing in the visible when it was exposed to UV light and it returned to transparency after switching the light off. In addition, scattering

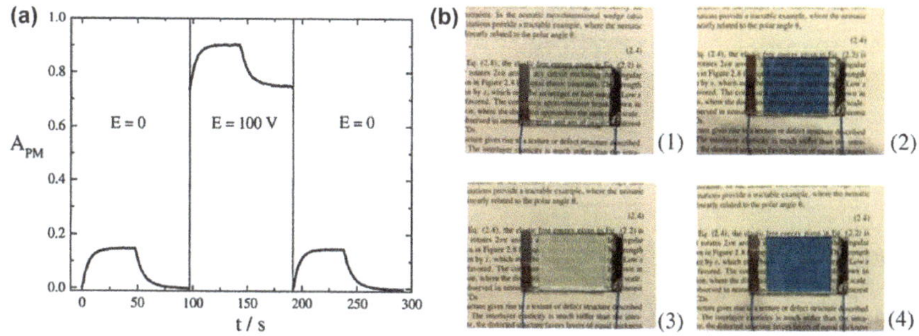

Figure 5.12 Behavior of a reverse mode PDLC containing a photochromic dye. (a) Absorbance of the colored form of a spiro-oxazine dopant in a PDLC as a function of time with and without UV illumination or electric field. UV light was on from 0 to 50 s, 95 to 145 s, and 190 to 235 s. Reprinted with permission from A. Romani, G. Chidichimo, P. Formoso, S. Manfredi, G. Favaro and U. Mazzucato, *J. Phys. Chem. B*, 2002, **106**, 9490. Copyright 2002 American Chemical Society. (b) Image of a photoresponsive dye-doped PDLC device. The device is in the off state in (1). The electric field is on for cases (3) and (4). UV light is on only for cases (2) and (4). Reprinted with permission from G. De Filpo, P. Formoso, S. Manfredi, A. I. Mashin and F. P. Nicoletta, *Liq. Cryst.*, 2017, **44**, 1607, Taylor & Francis Ltd. (www.tandfonline.com).

was turned on by an electric field (Figure 5.12a). Although a relatively large fraction of the photochromic molecules remained in the polymer matrix and only some were in the LC droplets, the photochromic reaction was activated in both. A device with similar operation was obtained using a formulation based on a mesogenic diacrylate monomer.[58] In this case, homeotropic alignment of the mixture before polymerization was achieved using glass substrates with rough surfaces. As shown in Figure 5.12b, the device could be switched between four states: (1) clear and uncolored (no field, no UV light), (2) clear and colored (no field, with UV irradiation), (3) opaque and uncolored (field on, no UV light), and (4) opaque and colored (field on, with UV irradiation).

Another interesting application is that of optically writable displays reported by Fuh *et al.*[59] In this study, a PDLC was prepared by PIPS from a mixture of E7, dipentaerythritol, and 1-vinyl-2-pyrrolidinone doped with a small amount of the azo-dye methyl red. The work exploits the surface assisted photoinduced alignment obtainable with methyl red. If a writing beam at 532 nm (in the absorption band of methyl red) was used to isomerize methyl red when the device was switched on, the alignment in the exposed regions could be locked in. The written parts of the film thus showed less scattering than the surrounding regions even after the writing beam and electric fields were turned off. The pattern could be erased by thermal treatment to reverse the isomerization process.

Bandwidth narrowing and random lasing have been observed in certain cases when the dyes used to dope the PDLC films are emissive chromophores.[60–62] For example, Harada *et al.* have shown that in a PDLC consisting of 5CB droplets in polystyrene doped with 1% 2-(4-biphenylyl)-6-phenylbenzoxazole (PBBO), the width of the PBBO emission decreased significantly with increasing pump energy and the emission intensity exhibited a threshold behavior (Figure 5.13).[61] Liu *et al.* observed bandwidth narrowing and threshold emission in an acrylate-based PDLC with nanosized droplets of E7 doped with the laser dye DCM.[62] It was also shown that bandwidth narrowing was absent, or present to a smaller extent, when the samples were heated above the nematic-to-isotropic transition temperature.[60,61] In these examples, the random lasing action was a result of the multiple scattering events that a light beam undergoes while propagating through a sample with the droplet morphology. The emission is mostly observed from the edge of the sample and it can be polarized.

Xiong *et al.* used doping of PDLCs with 4-butyl-4-methoxyazobenzene to demonstrate a photoactivated microlens array.[63] In their implementation, each microlens was a spherical PDLC bead (consisting of a polystyrene matrix with E7 droplets), created *via* SIPS in a microfluidics apparatus. A set of beads were then arranged on a plane in a close-packed array. The materials selection was such that the refractive index of the polymer was close to that of the LC in the isotropic phase. As prepared and at room temperature, the beads were opaque (Figure 5.14a). The beads became transparent when heated above the nematic-to-isotropic transition temperature of E7, where the refractive index mismatch was minimized. In that situation, given the shape, each bead (microlens) produced an image of an object put behind the array. The transition between the opaque and focusing states could also be achieved by exposure of the device to UV light, which leads to isomerization of the azobenzene dopant and a loss of the order of the LC molecules in the droplets (Figure 5.14b).

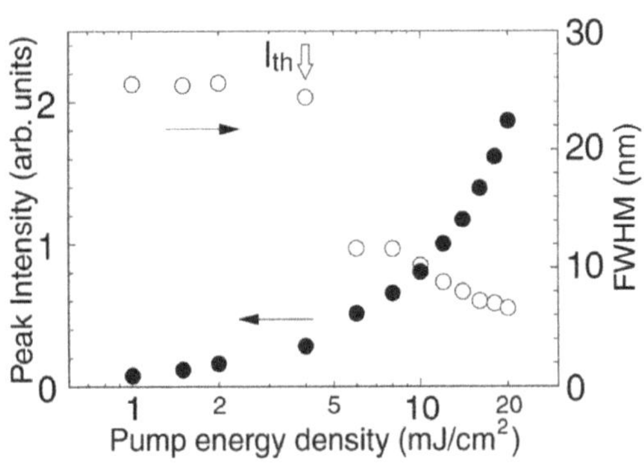

Figure 5.13 Peak intensity (filled circles) and full width at half maximum (FWHM, open circles) of the emission from the dye PBBO in a doped PDLC as a function of pumping energy density at 337 nm.
Reproduced from ref. 61 with permission from The Japan Society of Applied Physics, Copyright 2005.

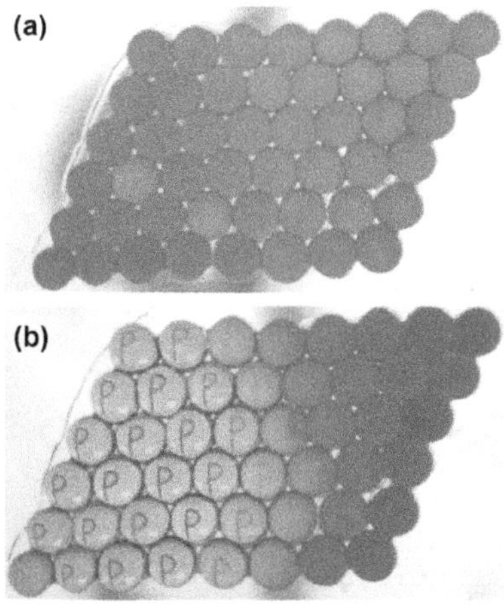

Figure 5.14 Appearance of a microlens array formed by PDLC beads doped with an azobenzene derivative. (a) If the device is not exposed to UV light, the beads are opaque. (b) When the left part of the device is exposed to UV light, the microlenses become transparent and focus light. Reproduced from ref. 63 with permission from John Wiley and Sons, Copyright © 2009 WILEY-VCH Verlag GmbH & Co. KGaA, Weinheim.

5.2.5 Other Liquid Crystal–Polymer Composites

Confinement of LCs in nanosized domains has also been used in other classes of LC–polymer composites and exploited to demonstrate EO responses not achievable in homogeneous systems. The presence of nanodroplets is thought to be responsible for the tuning behavior of a cholesteric liquid crystal device reported by Inoue *et al.*[64] It was found that using a photopolymerizable liquid crystal monomer, a structure with phase-separated nanodomains was obtained at 26% monomer content, lower than when using a non-mesogenic monomer (Figure 5.15a). Because the monomer is mesogenic and the polymerization was conducted in an ordered phase, the order and alignment were retained during the polymerization. The resulting polymer matrix was aligned and the same alignment was transferred to the small-molecule components of the mixtures, as is often the case in PSLCs.[65] The field-induced reorientation of the small-molecule LC component leads to an effective change in the refractive index of the device (the polymer content in the system is high enough that the polymer matrix does not respond to electric fields). For a system in the nematic phase, this can be exploited directly for phase modulation, whereas for devices constructed using a chiral dopant or chiral nematic monomer, the refractive index change manifests itself as a shift in the red edge of the Bragg reflection mode as a function of applied field, which can be exploited in an optical amplitude modulator (Figure 5.15b). In either case, the response times are very fast (20–150 μs for the rise side, depending on the applied

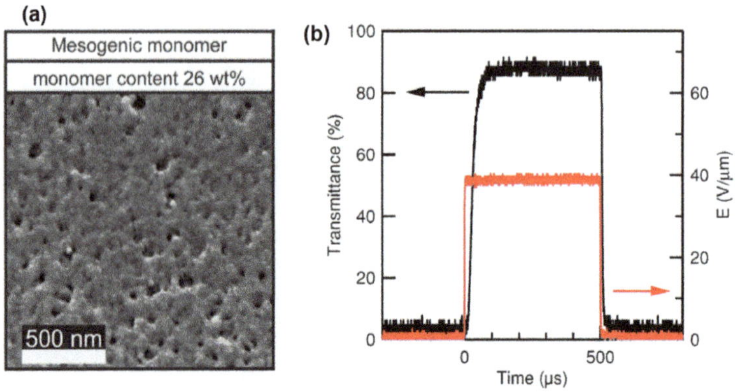

Figure 5.15 (a) SEM image of a polymerized LC composite, revealing the presence of nano-sized domains. Sample composition: 26% RMS03-008, 60% BL-011, and 14% R811. (b) Transient change in transmittance (black trace, left axis) when a pulsed electric field (red trace, right axis) is applied to the sample. The probing wavelength (633 nm) is within the reflection band in the *off*-state; the change in transmittance is due to a shift in the edge of the reflection band to below 633 in the *on*-state. The rise and decay times are reported to be 34 and 6 μs, respectively. Reproduced from ref. 64 with permission from John Wiley and Sons, © 2013 WILEY-VCH Verlag GmbH & Co. KGaA, Weinheim.

field and about 5 μs on the decay side), much faster than in bulk LCs, again due to the confinement of LCs in small pores.

Fast electric field-induced changes in birefringence were also observed in another composite of a mesogenic monomer (RM257) and BL-011 at a weight ratio 3 : 7.[66] In this case, no clear boundaries between the polymer and low molar mass LC components were visible by SEM and it was inferred that small nanodroplets were formed. The maximum achievable Δn in this device was 0.03, observed at 40 V μm^{-1}. The device could be driven by fields up to 10 kHz in frequency. In contrast to the optically isotropic nano-PDLC devices discussed earlier in Section 5.2.2, in the present case, the device response is polarization dependent. When the polymer precursor has liquid crystal phases by itself, devices with different operation characteristics can be obtained. Ouskova *et al.*[67] used a liquid crystalline acrylate monomer (PLC-20-14C, developed by Beam Co.) mixed at 40% level in the nematic 6CHBT. When a cell with substrates treated for planar alignment was filled with this mixture, the mixture assumed a homogeneous alignment, which was retained after photopolymerization. The polymerized system was found to have bicontinuous polymer/LC domains and it was transparent in the *off*-state. The device functioned as a phase modulator when an electric field was applied through the film thickness, as the molecules in the LC-rich phase followed the field. As is the case for bulk LC devices, the response was polarization dependent. Fast response times were achieved (0.18 and 0.24 ms for the rise and decay, respectively) due to the strong restoring forces provided by the polymer-rich domains. The same research group reported a device prepared in a similar manner but including mesogenic azobenzene derivatives instead of 6CHBT and exhibiting photoinduced birefringence changes (of the order of 0.01) as a result of order decrease when the azo-dye photoisomerized (no LC reorientation was involved in the operation in this case).[68] The photoinduced changes occurred on the timescale of a few seconds and were reversible when illumination was removed.

5.3 Periodic Polymer Dispersed Liquid Crystals

If a film of a mixture that operates by photo-induced PIPS is exposed in a non-uniform light profile, the polymerization is initiated at different rates in different parts of the sample, resulting in modulation of the phase separation process across the sample. If holographic illumination, such as from two or more interfering laser beams, is used, the periodicity and pattern of the intensity field will translate into a sample with a periodic phase-separated structure, as polymer forms at different rates in the high intensity and low intensity regions.[69] The resulting structures exhibit an alternation of regions rich in polymer with no or very limited signs of phase separation and regions with a high density of LC droplets separated by thin polymer walls. These structures have been called *h*olographic *p*olymer *d*ispersed *l*iquid *c*rystals (HPDLCs) and they operate as gratings. The large majority of the literature has described HPDLCs with periodicity in one dimension and they will be the focus of this section, but higher dimension gratings have also been reported.[70–73]

The typical morphology of a one-dimensional HPDLC film acquired by high resolution SEM is shown in Figure 5.16. Periodic regions of phase-separated LC domains exist between regions of pure polymer and the resultant index contrast leads to an optical grating.

A diffraction grating is a device able to scatter the impinging light in an 'ordered' manner, that is, in a limited number of directions. This behavior is due to the fact that a diffraction grating exhibits a periodic spatial modulation of its dielectric constant. The operation of diffraction gratings is often classified in two categories: the Raman-Nath regime and the Bragg regime. The first one refers to gratings which diffract light into multiple orders; gratings of this type are commonly called *thin gratings*. In the case of the Bragg regime, exhibited by "volume" or *thick gratings*, the impinging beam is redirected into a single diffracted beam in addition to a partially transmitted one; in the Bragg regime, the grating wave vector satisfies the Bragg condition.[74] A quantitative distinction between *thin* and *thick* gratings can be found in the work by Gaylord and Moharam.[75] One of the optical parameters used to characterize a diffraction grating is the diffraction efficiency (η_d), defined as the ratio between the intensity of the diffracted beam of interest and the intensity of the incident beam. In the case of a phase grating, where the grating's diffraction properties stem from a periodic variation of the refractive index of the material, the diffraction efficiency is a function of the

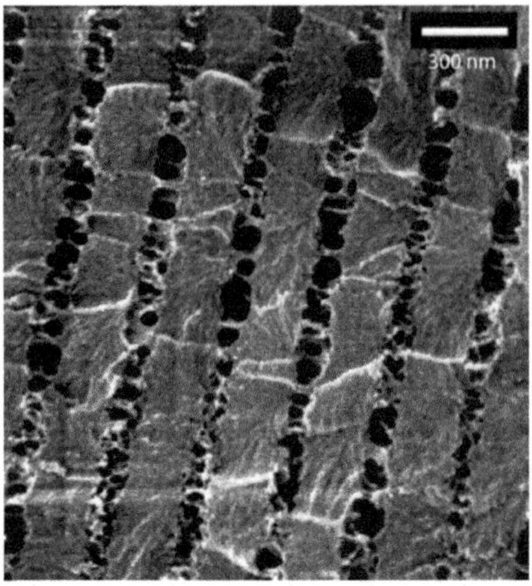

Figure 5.16 SEM image of a typical HPDLC transmission grating.

incidence angle θ_{inc}, probe wavelength λ, grating depth d, and refractive index modulation Δn.

Volume phase holographic diffraction gratings (HDGs) or simply Bragg gratings can be realized through a holographic polymerization (HP) process. In HPDLCs the HDG is partially composed of a reconfigurable material (the liquid crystal component) and thus, the optical properties can be modulated and a dynamic photonic behavior is enabled. HPDLCs possess a periodic variation of the index of refraction, n, which is typically small ($\Delta n \approx 0.05$) relative to the average index across the grating (≈ 1.6).

In an HPDLC, if no external electric field is applied, the incident radiation is diffracted in a single order beam with a given efficiency η_d (Figure 5.17a). If an electric field of appropriate magnitude is applied through the sample, the LC molecules in the droplets reorient along the field and, if $n_o \sim n_p$, the periodic refractive index modulation is washed out and the incident light is transmitted. To minimize scattering, the droplet sizes need to be smaller than the wavelength of the probing light, as discussed in Section 5.2.2. HPDLC gratings can be divided into two classes according to the holographic recording geometries used to generate them: *transmission* and *reflection* gratings (Figure 5.17b). Transmission gratings are realized by using interference fringes with a periodicity in a direction parallel to the substrates. In reflection gratings the periodicity of the fringes is perpendicular to the substrates and the diffracted beam is on the same side of the grating as the incident beam. From an application standpoint, transmission gratings have been used for multiplexing or demultiplexing light beams[76] while reflective structures have been employed for the realization of agile light filtering. From a practical perspective, the length scale of the phase separation process is fundamentally different for transmission and reflection gratings. For typical transmission HPDLC systems, LC inclusions of the order of several

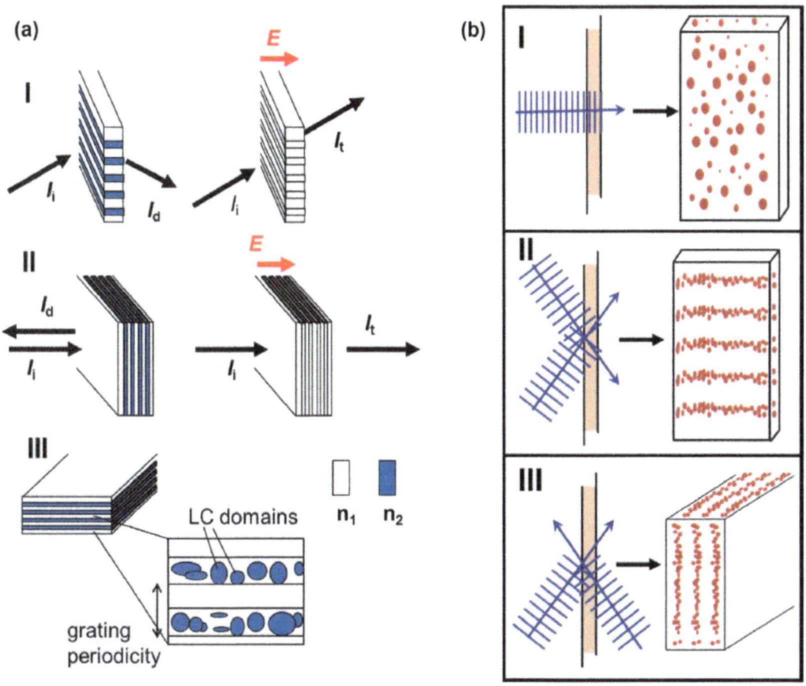

Figure 5.17 (a) Schematic of the working principle of a HPDLC transmission grating (I) and reflection grating (II). I_i, I_d, and I_t are the incident, diffracted, and transmitted beams, respectively; n_1 and n_2 are the refractive indices of the polymer-rich and LC-rich regions, respectively. The devices on the left side are in the *off*-state, the ones on the right are in the *on*-state, with an electric field directed along the red arrow. The internal structure of the LC-rich domains is illustrated in (III). (b) Schematic of the writing geometry (left) and arrangement of LC-rich domains (right; LC droplets shown in red) for (I) isotropic PDLCs, (II) transmission HPDLCs, and (III) reflection HPDLCs. The blue arrows represent the writing beams and the blue line their wavefronts.

hundred nm are formed while for reflection gratings (operating in the visible), the phase-separated domains are typically an order of magnitude smaller.

A typical HPDLC mixture contains 20–50wt% of LC and a photopolymerizable monomer. HPDLC structures can be fabricated in few seconds and the symmetry, size, and refractive index modulation can be easily controlled by the experimental parameters and the materials used. There are several papers which go into detail about the realization and characterization of such HPDLCs.[69,77–80] The type of photopolymerization, particularly the kinetics driven by the monomer nature (chain-growth *versus* step-growth polymerization), the length scale of the diffusion relative to the grating spacing, and the miscibility of the starting non-reactive materials with both the initial reactive fluid and final polymer matrix, all play a role in determining the final local 'grating' morphology. In turn, the morphology of the resulting grating is key to the optical and electro-optical performance of the structures formed. The relative refractive index difference between the LC-rich and polymer-rich regions is the main attribute in determining the diffraction efficiency of the grating; the heterogeneity and magnitude of the LC domain size are critical in determining the optical quality of the structure (amount of scattering); and the domain size and shape are key parameters in determining the initial relative director conformation.

5.3.1 Photo-polymerization Regimes and Materials

The fabrication of high quality HPDLCs depends on the balance between the polymerization process, phase separation conditions, and diffusion kinetics. If the photo-induced polymerization process is too fast, the non-reacting species are "trapped" in the high intensity regions of the curing interference pattern and they cannot phase separate, resulting in a random morphology. If the polymerization process is too slow, depending on the "affinity" of the non-reactive species with the growing polymer matrix, a random one- or two-phase morphology will exist. Only if the kinetics of the polymerization process and the photo-induced phase separation are appropriately balanced with the mass transport kinetics due to the size, shape, and periodicity of the pattern, can heterogeneous two-phase morphologies be obtained. HPDLCs have been mainly fabricated by using two classes of monomers: acrylate-based monomers and thiol-ene-based monomers.

As discussed in Section 5.2.1, multi-functional acrylate systems, in general, reach the gelation point very early in the reaction and the nature of the phase separation process is very complicated. The growing polymer chains phase separate out of the liquid monomer solution and the reaction proceeds to grow the polymer. The liquid domains become enriched with LC, which fixes the inherent local shape and distribution of these structures. All other things being equal, if the polymerization occurs too fast, little or no phase separation can take place, as one can trap the LC molecules in the polymer material much like a plasticizer. The rules governing the growth of HPDLCs are different from those of conventional PDLCs, however, because of the fixed boundary conditions established by the periodic intensity profile in the HPDLC case. This set of boundary conditions imparts directionality to the physical processes including diffusion and polymerization, as well as the structures that form. Phase separation is no longer driven by random thermal fluctuations; rather it occurs at different rates in different regions through the gradients in composition established across the film.

For material systems fabricated from acrylate-based monomers using standard writing conditions, well-defined transmission and reflection gratings can typically be formed if the period is less than 750 nm. Transmission grating structures larger than this can be fabricated, although enabling good confined phase separation of the LC domains becomes increasingly difficult. As the period becomes larger than several microns, the periodic two-phase morphological structure is typically smeared out and much less confinement of the phase-separated domains is observed. In the extreme, the long-period samples resemble classic PDLC structures, where there is an isotropic distribution of discontinuous LC domains residing in a continuous polymer structure. Usually, the amount of anisotropy of the LC droplets decreases as the Bragg spacing is increased. Figure 5.18a–b show typical scanning electron microscopy images of transmission gratings formed from free-radical chemistry systems. Local morphology differences are evident in Figure 5.18a and b, depending on the amount of LC present in the starting syrup and the grating period. The generic TEM image in Figure 5.18c clearly shows the non-uniform morphology consistent with this type of polymerization process.

For sub-micron gratings, the bulk of the characterization has been performed using electron microscopy techniques, although a small amount of probe microscopy has also been reported. In general, the literature is rich with images of 'surface' structures, particularly those of transmission gratings imaged with SEM techniques. It should be noted that interpretation of these images should proceed with caution as the internal, bulk phase-separated structures can be much different. Surface tension effects and preferential wetting of the fluid monomer mixture during the curing process typically lead to a surface morphology that can be considerably different. For reflection grating systems, where individual domain sizes can be of the order of tens of nm, low voltage SEM techniques using a minimal thickness of conductive coating should be employed. Transmission electron microscopy (TEM) imaging (including associated appropriate staining), although more labor intensive, if done correctly, gives much more insight into the true nature of the internal structure. The mounting and microtoming axes need to be in the proper orientation relative the Bragg period to ensure the correct viewing of the structures. Figure 5.19 shows images from the same chemistry

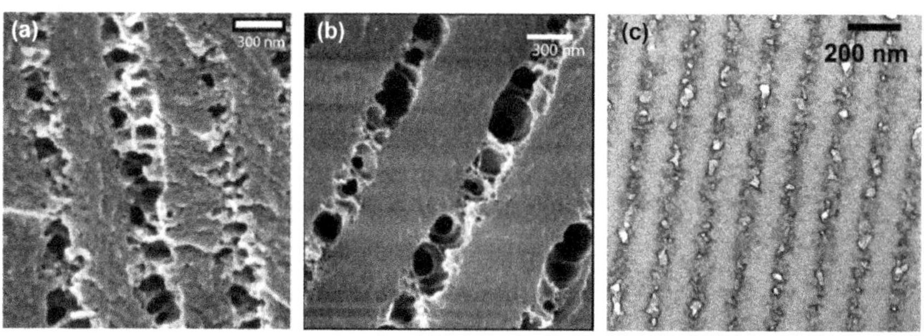

Figure 5.18 Morphologies of HPDLC transmission gratings formed from free-radical chemistry. (a) and (b) show SEM images of different types of periodically spaced non-uniform LC domains separated by pure polymer domains. Image (c) is a typical TEM image, again showing the irregular shapes of the LC domains caused by the free-radical polymerization and the relatively large periodicity.

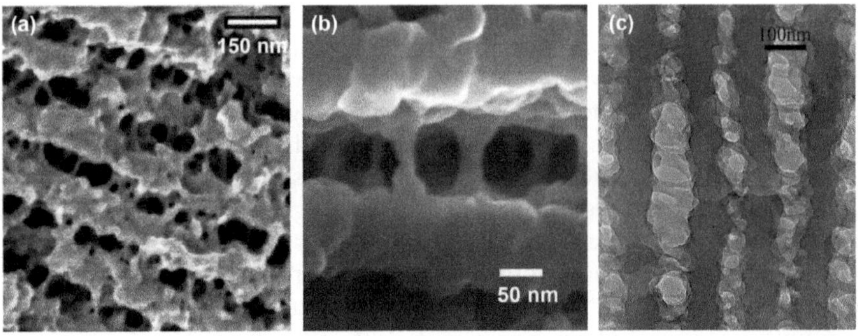

Figure 5.19 Morphologies of HPDLC reflection gratings formed from free-radical chemistry. The periodicity of such systems is typically much smaller than for transmission gratings and as such the domain sizes of the phase-separated structures are smaller. Images are typical SEM (a) and (b), and TEM (c) micrographs.

as in Figure 5.18 but in a reflection grating morphology, where the spacing and phase-separated structures are much smaller.

Two different volume fractions of LC domains are relevant within a periodic, heterogeneous two-phase morphology: the global volume fraction across many periods and the local volume fraction within the LC-rich region itself. Although both are related to the starting position on the phase diagram, the latter is greatly influenced by the complex play of kinetics during the grating formation. For systems that are on the verge of the two-phase window, a condition driven by the starting concentration of LC in the mixture and the periodicity, very small, nearly spherical domains can be formed; however, their confinement into a periodic structure is typically lessened. Both the global and the local volume fractions are decreased in this case, as more LC typically remains trapped in the polymer material (*i.e.*, there is a finite solubility of the LC in the polymer structure). In order to maximize the local volume fraction, a fine balance exists between the starting LC concentration in the syrup, the Bragg spacing, and the kinetics of photopolymerization. For the very small Bragg spacings necessary for reflection gratings in the visible, typically the fraction of the grating period consisting of the LC-rich regions is between 40 and 50% and highly interconnected domains are observed.

These interconnected LC-rich regions have been taken to the extreme in a transmission geometry by an Italian group which has used an elevated temperature, slow polymerization rate, and an active feedback process to form structures consisting of relatively pure slices of phase-separated nematic LC between polymer sheets (see Section 5.4). These structures, called POLICRYPS, have the advantage of lower switching voltage and reduced scatter relative to traditional HPDLCs, due to less interfacial matter. However, they have not been fully demonstrated to date in a reflection grating geometry and detailed high-resolution cross-sectional morphological characterization is still lacking.

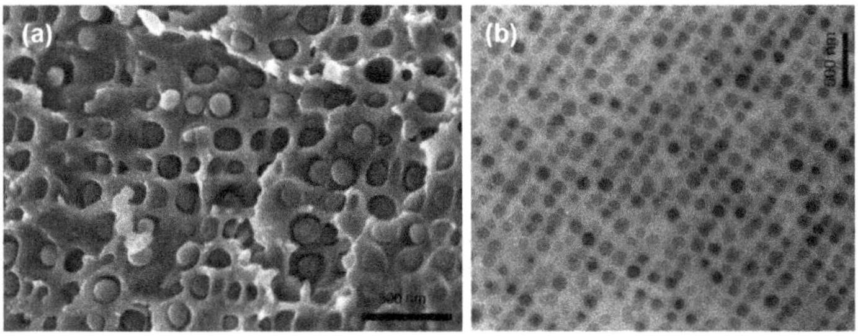

Figure 5.20 HPDLC reflection gratings formed from thiol-ene chemistry. Image (a) shows a cryo-SEM micrograph where the individual frozen LC domains are still present in many of the phase-separated LC domains. (b) is a TEM image showing the darker phase-separated LC domains.

For HPDLC gratings that are formed from a thiol-ene-based photochemistry system, the morphology is fundamentally different.[81,82] Because the polymerization reaction is step-growth, the system evolves as two immiscible fluid phases until the gelation point is reached in the polymer-rich phase. Under these conditions, individual near-spherical domains are typically formed due to surface tension effects in the systems, at least when the initial LC concentration is small or the Bragg spacing is large. Droplet coalescence can be observed in small spacing systems (reflection gratings) where the initial concentration of LC is high. The images in Figure 5.20 clearly indicate a difference in morphology relative to those discussed above, as a much more classic 'near spherical' domain morphology exists even though the droplet regions are periodic. Figure 5.20(a) was acquired cryogenically and, as such, the actual frozen LC droplets can be seen in some of the domains.

5.3.2 Electro-optical Properties of HPDLCs in Transmission Geometry

The diffractive properties of a transmissive Bragg grating can be described by the coupled wave theory of Kogelnik.[74] In this regime, the single fundamental diffracted beam has a diffraction efficiency given by:

$$\eta_d = \sin^2\gamma \tag{5.3}$$

with

$$\gamma = \frac{\pi\varepsilon_1 d \cos 2\theta}{2\lambda\sqrt{\varepsilon_0}\cos\theta} \tag{5.4}$$

where d is the thickness of the grating, ε_1 and ε_0 are the amplitude of the modulation and the average relative dielectric constants of the medium, respectively, and θ is the incidence angle of the probe radiation at the Bragg condition at wavelength λ. Typical electro-optic response of a transmission HPDLC structure is reported in Figure 5.21.[83]

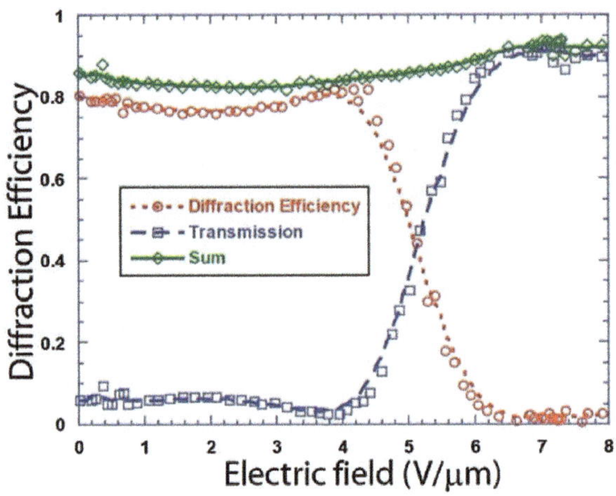

Figure 5.21 Electro-optical response of a HPDLC sample. Reproduced from ref. 83 with permission from The Authors.

The behavior of the first-order diffracted beam (Figure 5.21, red curve), zero-order transmitted beam (Figure 5.21, blue curve), and total transmission (Figure 5.21, green curve) are plotted *versus* the applied electric field (square wave, 1 KHz). At zero field, the beam is diffracted into the first order with large efficiency due to the mismatch in refractive index. With increasing field, the diffraction efficiency decreases and it approaches zero when the refractive index of the LC domains matches that of the surrounding polymer matrix (similar to the switching behavior of classic PDLCs described previously). The transmitted beam has the complementary behavior. It is worth pointing out that the total intensity (Figure 5.21, green curve) is only slightly less than 1 and remains almost constant for all the values of the applied external electric field. The high efficiency and small scattering losses are indicative of LC domains (droplets) of submicron size and a large droplet density gradient along the grating period direction. Sub-millisecond switching times (on and off) are typically observed and have been reported in several papers.[84,85]

The electro-optical properties of HPDLC structures can also be improved by adding monomers and additives into the initial composition. Fluorinated monomers have been shown to reduce the surface energy and thus the interaction of the polymer and LC molecules at the boundaries. Inert surfactant molecules have also been used to minimize these surface interactions.[29] Surfactants are like lubricants between the polymeric matrix and liquid crystals. During the phase separation process, the surfactant acts like an intermediate layer between the polymer and the LC by reducing the anchoring energy at the interface, thus allowing the LC molecules to reorient more easily upon the application of an electric field. HPDLCs with superior electro-optical properties have been obtained by doping the curing mixture with varying amounts of multiwalled carbon nanotubes[86] or plasmonic

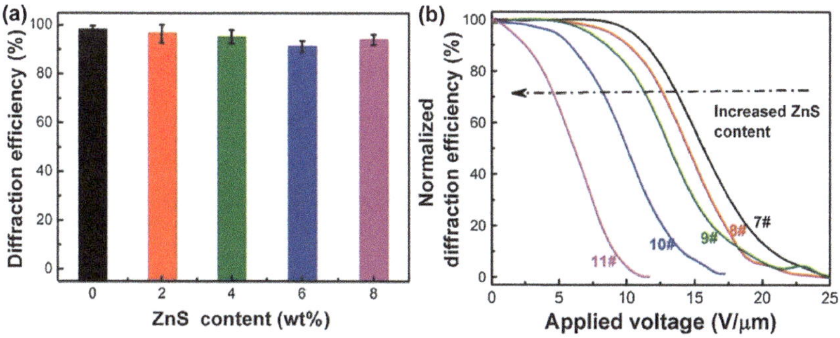

Figure 5.22 (a) P-polarized diffraction efficiency and (b) normalized diffraction efficiency *versus* applied voltage for HPDLC gratings with varied content of ZnS nanoparticles: 7#, zero; 8#, 2wt%; 9#, 4wt%; 10#, 6wt%; 11#, 8wt%; (a) and (b) reproduced from ref. 87 with permission from The Royal Society of Chemistry.

nanomaterials. For example, it has been reported that the diffraction efficiency and the electro-optical properties of HPDLCs are improved through the addition of Ag and Au NPs.

Very recently, ZnS nanoparticle doped HPDLC gratings with well-defined phase separation have been reported.[87] The realized structures exhibited very high diffraction efficiencies and low switching voltages. The presence of ZnS nanoparticles did not affect the diffraction efficiency of the HPDLC gratings (Figure 5.22a) whereas the threshold voltage decreased from 11.6 $V\,\mu m^{-1}$ to 2.5 $V\,\mu m^{-1}$ (Figure 5.22b). High resolution SEM characterization showed that there is no apparent change in the morphology of the gratings upon varying the ZnS concentration, and well-defined scaffold structures are obtained in all cases.

5.3.3 Electro-optical Properties of HPDLCs in Reflection Geometry

Switchable reflective gratings formed in HPDLCs are the subject of significant current technological and scientific investigation. Potential application for HPDLC materials include wavelength division multiplexing, reflective color display, photonic crystals, and novel narrow notch spectral filtering devices.[88,89]

A typical holographic setup for realizing HPDLC structures in reflection mode is shown in Figure 5.23. An expanded and collimated CW laser (for example with $\lambda = 532$ nm) is divided into two parts by a beam splitter. The two beams (1 and 2) are reflected by two mirrors (M1 and M2) and overlap in a counter-propagating geometry at the sample, giving rise to an interference pattern with spatial periodicity from 200 nm to several micrometers, allowing the fabrication of reflection gratings with a reflection notch across the visible and into the NIR region, as shown in Figure 5.24.

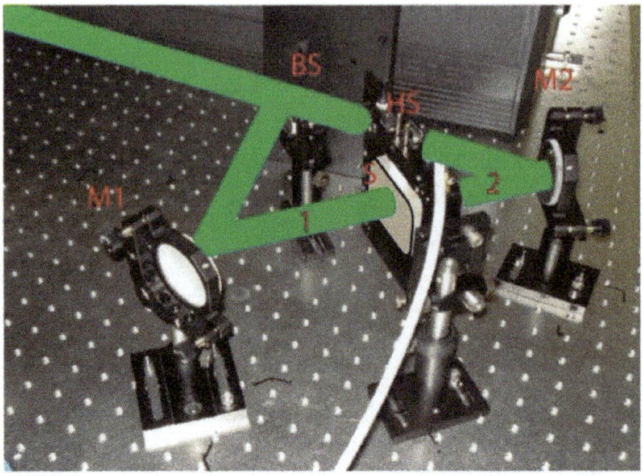

Figure 5.23 Setup for the realization of holographic reflective gratings: BS, beam splitter; M1, M2 mirrors; HS, hot stage; S, sample.
Reproduced from ref. 97 with permission from The Optical Society of America.

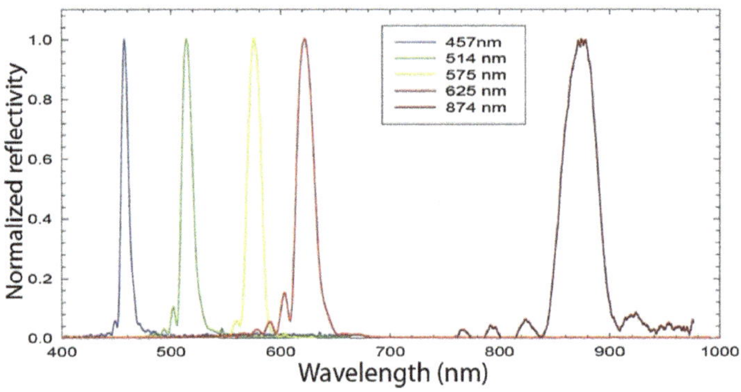

Figure 5.24 Spectral response of several reflection HPDLC samples with different grating periods.
Reproduced from ref. 83 with permission from The Authors.

Complex holographic structures (2D and 3D) in HPDLC geometries have been realized such as orthorhombic, face-centered cubic, transverse square, diamond-like, and Penrose structures.[70,72,73] Particularly, photonic crystals written in HPDLC materials show the typical correlation between the diffractive properties and the applied electric field. The index modulation between the polymer and the LC is much too low to be considered for applications that require a complete bandgap, but dynamic modulation (grey-scale) between an angle-dependent diffraction and a transparent *on*-state is achievable. Much larger index modulations are obtainable by removing the LC but then all dynamic behavior is lost.

Experimental realization of mirrorless lasers in the last decade has provided a compelling tool for the fabrication of on-chip photonic devices.

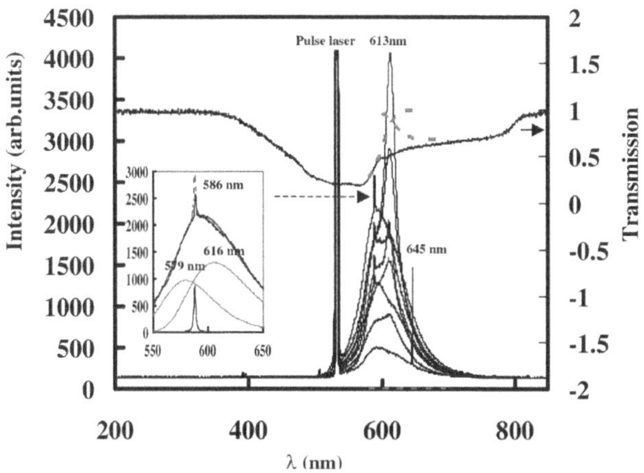

Figure 5.25 Emission spectra of an Nd:YAG laser-pumped HPDLC reflection grating. The inset shows the peak fitting of a lasing line at 586 nm with two broad reflections centered at 579 nm and 616 nm. The lasing peak is a Lorentzian shape with a full width at half maximum of ~3 nm. Reproduced from ref. 91 with permission from The Japan Society of Applied Physics, Copyright 2005.

To this end, HPDLCs have been exploited in several geometries and compositions for the realization of high quality and inexpensive lasers.[90,91] For example, it has been demonstrated that by adding a small amount of organic dye (~1wt% DCM, Exciton) to a reflective HPDLC composition, it was possible to achieve a lasing effect. To do so, the dye-doped HPDLC reflection grating was pumped using a Q-switched Nd:YAG pulsed laser (~8 ns). Figure 5.25 shows the amplified spontaneous emission (ASE) of the HPDLC with increasing pump energy from 1 mJ to 25 mJ per pulse, along with the transmission spectrum of the reflective structure. The inset of Figure 5.25 displays the peak fitting of a curve that shows lasing (at 586 nm) with two reflections.

5.4 POLICRYPS Gratings

The optical properties of HPDLC devices strongly depend on the presence of a periodic distribution of LC droplets inside the gratings. Because LC droplets need to be nano-sized in HPDLCs in order to limit the fraction of light scattered, the switching voltages are typically high. Several years ago, an attempt has been made to fabricate a new kind of holographic grating with new and intriguing morphologies. These gratings consist of polymer slices alternated with films of regularly aligned nematic liquid crystals, NLC (named POLICRYPS, acronym for POlymer LIquid CRYstal Polymer Slices).[92–94]

The basic idea for the realization of POLICRYPS is to avoid the formation of separate NLC droplets during the recording process and to obtain only a macroscopic phase separation due to the complete re-distribution of nematic and monomer components inside the POLICRYPS mixture during the

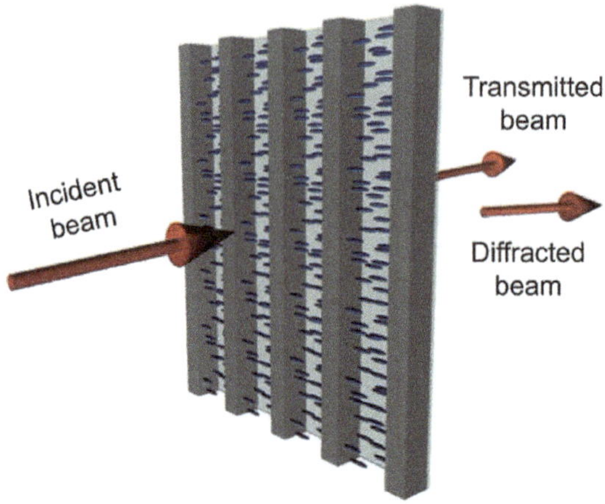

Figure 5.26 Schematic of the configuration of a transmission POLICRYPS structure. The grey regions represent the polymer slices and the blue ellipsoids are the aligned nematic liquid crystal molecules.
Reproduced from ref. 93 with permission from IOP Publishing.

writing of the grating (Figure 5.26). This result is obtained by utilizing the high diffusivity of the NLC molecules in the isotropic state. POLICRYPS gratings have been realized using a new technique called MPTIPS, or *m*ixed *p*olymerization *t*hermal *i*nduced *p*hase *s*eparation. The standard procedure can be summarized with the following steps: (a) heating of a blend made of monomer, photoinitiator and NLC above the nematic to isotropic phase transition; (b) illumination of the curing mixture (confined in a glass cell) with the interference pattern of a curing UV/visible radiation; (c) slow cooling to room temperature of the sample after the curing radiation has been switched off. High quality periodic structures with very well aligned LC stripes and no evidence of microscopic NLC droplets have been obtained from this method (Figure 5.27).

Electro-optical experiments show (Figure 5.28) that POLICRYPS gratings can be switched off with a relatively low external field (6 V μm^{-1}). At zero field the structure has a large diffraction efficiency. By applying an external voltage (square wave, 1 KHz) the applied electric field induces the reorientation of the NLC director and an incident probe light is completely transmitted due to the refractive index matching between the polymer and the NLC. POLICRYPS gratings are completely reversible when the applied field is turned off. Typical response times are about 1–2 ms. POLICRYPS structures have been used in devices of interest for several applications ranging from photonics[95] to plasmonics.[96]

Very recently, the first demonstration of reflective POLICRYPS structures has been reported, which could lead to the realization of smart Bragg mirrors with superior optical and electro-optical properties.[97] As sketched in the cartoon in Figure 5.29a, the reflective POLICRYPS is made of polymeric

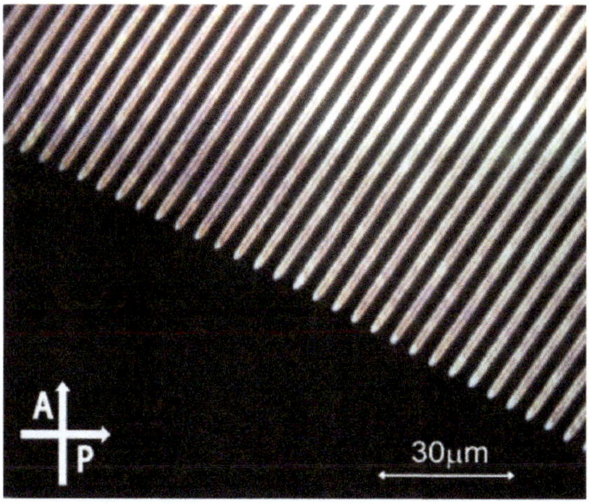

Figure 5.27 Polarized optical microscope view of a transmission POLICRYPS grating.
Reproduced from ref. 94 with permission from John Wiley and Sons, Copyright © 2013 Wiley Periodicals, Inc.

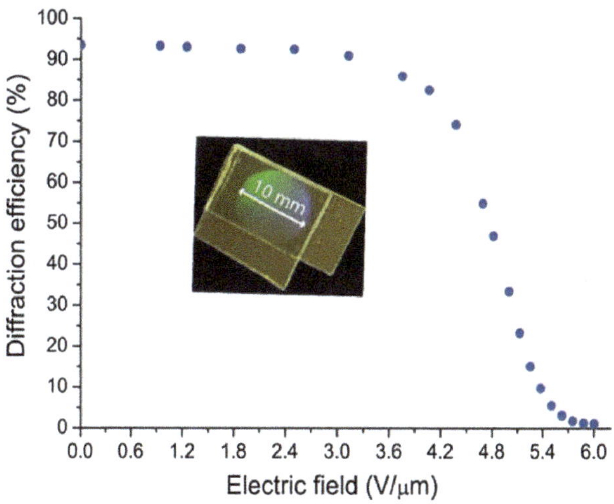

Figure 5.28 Electro-optical property of a transmission POLICRYPS grating.
Reproduced from ref. 94 with permission from John Wiley and Sons, Copyright © 2013 Wiley Periodicals, Inc.

layers alternated to very well phase-separated, vertically aligned NLC layers. The geometry was confirmed by analyzing the realized sample with a polarized optical microscope in conoscopic mode. The observed dark Maltese cross (Figure 5.29b) confirmed that the NLC molecules were homeotropically aligned within the polymeric films. As a consequence, the sample is clear (Figure 5.29c) under ambient light due to the refractive index matching

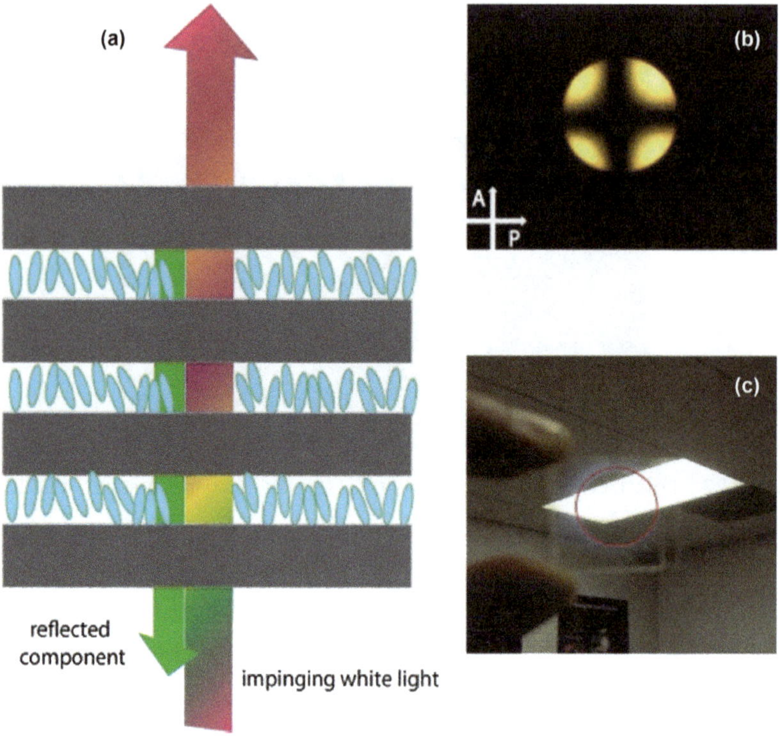

Figure 5.29 (a) Schematic of the NLC configuration in reflective POLICRYPS structures. (b) Conoscopic image and (c) photograph of a POLICRYPS sample. Reproduced from ref. 97 with permission from The Optical Society of America.

between the polymer ($n_p \approx 1.54$) and the ordinary index ($n_o \approx 1.50$) of the NLC molecules.

To summarize, the combination of low molar mass liquid crystals mixed in a polymer matrix leads to systems which have interesting dynamic optical properties. The utility of such 'isotropic' PDLC systems in dynamic scattering applications and periodic PDLC systems in dynamic diffractive and photonic applications is still relatively untapped. The major limitation to date has been the high voltages needed to induce modulation. However, as this chapter demonstrates, a clear understanding of the phase separation details and the associated wide variety of morphologies that can be obtained and optimized lead to continued optimism that significant improvements can still be obtained. The simple perspective of round LC droplets that are separated in a 'Swiss cheese'-like morphology is limiting. A wide variety of parameters can be controlled to generate a range of complex and nuanced morphologies, all having specific attributes depending on the application. A proper understanding of the phase separation kinetics and all the variables that control this is key. In the periodic PDLC case, this understanding has to be balanced with a sound understanding of diffusion kinetics and localization-dependent behavior.

Acknowledgements

The authors are thankful to Ms Pamela Lloyd for help with the preparation of figures with SEM and TEM images. Financial support for this work was provided in part by the Air Force Office of Scientific Research (Grant 15RXCOR179).

References

1. J. L. West, in *Liquid-Crystalline Polymers*, ACS Symposium Series, American Chemical Society, 1990, vol. 435, pp. 475–495.
2. P. S. Drzaic, *Liquid Crystal Dispersions*, World Scientific, Singapore, 1995.
3. D. Coates, *J. Mater. Chem.*, 1995, **5**, 2063.
4. M. Mucha, *Prog. Polym. Sci.*, 2003, **28**, 837.
5. S. Bronnikov, S. Kostromin and V. Zuev, *J. Macromol. Sci., Part B: Phys.*, 2013, **52**, 1718.
6. R. Ondris-Crawford, E. P. Boyko, B. G. Wagner, J. H. Erdmann, S. Žumer and J. W. Doane, *J. Appl. Phys.*, 1991, **69**, 6380.
7. J. L. Fergason, *US Pat.* 4,435,047, 1984.
8. P. S. Drzaic, *J. Appl. Phys.*, 1986, **60**, 2142.
9. P. Drzaic, *Liq. Cryst.*, 2006, **33**, 1281.
10. J. W. Doane, N. A. Vaz, B.-G. Wu and S. Žumer, *Appl. Phys. Lett.*, 1986, **48**, 269.
11. Y. Hirai, S. Niiyama, H. Kumai and T. Gunjima, *Proc. SPIE*, 1990, **1257**, 2.
12. F. G. Yamagishi, L. J. Miller and C. I. van Ast, *Proc. SPIE*, 1989, **1080**, 24.
13. R. A. Vaia, D. W. Tomlin, M. D. Schulte and T. J. Bunning, *Polymer*, 2001, **42**, 1055.
14. R. S. Justice, D. W. Schaefer, R. A. Vaia, D. W. Tomlin and T. J. Bunning, *Polymer*, 2005, **46**, 4465.
15. N. J. Crawford, M. D. Dadmun, T. J. Bunning and L. V. Natarajan, *Polymer*, 2006, **47**, 6311.
16. A. J. Lovinger, K. R. Amundson and D. D. Davis, *Chem. Mater.*, 1994, **6**, 1726.
17. R. Bhargava, S.-Q. Wang and J. L. Koenig, *Macromolecules*, 1999, **32**, 8989.
18. D. Nwabunma, H.-W. Chiu and T. Kyu, *Macromolecules*, 2000, **33**, 1416.
19. G. P. Montgomery, J. L. West and W. Tamura-Lis, *J. Appl. Phys.*, 1991, **69**, 1605.
20. T. J. White, L. V. Natarajan, V. P. Tondiglia, T. J. Bunning and C. A. Guymon, *Macromolecules*, 2007, **40**, 1112.

21. T. J. White, L. V. Natarajan, T. J. Bunning and A. Guymon, *Liq. Cryst.*, 2007, **34**, 1377.
22. P. S. Drzaic, *Liq. Cryst.*, 1988, **3**, 1543.
23. M. Mucha, *J. Appl. Polym. Sci.*, 1991, **43**, 175.
24. S. C. Jain and D. K. Rout, *J. Appl. Phys.*, 1991, **70**, 6988.
25. G.-M. Zhang, Z. Hong, C. Zhou, B. G. Wu and J. W. Lin, *Proc. SPIE*, 1992, **1815**, 233.
26. S. C. Jain, R. S. Thakur and S. T. Lakshmikumar, *J. Appl. Phys.*, 1993, **73**, 3744.
27. Z. J. Lu and D. K. Yang, *Appl. Phys. Lett.*, 1994, **65**, 505.
28. V. P. Tondiglia, L. V. Natarajan, R. M. Meal, R. L. Sutherland and T. J. Bunning, *Mater. Res. Soc. Symp. Proc.*, 1997, **479**, 235.
29. J. Klosterman, L. V. Natarajan, V. P. Tondiglia, R. L. Sutherland, T. J. White, C. A. Guymon and T. J. Bunning, *Polymer*, 2004, **45**, 7213.
30. D. Coates, *Displays*, 1993, **14**, 94.
31. Y.-D. Ma, B. G. Wu and G. Xu, *Proc. SPIE*, 1990, **1257**, 46.
32. F. P. Nicoletta, G. De Filpo, J. Lanzo and G. Chidichimo, *Appl. Phys. Lett.*, 1999, **74**, 3945.
33. T. Gotoh and H. Murai, *Appl. Phys. Lett.*, 1992, **60**, 392.
34. M. J. Sansone, G. Khanarian, T. M. Leslie, M. Stiller, J. Altman and P. Elizondo, *J. Appl. Phys.*, 1990, **67**, 4253.
35. M. J. Sansone, G. Khanarian and M. S. Kwiatek, *J. Appl. Phys.*, 1994, **75**, 1715.
36. S. Matsumoto, M. Houlbert, T. Hayashi and K.-i. Kubodera, *Appl. Phys. Lett.*, 1996, **69**, 1044.
37. S. Matsumoto, Y. Sugiyama, S. Sakata and T. Hayashi, *Liq. Cryst.*, 2000, **27**, 649.
38. P. J. W. Hands, A. K. Kirby and G. D. Love, *Proc. SPIE*, 2005, **5894**, 58940L.
39. D. E. Lucchetta, R. Karapinar, A. Manni and F. Simoni, *J. Appl. Phys.*, 2002, **91**, 6060.
40. S. Aya, K. V. Le, F. Araoka, K. Ishikawa and H. Takezoe, *Jpn. J. Appl. Phys.*, 2011, **50**, 051703.
41. H. Ren, Y.-H. Fan, Y.-H. Lin and S.-T. Wu, *Opt. Commun.*, 2005, **247**, 101.
42. A. Hinojosa and S. C. Sharma, *Appl. Phys. Lett.*, 2010, **97**, 081114.
43. C. C. Hsu, Y. X. Chen, H. W. Li and J. S. Hsu, *Opt. Express*, 2016, **24**, 7063.
44. W. B. Li, M. J. Zhu, X. K. Ding, B. F. Li, W. Huang, H. Cao, Z. Yang and H. Yang, *J. Appl. Polym. Sci.*, 2009, **111**, 1449.
45. S. V. Kalashnikov, N. A. Romanov and A. V. Nomoev, *J. Appl. Phys.*, 2016, **119**, 094304.
46. H. Shim, H. K. Lyu, B. Allabergenov, Y. Garbovskiy, A. Glushchenko and B. Choi, *Liq. Cryst.*, 2016, **43**, 1390.
47. V. Rachet, K. Lahlil, M. Berard, T. Gacoin and J. P. Boilot, *J. Am. Chem. Soc.*, 2007, **129**, 9274.
48. P. S. Drzaic, R. Wiley and J. McCoy, *Proc. SPIE*, 1989, **1080**, 41.
49. P. S. Drzaic, *Displays*, 1991, **12**, 2.
50. J. L. West, R. Ondris and M. Erdmann, *Proc. SPIE*, 1990, **1257**, 76.

51. J. L. West and R. Ondris-Crawford, *J. Appl. Phys.*, 1991, **70**, 3785.
52. P. Malik and K. K. Raina, *Phys. B*, 2010, **405**, 161.
53. K. J. Yang, S. C. Lee and B. D. Choi, *Jpn. J. Appl. Phys.*, 2010, **49**, 05EA05.
54. R. R. Deshmukh and A. K. Jain, *Liq. Cryst.*, 2014, **41**, 960.
55. P. Kumar, V. Sharma, C. Jaggi and K. K. Raina, *Liq. Cryst.*, 2017, **44**, 757.
56. A. Romani, G. Chidichimo, P. Formoso, S. Manfredi, G. Favaro and U. Mazzucato, *J. Phys. Chem. B*, 2002, **106**, 9490.
57. M. Macchione, G. De Filpo, F. P. Nicoletta and G. Chidichimo, *Liq. Cryst.*, 2005, **32**, 315.
58. G. De Filpo, P. Formoso, S. Manfredi, A. I. Mashin and F. P. Nicoletta, *Liq. Cryst.*, 2017, **44**, 1607.
59. A. Y. G. Fuh, C. C. Chen, C. K. Liu and K. T. Cheng, *Opt. Express*, 2009, **17**, 7088.
60. S. Gottardo, S. Cavalieri, O. Yaroshchuk and D. S. Wiersma, *Phys. Rev. Lett.*, 2004, **93**, 263901.
61. H. Harada, M. Nakatsu and H. Naito, *Jpn. J. Appl. Phys.*, 2005, **44**, L915.
62. Y. J. Liu, X. W. Sun, H. I Elim and W. Ji, *Appl. Phys. Lett.*, 2006, **89**, 011111.
63. G. R. Xiong, G. Z. Han, C. Sun, H. Xu, H. M. Wei and Z. Z. Gu, *Adv. Funct. Mater.*, 2009, **19**, 1082.
64. Y. Inoue, H. Yoshida, H. Kubo and M. Ozaki, *Adv. Opt. Mater.*, 2013, **1**, 256.
65. R. A. M. Hikmet, in *Liquid Crystals in Complex Geometries Formed by Polymer and Porous Networks*, ed. G. P. Crawford and S. Žumer, Taylor & Francis, London, 1996, pp. 53–82.
66. Y. Inoue, H. Yoshida and M. Ozaki, *Opt. Mater. Express*, 2014, **4**, 916.
67. E. Ouskova, L. De Sio, R. Vergara, T. J. White, N. Tabiryan and T. J. Bunning, *Appl. Phys. Lett.*, 2014, **105**, 231122.
68. L. De Sio, E. Ouskova, P. Lloyd, R. Vergara, N. Tabiryan and T. J. Bunning, *Opt. Mater. Express*, 2017, **7**, 1581.
69. T. J. Bunning, L. V. Natarajan, V. P. Tondiglia and R. L. Sutherland, *Annu. Rev. Mater. Sci.*, 2000, **30**, 83.
70. V. P. Tondiglia, L. V. Natarajan, R. L. Sutherland, D. Tomlin and T. J. Bunning, *Adv. Mater.*, 2002, **14**, 187.
71. R. L. Sutherland, V. P. Tondiglia, L. V. Natarajan, S. Chandra, D. Tomlin and T. J. Bunning, *Opt. Express*, 2002, **10**, 1074.
72. M. J. Escuti, J. Qi and G. P. Crawford, *Appl. Phys. Lett.*, 2003, **83**, 1331.
73. Y. J. Liu and X. W. Sun, *Jpn. J. Appl. Phys.*, 2007, **46**, 6634.
74. H. Kogelnik, *Bell Syst. Tech. J.*, 1969, **48**, 2909.
75. M. G. Moharam and T. K. Gaylord, *J. Opt. Soc. Am.*, 1981, **71**, 811.
76. S. Li and J. Wang, *Sci. Rep.*, 2015, **5**, 15406.
77. D. Munekazu, T. Yoshie and K. Kinya, *J. Phys. D: Appl. Phys.*, 1998, **31**, 2225.
78. Y. J. Liu and X. W. Sun, *Adv. OptoElectron.*, 2008, **2008**, 684349.
79. E. Kim, J. Woo and B. Kim, *Liq. Cryst.*, 2007, **34**, 79.
80. M. Jazbinšek, I. Drevenšek Olenik, M. Zgonik, A. K. Fontecchio and G. P. Crawford, *J. Appl. Phys.*, 2001, **90**, 3831.

81. L. V. Natarajan, D. P. Brown, J. M. Wofford, V. P. Tondiglia, R. L. Sutherland, P. F. Lloyd and T. J. Bunning, *Polymer*, 2006, **47**, 4411.
82. A. F. Senyurt, G. Warren, J. B. Whitehead and C. E. Hoyle, *Polymer*, 2006, **47**, 2741.
83. L. De Sio, N. Tabiryan, T. Bunning, B. R. Kimball and C. Umeton, in *Progress in Optics*, ed. E. Wolf, Elsevier, 2013, vol. 58, pp. 1–64.
84. L. V. Natarajan, R. L. Sutherland, V. P. Tondiglia, T. J. Bunning and W. W. Adams, *J. Nonlinear Opt. Phys. Mater.*, 1996, **05**, 89.
85. L. V. Natarajan, C. K. Shepherd, D. M. Brandelik, R. L. Sutherland, S. Chandra, V. P. Tondiglia, D. Tomlin and T. J. Bunning, *Chem. Mater.*, 2003, **15**, 2477.
86. K. R. Sun and B. K. Kim, *Polym. Adv. Technol.*, 2011, **22**, 1993.
87. M. Ni, G. Chen, H. Sun, H. Peng, Z. Yang, Y. Liao, Y. Ye, Y. Yang and X. Xie, *Mater. Chem. Front.*, 2017, **1**, 294.
88. R. L. Sutherland, V. P. Tondiglia, L. V. Natarajan and T. J. Bunning, *Appl. Phys. Lett.*, 2001, **79**, 1420.
89. L. V. Natarajan, R. L. Sutherland, V. P. Tondiglia, S. Siwecki, R. Pogue, M. Schmitt, D. Brandelik, B. Epling, G. Berman, C. Wendel, M. Ritter, M. Stallings and T. J. Bunning, *MRS Online Proc. Libr.*, 1999, **559**, 109.
90. M. Liu, Y. Liu, G. Zhang, Z. Peng, D. Li, J. Ma and L. Xuan, *J. Phys. D: Appl. Phys.*, 2016, **49**, 465102.
91. S.-T. Wu and A. Y.-G. Fuh, *Jpn. J. Appl. Phys.*, 2005, **44**, 977.
92. R. Caputo, L. De Sio, A. Veltri, C. Umeton and A. V. Sukhov, *Opt. Lett.*, 2004, **29**, 1261.
93. R. Caputo, A. DeLuca, L. De Sio, L. Pezzi, G. Strangi, C. Umeton, A. Veltri, R. Asquini, A. d'Alessandro, D. Donisi, R. Beccherelli, A. V. Sukhov and N. V. Tabiryan, *J. Opt. A: Pure Appl. Opt.*, 2009, **11**, 024017.
94. L. De Sio and N. Tabiryan, *J. Polym. Sci., Part B: Polym. Phys.*, 2014, **52**, 158.
95. L. De Sio, N. Tabiryan, R. Caputo, A. Veltri and C. Umeton, *Opt. Express*, 2008, **16**, 7619.
96. L. De Sio, R. Caputo, U. Cataldi and C. Umeton, *J. Mater. Chem.*, 2011, **21**, 18967.
97. L. De Sio, N. Tabiryan and T. J. Bunning, *Opt. Express*, 2015, **23**, 32696.

CHAPTER 6

Introduction to Polymer Stabilized Liquid Crystals

INGO DIERKING

School of Physics and Astronomy, University of Manchester, Oxford Road, Manchester M13 9PL, UK
Email: Ingo.Dierking@manchester.ac.uk

6.1 General Sample Preparation

Polymer stabilized liquid crystals[1–3] are obtained towards the small monomer/polymer concentration side of the polymer–liquid crystal phase diagram, generally at polymer concentrations of less than approximately 10%. Typical bifunctional, photo-reactive monomers, often mesogenic by themselves, are shown in Figure 6.1. Pioneering work on such materials was done by the Broer and the Philips group.[4–6] These are mixed into a liquid crystal host, together with a small amount of a photo-initiator, often benzoin methyl ether (BME) or Irgacure 184, a hydroxyketone, at about 1% of the monomer concentration (see Figure 6.1). The photo-initiator starts the polymerization reaction but is not used up during the process. To achieve a molecularly uniform dispersion, the liquid crystal, monomer and photo-initiator are generally dissolved in an organic solvent like dimethylether and mixed while the solvent is slowly evaporated. Care has to be taken that the solvent evaporates completely. This has to be carried out under red light conditions, because otherwise the polymerization reaction occurs.

After preparation of the dispersion, it then fills the sandwich cells by capillary action and the liquid crystal is oriented in the desired configuration, which can be achieved by using boundary conditions, orientation

Soft Matter Series No. 8
Polymer-modified Liquid Crystals
Edited by Ingo Dierking

Published by the Royal Society of Chemistry, www.rsc.org

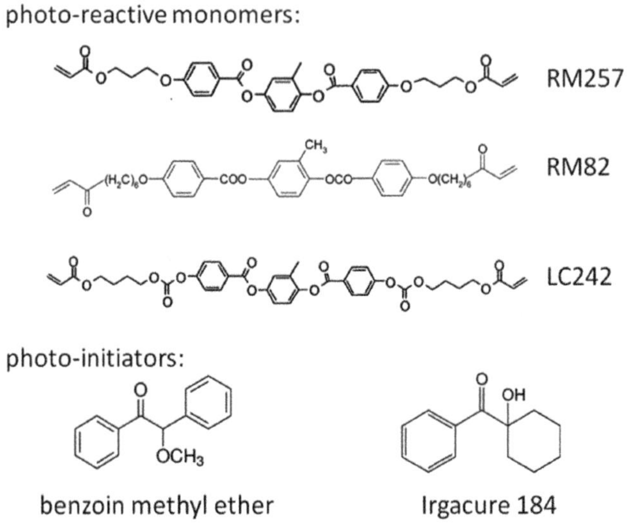

Figure 6.1 Some commonly used, commercially available, bifunctional photo-reactive monomers, and respective photo-initiators to promote the UV induced polymerization process within a liquid crystal host.

layers, or electric and magnetic field treatments. The sample is then ready for the creation of the stabilizing polymer network, which is generally formed at constant temperature through illumination by UV light. During the polymerization, a cross-linked network is formed that phase separates from the liquid crystal, as can be confirmed by infrared spectroscopy or neutron scattering.[7] As a rule of thumb, as we will see below, it is usually desirable to form the polymer network at (i) temperatures well below the phase transition temperature, (ii) by using relatively small UV intensities (otherwise the temperature may also change), and (iii) carrying out the polymerization until full monomer conversion is achieved, which is normally the case after about 60–90 minutes. This will ensure a slow continuous polymerization process, which leads to well established networks and reproducible samples. The whole process has to be carried out in a darkroom under red light conditions. The temperature can be controlled by a standard temperature controller with hotstage, and the UV illumination can be applied for example with a standard hand held UV lamp (~10 mW), available from chemical lab supply companies or UV curing lamps from dental suppliers or nail studio suppliers.

6.2 Polymer Networks Templating Liquid Crystalline Order

The general idea behind the formation of polymer stabilized liquid crystals is to stabilize a macroscopic liquid crystalline structure, which includes the smectic layer arrangement if present, as well as the director field of orientational order. The polymer network thus forms a template of the liquid crystalline order in which it was formed. At the same time, a large internal surface area is created, which is much larger than that of the bounding substrates. This implies that the elastic interaction between the liquid crystal and the polymer network is large, which is exploited to achieve faster response times back to the field-off state of a device, than achievable *via* substrate surfaces alone. A classic example would be a polymer stabilized cholesteric phase, driven from a uniform Grandjean texture to a focal conic texture by electric field application. After turning the field off, the re-orientation back to the Grandjean state normally takes minutes, hours or days for a non-stabilized cholesteric, if it is achieved at all. In a polymer stabilized cholesteric, this re-orientation is achieved in the time period of milliseconds. But of course, many other applications can also be envisioned through the possibility of stabilizing certain structures and configurations, which will be discussed in more detail in the chapters below.

As the basic intention of polymer stabilized liquid crystals (PSLCs) is to template the liquid crystal structure,[8–10] several examples will be presented, discussing different phases, and different configurations. The simplest, and possibly the most illustrative example is the nematic phase under planar and homeotropic boundary conditions.[11,12] This is demonstrated as a scanning electron microscopy (SEM) top view of the planar nematic network after removal of the liquid crystal (Figure 6.2(a)), and a side view of a homeotropic nematic in Figure 6.2(b). For the planar configuration a clear directionality of the network is already observed by visual inspection, and it is found that the orientational order parameter of the network corresponds roughly to that of the nematic liquid crystal, S ~0.7. In the homeotropic case, the side view of the cell clearly demonstrates that the network is formed from the bottom of the cell to the top, connecting both substrates and being perpendicular to the substrate plane. In Figure 6.2(b) it is also visible that the network must have been formed by UV illumination of the cell from the

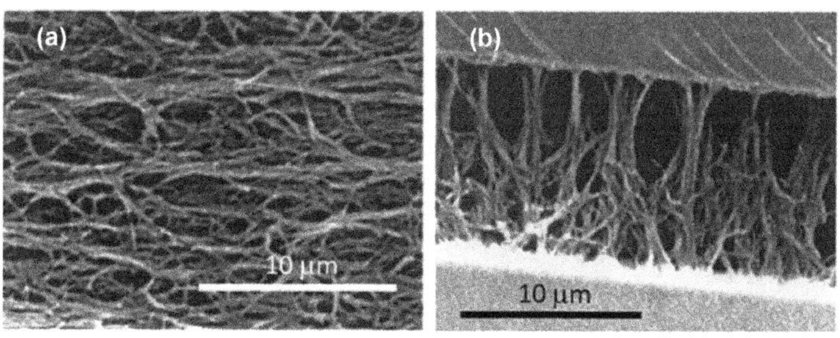

Figure 6.2 Electron scanning microscopy (SEM) images of a polymer network formed in a nematic liquid crystal under (a) planar and (b) homeotropic boundary conditions. For the imaging process, the polymer network strands are coated with a very thin layer of gold, after the liquid crystal is removed by a suitable organic solvent.
Reproduced from ref. 11 with permission from Taylor and Francis.

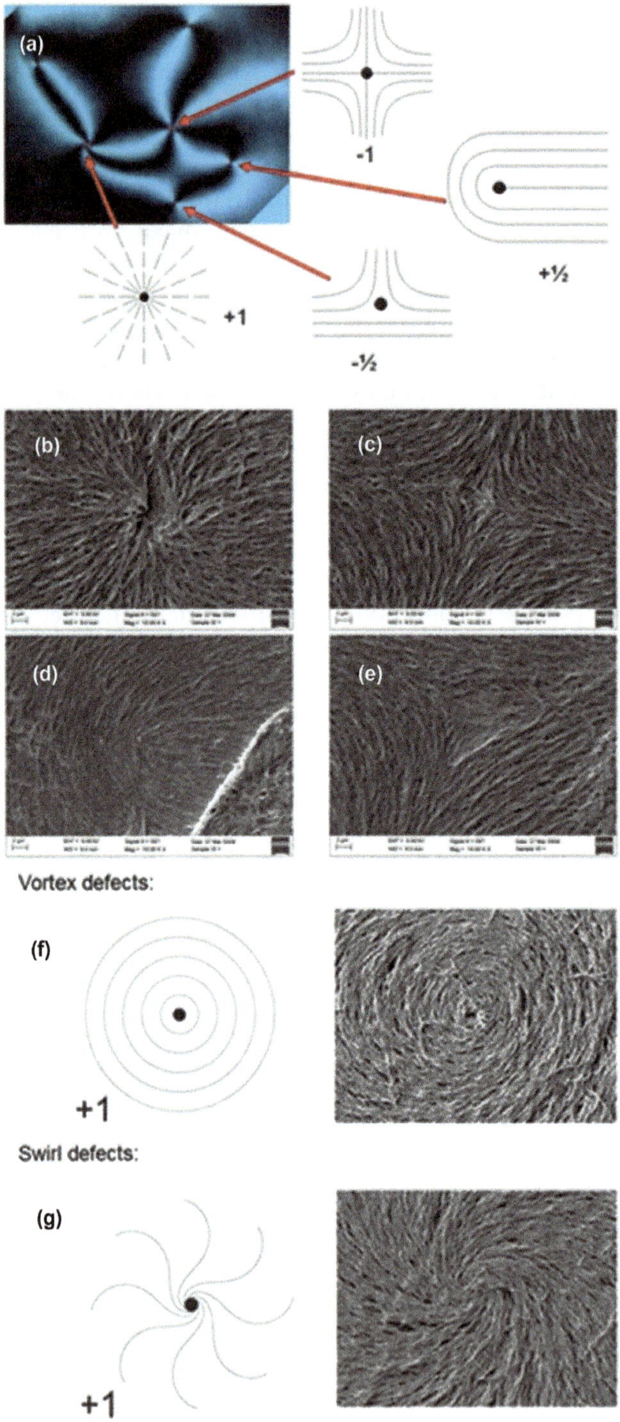
(a)
-1
+½
+1
-½
(b)
(c)
(d)
(e)
Vortex defects:
(f)
+1
Swirl defects:
(g)
+1

bottom substrate, as the network density changes through the cell from higher to lower UV intensity illumination.

It is well known that the nematic phase exhibits topological defects of different sign and strength.[13,14] These can for example appear after a temperature quench across the isotropic-nematic transition, where defects of strength $s = \pm 1/2$ and $s = \pm 1$ are observed. A texture showing all four types of topological defects is shown in Figure 6.3(a), together with the respective associated director fields indicated as lines. The singularities are indicated by a black dot. Very similar defects, but exclusively of strength $s = \pm 1$, can be generated by applying an electric field to a homeotropically oriented nematic with negative dielectric anisotropy, $\Delta\varepsilon = \varepsilon_{\parallel} - \varepsilon_{\perp} < 0$. In this case the size of the defect core is finite, and one speaks of umbilical defects.[15]

Figure 6.3(b)–(e) demonstrate that the formed polymer network not only allows imaging of the director fields of all the discussed defects by following the macroscopic directions of the director, but indeed templates the defect structure. Other, more exotic $s = +1$ defect structures can also be identified, like the vortex defect (Figure 6.3(f)) and the swirl defect (Figure 6.3(g)), as reported in reference.[16] Without the polymer network it would be impossible to distinguish these from the $s = +1$ defect of Figure 6.3(b). Polarized optical microscopy would show the same four-fold Schlieren structure around the defect core, and sample rotation would lead to the positive sign of the defect, but the differing director field could not be made visible.

Besides defects, other structures of liquid crystal phases can also be made visible by polymer stabilization, which is used extensively to increase the temperature interval of existance of the Blue Phases, for example. Normally, the cubic Blue Phase defect lattice phases, BP, only exist over temperature regions of approximately 1 K or even less, but as shown by Kikuchi *et al.*,[17] this can be increased to several tens of degrees by employing carefully designed polymer network stabilization. This stabilization enables the exploitation of the Kerr effect of Blue Phases in display applications without alignment layers and at a much faster switching speed than the standard nematics. This implies better performance at lower production costs, a topic which is discussed in much more detail below.

The fact that Blue Phases exhibit a cubic structure of defects, which can be local (macroscopically amorphous) as in BPIII, simple cubic as in BPII or

Figure 6.3 (a) Nematic texture with the commonly observed topological defects of strength $s = \pm 1/2$ and $s = \pm 1$. The corresponding director field configurations are shown schematically as lines. SEM images of the polymer networks following the director field of the four standard topological defects, (b) $s = +1$, (c) $s = -1$, (d) $s = +1/2$ and (e) $s = -1/2$. Non-standard $s = +1$ defects may also give images, like (f) the vortex defect and (g) the swirl defect. The latter two topological defects cannot be distinguished from that of part (b) by polarized optical microscopy. Reproduced from ref. 16 with permission from the Royal Society of Chemistry.

body centred cubic as in BPI, can be shown using Kossel diagrams. Polymer stabilization, the formation of a polymer network, which appears to be formed within the defects, can also be used to image different lattice planes, with an example given in Figure 6.4, showing the (1 0 0)-plane. Images of further lattice planes have been reported in ref. 17. It is believed that firstly, while it is formed, the polymer network is accumulated within the defect regions to lower the free energy of the BP, which would correspond to an expansion of the existence region. Secondly, it appears that as the polymer grows and crosslinks, it is also expelled from highly twisted regions, *i.e.* the double twist cylinders which make up the Blue Phases. This point is discussed below for other phases.

The cholesteric phase is the one observed when cooling a chiral liquid crystal from the isotropic liquid or the Blue Phase. It is the helical analogy of the nematic phase, with the director twisted to form a helical superstructure with helix axis perpendicular to the long axis of the mesogen. Confined to sandwich cells of a specific gap and with planar boundary conditions, the cholesteric phase forms a helix with helical axis perpendicular to the substrates, which is called the Grandjean orientation. Due to the boundary conditions, the pitch is adjusted so that an integer number of π-twists fits within the gap. Such a structure is also quite often found in biological systems, as realized by Yves Bouligand, who demonstrated the helix through the arcs visible when cutting a helix at an oblique angle, a so-called Bouligand cut,[18] as schematically shown in Figure 6.5(a). For polymer stabilized cholesterics one can employ the same method to demonstrate that the network is templating the helical superstructure, which is shown as an oblique cut through a sandwich cell (Figure 6.5(b)). For the investigation the helical pitch was adjusted to 10 μm, confined to a gap of 15 μm. One would thus expect three π-twists. The resulting SEM image of Figure 6.5(c) clearly shows

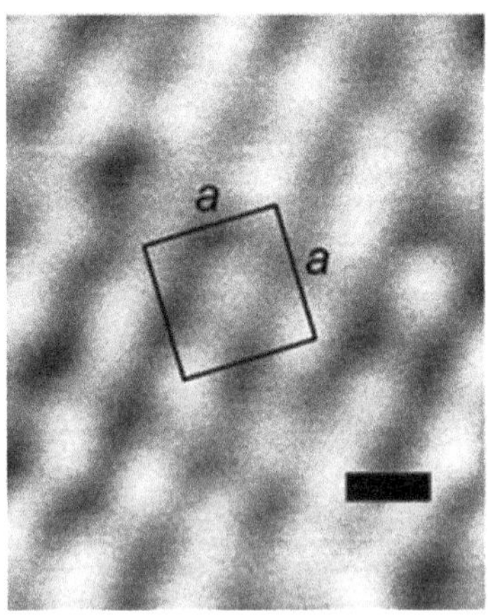

Figure 6.4 SEM image illustrating the cubic structure of the Blue Phase. Reproduced from ref. 17 with permission from Springer Nature, Copyright 2002.

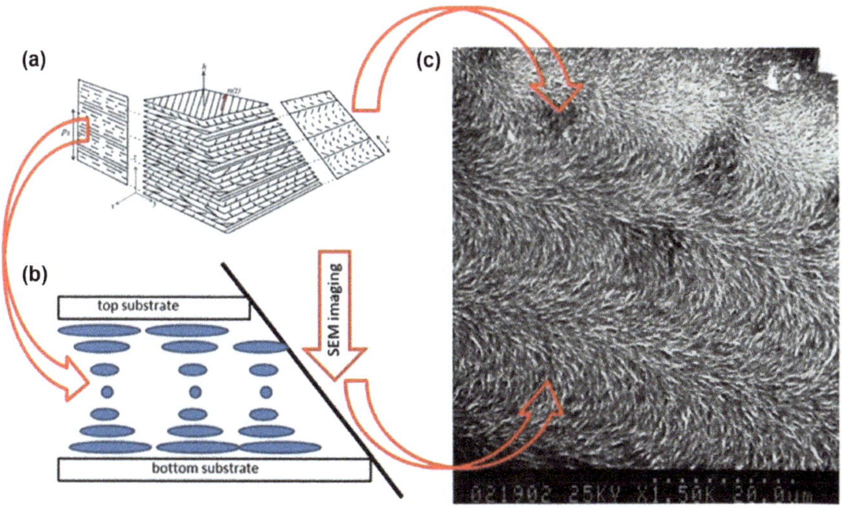

Figure 6.5 (a) Bouligand cut illustrating the arc pattern observed for an oblique cut through a helical structure. Reproduced from ref. 18 with permission from the Royal Society of Chemistry. P_0 is the pitch of the helical structure, **n**(**z**) the director, **h** the normal to a quasi-nematic layer, and L the size of the arc pattern. (b) Schematic illustration showing how such a Bouligand cut is performed on a liquid crystal sandwich cell, and (c) resulting SEM image of the polymer network structure formed in a cholesteric phase of pitch 10 μm confined to a 15 μm thick cell. The three arcs observed correspond to the three π-twists. Reproduced from ref. 19 with permission from AIP Publishing.

the expected three arcs, with the top and the bottom substrate corresponding to the top and the bottom of the image.[19]

Above it was mentioned that the polymer networks appear to be expelled from regions of strong deformations. Another example of such behaviour can be found in cholesteric liquid crystals which exhibit characteristic oily streaks defects. The director structure within those line defects can be quite complicated,[20] as shown in Figure 6.6(a), but is always highly deformed in comparison to the bulk cholesteric phase. The SEM images in Figure 6.6(b) show the polymer network of a macroscopic sample with oily streaks originating at spacer beads that have been sprayed onto the substrate surface to assure a constant cell thickness. The insets in the figure represent a close up of the original image and it can clearly be seen that the polymer network is not formed in the deformed regions of the oily streaks. Also, in the regions around an individual spacer bead the director configuration appears to be highly deformed, as expected. This observed behaviour can be attributed to the increase in elastic energy when the forming polymer network increases in length and crosslinking during photo-polymerization. Once the elastic energy term becomes unsustainable, the polymer network is expelled from the deformed director field to continue forming in structures of smaller splay, twist or bend.

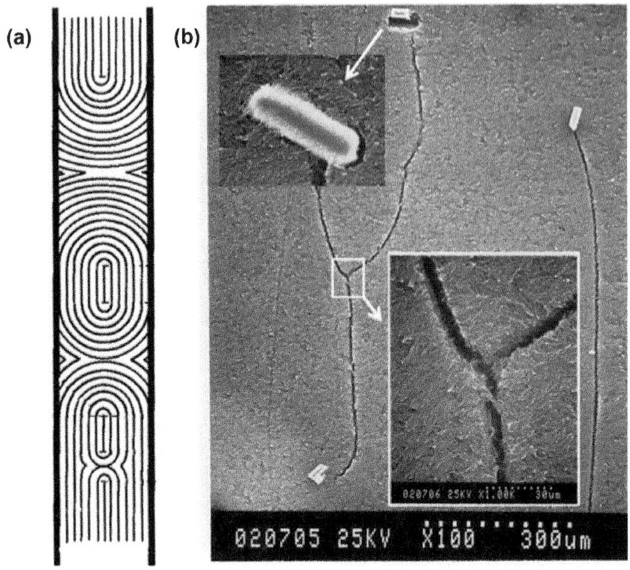

Figure 6.6 (a) Schematic diagram of possible deformations within oily streaks structures. Reproduced from ref. 20 with permission from the American Physical Society, Copyright 1998. (b) SEM image at low resolution, illustrating the absence of the polymer network within the former oily streaks defects. The insets show close-ups of oily streaks meeting, and originating at spacer beads.

Another example of a polymer network templating the structure of the phase it was formed in, while also illustrating that such networks are often expelled from highly deformed regions, can be found in the frustrated Twist Grain Boundary (TGB) phases.[21] These phases occur due to a competition between cholesteric twist and smectic layer formation, thus only for highly chiral materials and between those mentioned phases in a small temperature interval. Figure 6.7(a) schematically depicts the structure of the TGB A* phase, which is formed of smectic A slabs, which are twisted with respect to each neighbour, while the slabs are separated by grain boundaries of arrays of screw dislocations. The helical axis lies in the smectic layer plane and the local director in each smectic block is perpendicular to the helix axis. Figure 6.7(b) shows the polymer network formed at the boundary between the normal smectic A* phase at the bottom left, and the TGB A* phase at the top right of the SEM image. It can clearly be seen that the smectic A* phase network only exhibits one preferred direction of the network, that along the director, *i.e.* parallel to the smectic layer normal, while the TGB A* phase shows several distinct directions.

From Figure 6.7(b) it becomes clear that the polymer network is only formed within the structure of the smectic blocks, being expelled from the grain boundaries of screw dislocations. This is yet another example where

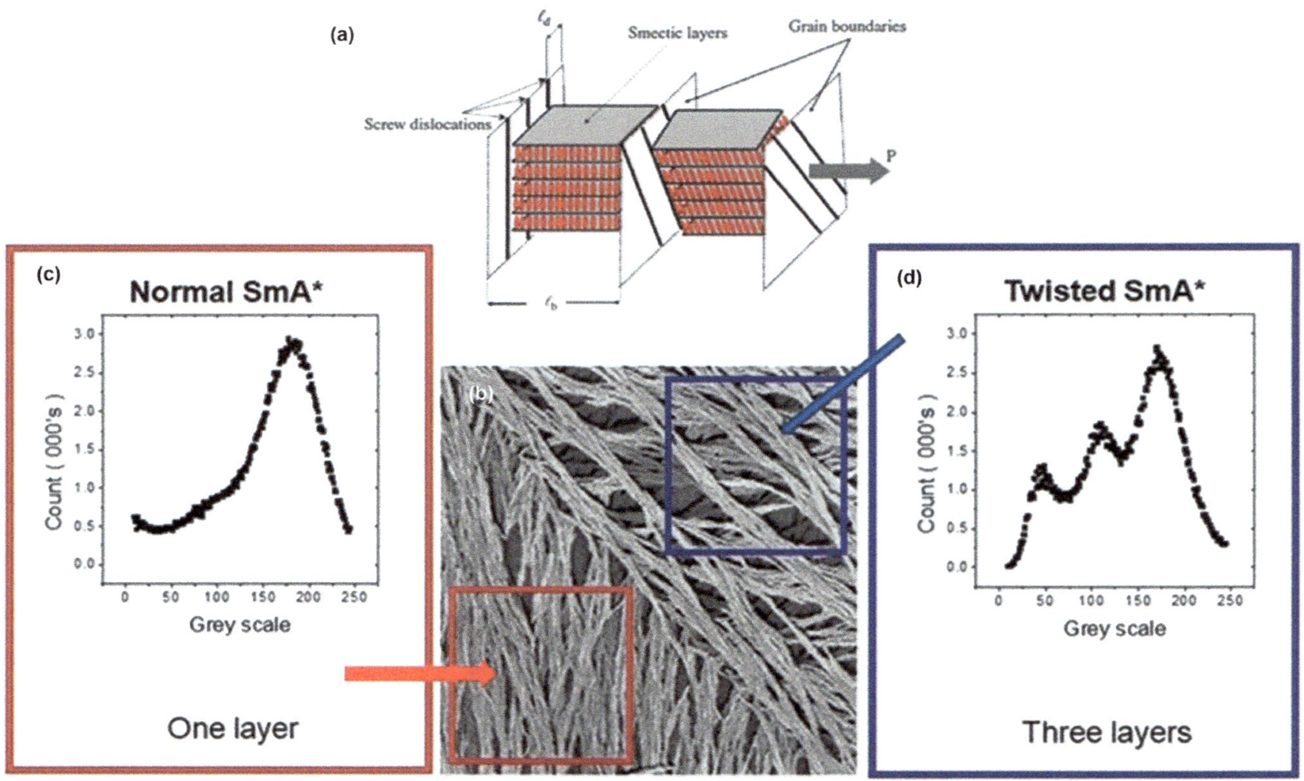

Figure 6.7 (a) Schematic of the structure of the helical Twist Grain Boundary (TGB A*) phase. The helix axis is perpendicular to the local director, like in the cholesteric phase, only with smectic layers being formed at the same time. For cells with planar boundary conditions, this implies a helix axis perpendicular to the substrates, similar to the Grandjean orientation. Reproduced from ref. 22 with permission from the Royal Society of Chemistry. (b) SEM image of the regular smectic A* phase at the bottom left and the TGB A* phase at the top right in Grandjean orientation, thus looking onto the helix axis for the TGB phase. Figure parts (c) and (d) illustrate a single preferred direction for the network formed in the standard smectic A* phase and discontinuously twisted smectic blocks for the TGB A* phase, respectively. The twist angle between adjacent blocks is constant and approximately 60 degrees. Reproduced from ref. 21 with permission from the Royal Society of Chemistry.

the polymer network does not form in highly deformed director fields, most likely due to reasons of elasticity, as discussed above.

Plotting the intensity of grey scale values of the SEM image for part of the standard smectic A* phase (Figure 6.7(c)), as well as the TGB A* phase (Figure 6.7(d)), one can infer that there is only one single preferred direction for the former, and, within the cell gap used, three separate, individual preferred directions for the latter phase, which verifies the discrete block structure of the TGB phase. The twist angle between adjacent smectic blocks is approximately 60 degrees, the phase is thus most likely an example of a so-called giant block Twist Grain Boundary phase.[23] The orientational order parameter in both phases and each smectic block is equivalent to the values observed for bulk samples, about S ~0.8.

6.3 Polymer Network Morphology and Electro-optic Performance

Quantitatively relating the electro-optic performance of polymer stabilized liquid crystals to the polymer network morphology is by no means a trivial task. The reason for this is that the local polymer morphology depends on almost every conceivable parameter that may be varied for the polymerization process. These can be the monomer material itself and its chemical constitution, the solubility within the host, curing temperature, curing time, UV intensity, the amount of photo-initiator, monomer diffusion dynamics within the host, and many more parameters that influence the UV polymerization dynamics and conversion. This in turn will affect the polymer network morphology and thus the electro-optic performance, mainly the threshold voltage, intensity modulation, scattering, reflectivity, and response times. It is thus clear that all of these parameters are somehow interrelated in a rather complicated way.

Nevertheless, there appears to be one major factor influencing the morphology and performance, namely the solubility of the photo-reactive, bifunctional monomer within the liquid crystalline host.[11,24] The former of course in turn depends on many of the above-mentioned conditions. It has become apparent that the best and most predictable performance of PSLCs is obtained from such devices which are stabilized by a smooth polymer network. These are obtained from soluble monomers during a slow polymerization process with good monomer conversion at moderate UV intensities, well away from any phase transitions. The general scheme is schematically depicted in Figure 6.8. We will discuss individual factors separately below.

Highly soluble monomers, or monomers well below the solubility limit, are most likely to be homogeneously dispersed throughout the liquid crystal host. UV illumination induces the polymerization, which will proceed more slowly to allow time for the formation of a smooth polymer network as depicted in Figure 6.8(a). Less soluble monomers, or monomers at concentrations above the solubility limit, will be inhomogeneously dispersed.[25,26] UV illumination will cause a rapid polymerization in regions of high monomer concentration, and the formation of polymer grains, which will subsequently slowly crosslink to form a more irregular and weaker structure than the smooth polymer networks (Figure 6.8(c)). For intermediate concentrations a continuous transition from a smooth to a grainy structure is observed (Figure 6.8(b)). As we shall see below, the smooth polymer structures of Figure 6.8(a) will lead to a more predictable, continuous and regular electro-optic performance. This will admittedly happen at the cost of a slightly higher threshold field, but with the advantage of a much decreased switching-off time and larger electro-optic modulation.

The effect of a smooth polymer network embedded in a continuous matrix of a liquid crystal is mainly of elastic nature. Compared to the sole surfaces of the substrates, an oriented polymer network, consisting of many polymer network strands, provides a much larger surface area, and thus a larger elastic surface interaction. The effective elastic constant is drastically increased, which accounts for the increased threshold electric field, E_{th}, for the deformation of the director field, as well as the much-enhanced decay times,

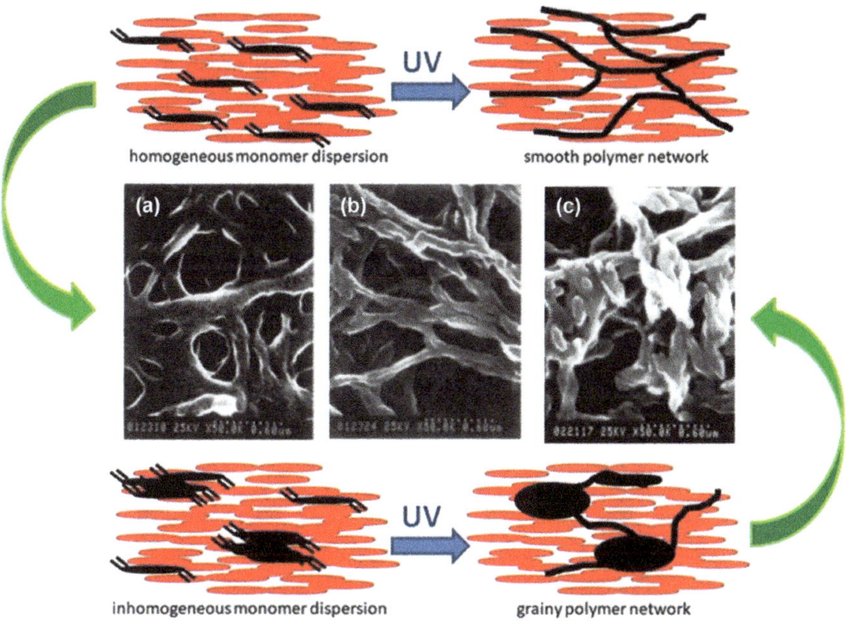

Figure 6.8 Smooth polymer networks are formed for homogeneously dispersed monomers, well below the solubility limit (top), while grainy networks form from inhomogeneously dispersed monomers above the solubility limit (bottom). A crossover from smooth to grainy networks is observed (a)–(c) as the monomer concentration is increased. The solubility limit itself depends on a variety of parameters, such as temperature, chemical constitution and many others.
SEM images reproduced from ref. 24 with permission from AIP Publishing.

τ_{off}. The effect of the polymer network on the liquid crystal can nicely be demonstrated *via* the residual birefringence,[27] which at the same time is a method to image the network without damaging the sample and removing the liquid crystal, as is necessary for SEM imaging. At the same time, the method provides less resolution than SEM, as it is done simply in a polarized optical microscope. The sample is heated just above the clearing temperature, and the residual birefringence is still visible between crossed polarizers, indicating the direction of the oriented polymer network director field (Figure 6. 9(a)).

The residual birefringence is an induced birefringence due to the orientation of liquid crystal molecules in the close vicinity of polymer network strands, due to elastic interactions. The elastic energy involved is in the order of $W_p \sim 10^3$ Jm^{-3}, increasing linearly with polymer concentration, for small concentrations.[28–30] The elastic penetration depth into the bulk of the liquid crystal is of the order of several hundred nanometres.[29] Therefore, the amount of liquid crystal being influenced by the network increases with increasing polymer concentration and is visible deeper into the isotropic phase at elevated temperatures (Figure 6.9(b)). The situation is schematically illustrated in Figure 6.9(c) for smaller and larger polymer concentrations.

This effect can also be observed in the electro-optic curves, for example of a reverse mode PSCT.[1] In Figure 6.10(a) the static diffuse reflectivity, *i.e.* the back scattered percentage of incoming light, is shown as a function of applied voltage for two PSCTs prepared at different temperatures, 14 °C and close to the clearing temperature, at 51 °C. The polymer network which was cured at lower temperature exhibited much smaller voids, at the same monomer concentration of 6% and equal polymerization conditions, than the network prepared at higher temperature, which exhibited small as well as large voids, as can be seen in Figure 6.10(c). While the former sample is composed solely of network dominated regions, the latter exhibits bulk and confined switching, which can be seen not only in the static experiment of Figure 6.10(a), but also in the dynamic one for the director reorientation back to the stabilized state after electric field removal (Figure 6.10(b)).

We shall now discuss the influence of sample preparation conditions and factors of UV illumination separately, while keeping all other parameters unchanged. Again, for convenience we will use experimental measurements on reverse mode PSCTs,[28] although an analogous behaviour also holds for other polymer stabilized liquid crystals, despite the fact that the morphology to some extent depends on the liquid crystal host.[12,31,32] The preparation conditions varied include the polymer concentration (in wt%), the curing temperature, T_{cure}, the time of UV illumination, t, and its intensity. Their effect on the most fundamental electro-optic parameters, namely diffuse reflectivity, R, transmission, T, threshold field, E_{th}, and decay time, τ_{off}, are discussed. Effects of photo-initiator concentration, cell gap, d, and measuring wavelength, λ, will also be pointed out.

The time of curing, *i.e.* time of UV illumination, is one of the most important parameters in sample preparation and in obtaining well established,

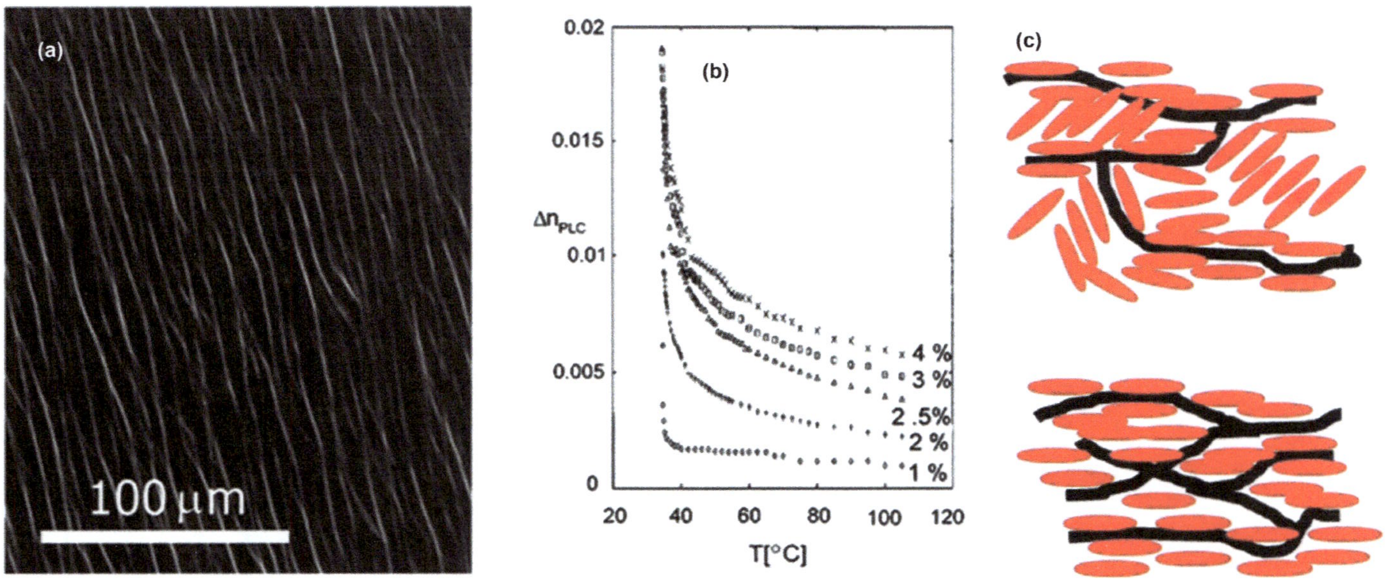

Figure 6.9 (a) Polarized optical microscopy image of a sample heated just above the clearing temperature into the isotropic phase between crossed polarizers. The polymer network is visible due to the residual birefringence, which is caused by the elastic interaction between mesogens and the surface of network strands. (b) The residual birefringence, Δn_{PLC}, increases with increasing polymer concentration due to a strongly increased surface area. Reproduced from ref. 27 with permission from the American Physical Society, Copyright 1997. (c) Schematic illustration of the influence of the polymer for a weakly and a strongly network-dominated liquid crystal, *i.e.* smaller and larger polymer concentration.

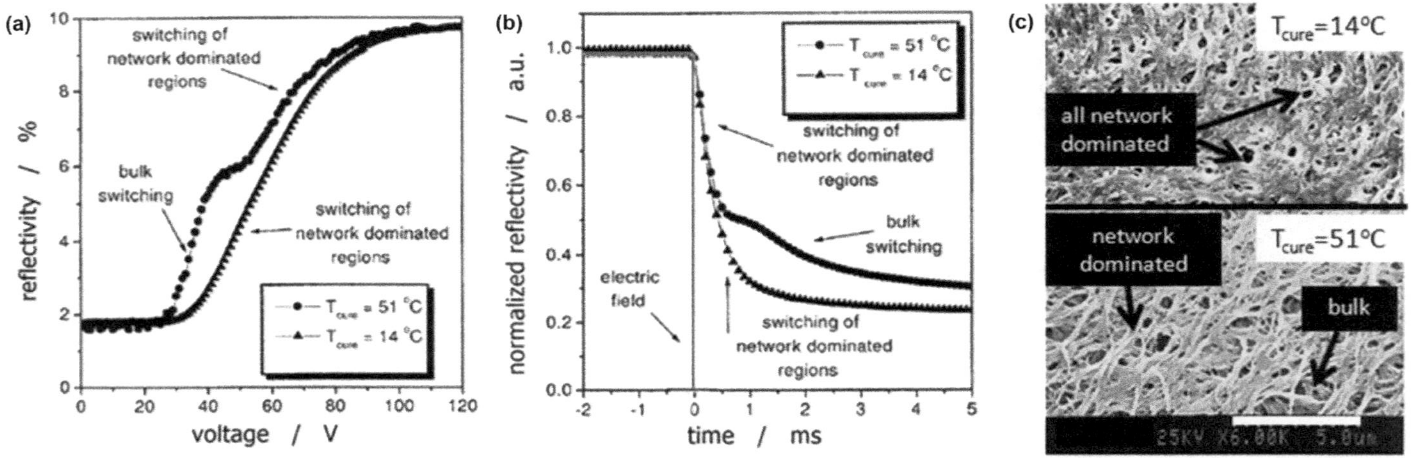

Figure 6.10 (a) Static reflectivity as a function of applied voltage for two samples prepared at low and high temperatures at equal monomer concentrations of 6%. The former exhibits a dense network, with the liquid crystal being wholly dominated by the polymer. Polymerization at higher temperatures leads to voids in the polymer network, which allow for bulk liquid crystal switching, which is indicated by the two-step switching process. (b) The equivalent behaviour can also be observed for the dynamic experiment considering the decay time. (c) The respective polymer networks are shown in the SEM images, indicating that voids of a size smaller than approximately $(1\ \mu m)^3$ are fully dominated by the network, while voids larger than about $(2\ \mu m)^3$ exhibit bulk reorientation at lower thresholds, smaller reflectivity and longer decay times, in addition to the polymer dominated switching.
Parts (a) and (b) reproduced from ref. 19 with permission from AIP Publishing.

reproducible polymer networks at high monomer conversion rate.[33] As a rule of thumb, samples should be cured for at least one hour, which is also apparent from the measurements shown in Figure 6.11. All of the important electro-optic quantities, such as the reflectivity (Figure 6.11(a)), transmission (Figure 6.11(b)), threshold field (Figure 6.11(c)) and response times (Figure 6.11(d)), clearly vary for networks cured for less than 30 minutes. Also, the network morphology is changing for polymerization times less than about one hour. Only after this time can no further changes in the sizes of polymer network strands and structural voids be observed (see SEM images in Figure 6.11(e)).

The reflectivity strongly increases with increasing curing time, as the polymer network forms and becomes increasingly dense. When after about one hour all monomer material is used up for the polymerization process and the network is completely formed and doesn't change its morphology any further, the reflectivity reaches a steady saturation state (Figure 6.11(a)). It is at this point when the behaviour of the polymer stabilized liquid crystal becomes reproducible. As the reflectivity increases, the transmission naturally decreases, also reaching a saturation state when the polymerization is complete (Figure 6.11(b)). During the proceeding network formation process, the threshold field also increases (Figure 6.11(c)), which can be understood in terms of the discussion above relating to Figure 6.9(c) and an increasingly dense polymer network which elastically dominates the liquid crystal completely.[34,35] As the polymer density increases, so does the elastic interaction between network and liquid crystal. This leads to faster decay times, which reach saturation when all of the liquid crystal is influenced by the polymer (Figure 6.11(d)). The discussed behaviour of the electro-optic properties is clearly qualitatively related to the morphology changes of the polymer network, as outlined in Figure 6.11(e) for the early stages of network formation. At longer polymerization times, larger than about one hour, the network morphology does not change any further.

The temperature at which the polymerization is carried out also has an effect on the morphology of the network structure, and with it, on the electro-optic performance of polymer stabilized liquid crystals.[25,36] This is not only due to an increased order parameter at lower temperatures, because the latter relatively quickly approaches quite constant values of usually $S \sim 0.6$–0.7 for nematic and cholesteric, or $S \sim 0.7$–0.8 for smectic phases. It is believed that the most pronounced contribution is due to the dynamics of polymerization, which is much slower at low temperatures than at higher ones. The slower the polymerization, the more uniformly it takes place, due to more even diffusion of the unreacted monomers. For lower curing temperatures, networks of smoother morphology and with smaller voids are observed, which become larger as T_{cure} approaches the clearing temperature T_C. This is qualitatively obvious from the SEM images of networks formed at different temperatures as indicated in Figure 6.12(e). Hence, the reflectivity decreases for increasing curing temperature, while the transmission increases (Figure 6.12(a) and (b),

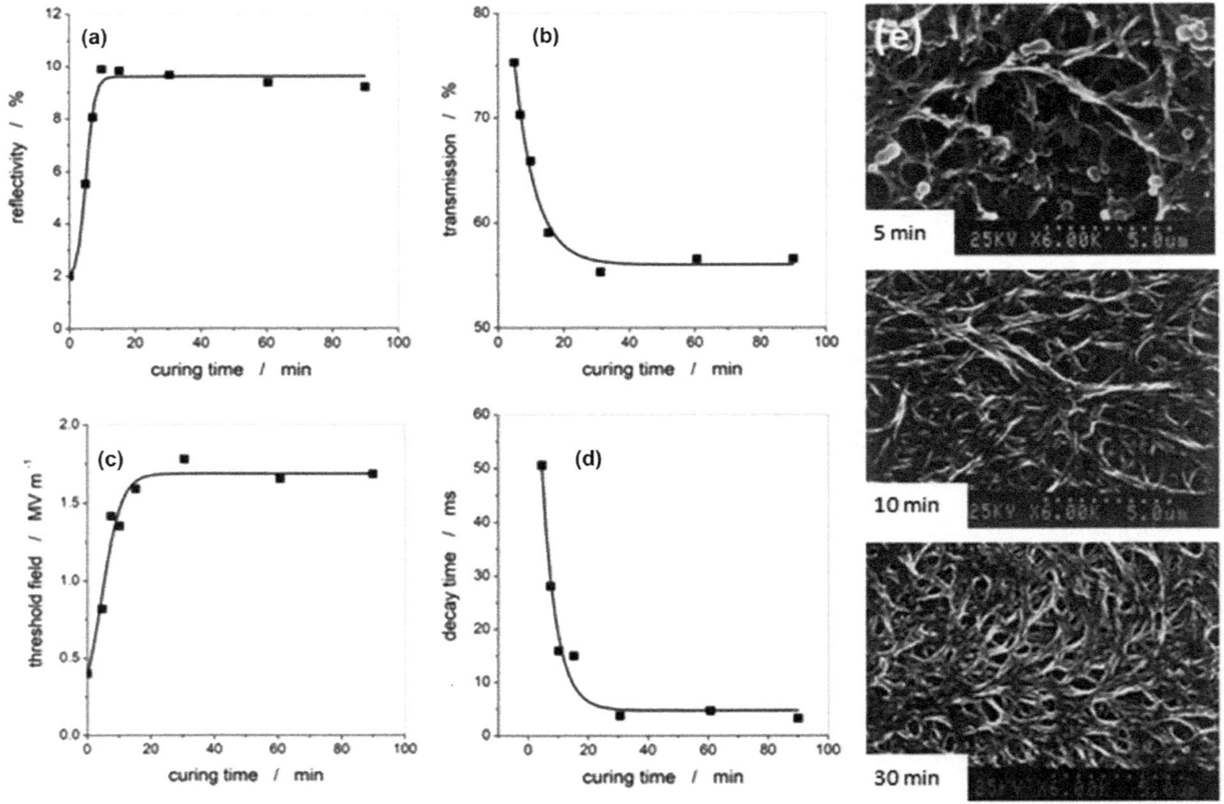

Figure 6.11 Electro-optic properties (a) diffuse reflectivity, (b) transmission, (c) threshold field and (d) decay time as a function of curing time, during which the network polymerization proceeds. (e) As the polymerization proceeds with time, increasingly dense networks are formed, which favour higher reflectivities and faster decay times at the cost of a slightly increased threshold field. After approximately one hour, a network has been formed in which all of the liquid crystal properties are dominated by the polymer. Solid lines are a guide to the eye.
Graph data reproduced from ref. 33 with permission from Taylor and Francis.

respectively). As expected from the discussion above, the threshold field decreases as the network becomes looser (Figure 6.12(c)) and the decay times increase (Figure 6.12(d)), a behaviour which can again be attributed to the morphology and its implications for the elastic coupling between network and liquid crystal.

A similar behaviour, although somewhat less pronounced, can be observed for a variation of the illuminating UV intensity during polymerization.[11,33] The reflectivity slightly decreases and the transmission increases with increasing UV intensity, while the threshold field decreases and the decay times increase, as depicted in Figures 6.13(a)–(d), respectively. Again, this behaviour can qualitatively be attributed to the changing polymer network morphology, which is shown in the series of SEM images in Figure 6.13(e). Hand-in-hand with the less pronounced changes in electro-optic parameters, the morphology changes are also more subtle, though still recognizable. Small UV intensities lead to a slower, more even polymerization, which produces smoother and more uniform network structures, while large intensities cause a more rapid polymerization, which results in a morphology of larger voids with a larger distribution of sizes.

The same qualitative effect as a prolonged polymerization time is obtained when increasing the concentration of photo-reactive monomer.[11,19,24,37] Naturally, a larger monomer concentration leads to denser networks with smaller voids and thus larger surface elastic interaction between polymer and liquid crystal[19,38] (SEM images in Figure 6.14(e)). This implies an increase in diffuse reflectivity, a decrease in transmission, higher threshold fields, but faster decay times, as shown in Figure 6.14(a)–(d), respectively.

The electro-optic performance, in particular the diffuse reflectivity, the threshold voltage and the decay time back to the original orientation after electric field removal, is consistent across all variations of polymerization conditions, namely the monomer concentration, the curing time and curing temperature, as well as UV intensity. This in turn can be understood in relation to the changing polymer network morphology, as the above parameters are varied. As a general observation, one realizes that the decay times can be influenced over a large range of values covering about one order of magnitude from the low millisecond regime to about one hundred milliseconds, at only slight cost in an increased threshold field of about a factor of three. Polymer concentration and curing time, thus the amount of polymer network present, have the largest impact on the reflectivity of the device, when keeping all other parameters, especially the cell gap, constant. The smallest influences on the electro-optics, in absolute terms as well as on the network morphology, appear to be generated from the curing temperature and especially the UV intensity.

In summary, one can conclude that the most effective, yet reproducible, stabilization is achieved by monomers which form smooth polymer strands, at moderately high concentrations but below the solubility limit (depending on the chemical constitution of monomer and liquid crystal host). The polymerization process should be carried out at slow pace, at low temperatures (room-temperature is generally sufficient) and small UV intensities

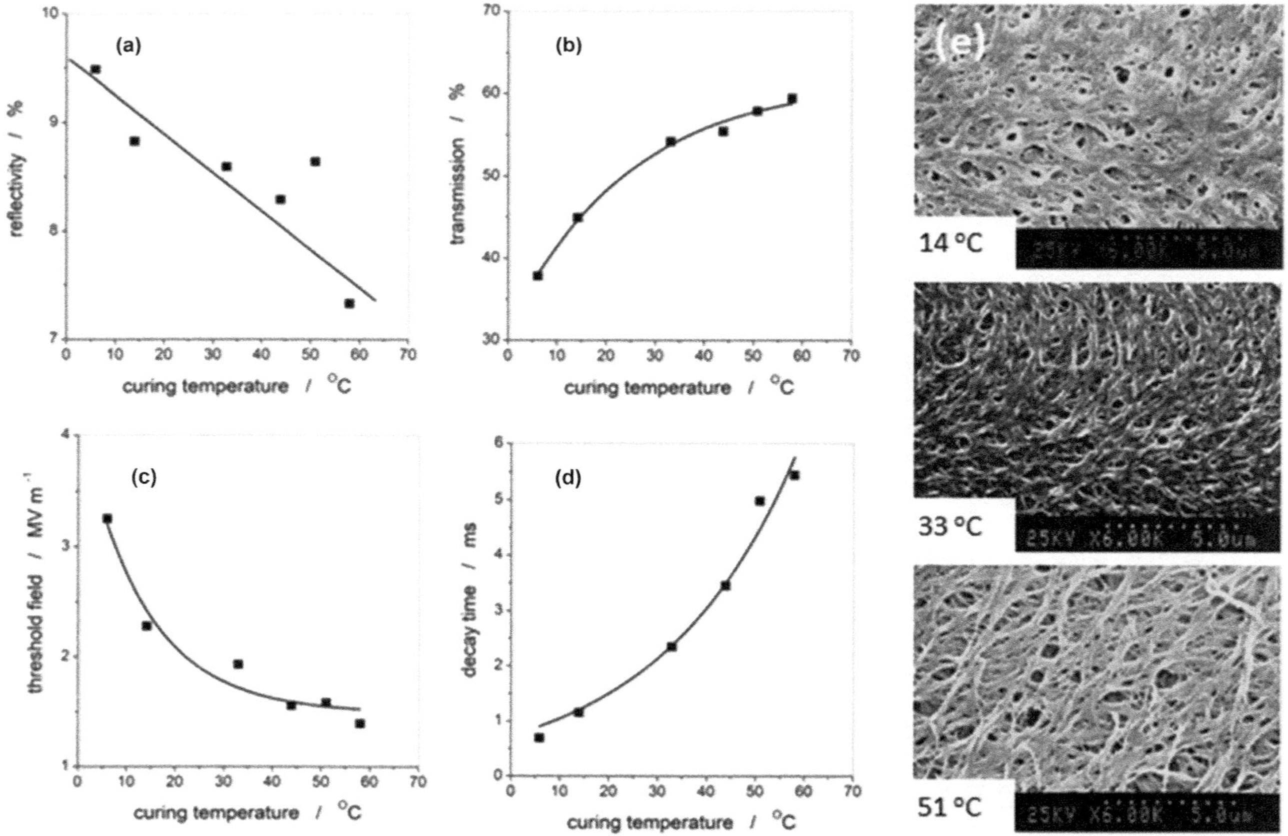

Figure 6.12 Electro-optic properties (a) diffuse reflectivity, (b) transmission, (c) threshold field and (d) decay time as a function of curing temperature, at which the network polymerization proceeds. (e) As the polymerization proceeds at higher temperatures, more open and less dense networks are formed, which lead to lower reflectivities and slower decay times, while the threshold field slightly decreases. Solid lines are a guide to the eye.
Graph data reproduced from ref. 36 with permission from Taylor and Francis.

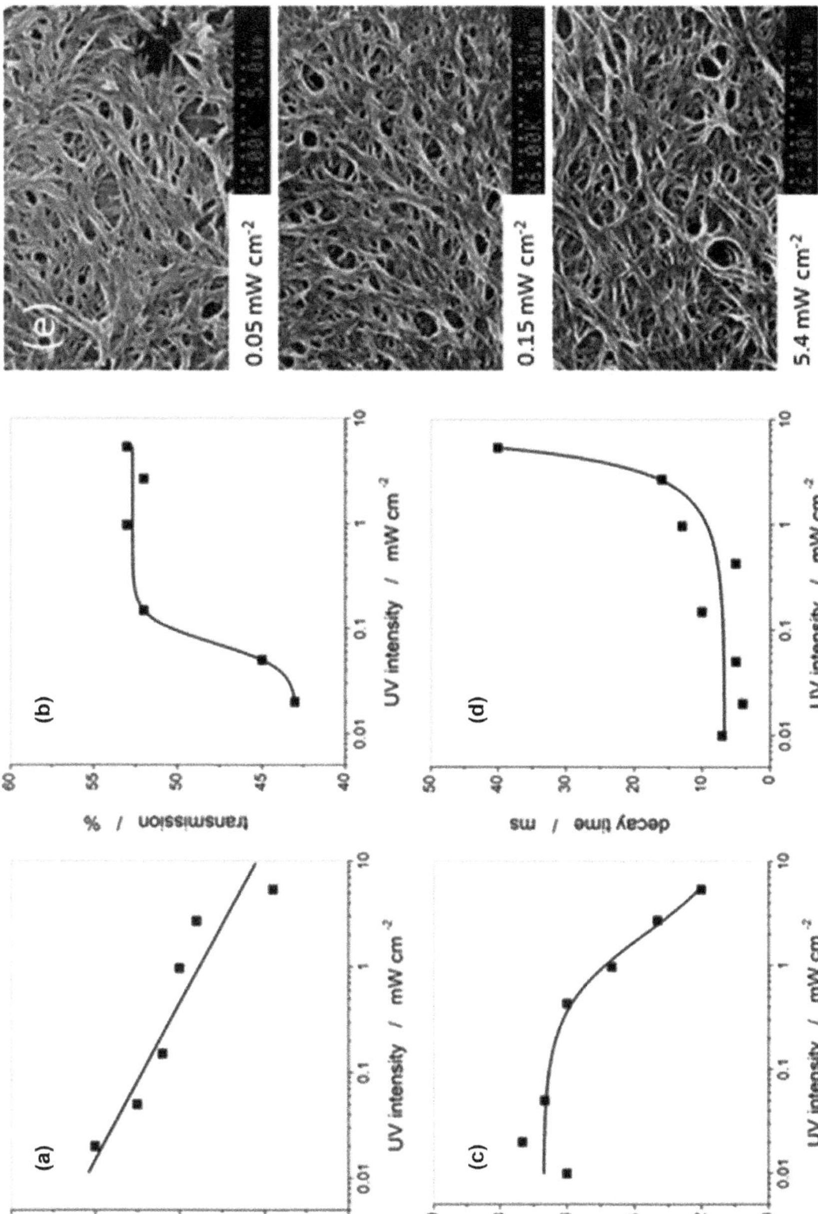

Figure 6.13 Electro-optic properties (a) diffuse reflectivity, (b) transmission, (c) threshold field and (d) decay time as a function of UV curing intensity, applied during the network formation. (e) UV intensity only has a slight influence on the polymerization process, and only at high intensities can one observe the formation of less dense networks with larger voids. These lead to lower reflectivities and slower decay times, while the threshold field slightly decreases. Solid lines are a guide to the eye. Graph data reproduced from ref. 33 with permission from Taylor and Francis.

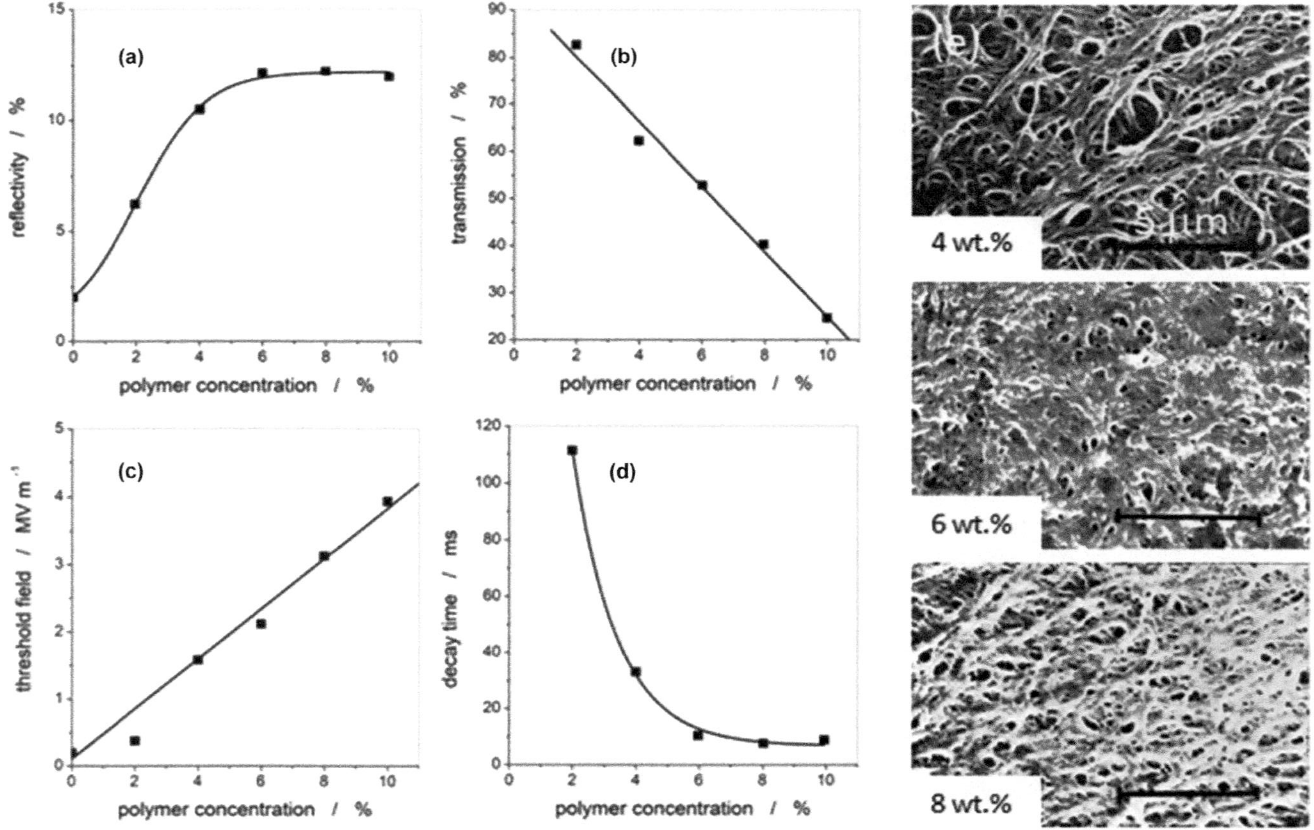

Figure 6.14 Electro-optic properties (a) diffuse reflectivity, (b) transmission, (c) threshold field and (d) decay time as a function of polymer concentration by weight percentage. (e) For completed polymerization a loose and open network can be observed for small concentrations, which leads to small reflectivities and long decay times at a somewhat reduced threshold field. On increasing the concentration, the network naturally becomes denser and the voids smaller, until the whole liquid crystal is polymer dominated, which can be inferred from the saturation of reflectivity and decay times. Solid lines are a guide to the eye. Graph data reproduced from ref. 19 with permission from AIP Publishing.

(0.1 mW cm^{-2} or smaller) and should last at least one hour to assure complete conversion of the monomers. Under these conditions it is most likely that a polymer network structure which is smooth, and with only small voids, will be obtained so that the behaviour of the liquid crystal is mainly governed by the elastic surface interactions with the polymer.

Nevertheless, there are a few more factors which may be varied and should briefly be mentioned for completeness. The amount of photo-initiator used does not seem to play any significant role[3] even up to high concentration of 15wt% of the monomer concentration (Figure 6.15(a)). As generally only 1–2wt% are used, this has no serious effect on sample preparation. As mentioned above, the main influence on the reflectivity appears to be due to the amount of stabilizing polymer formed within the liquid crystal. The employed cell gap is thus expected to have a major influence on the reflectivity, because the scattering volume is doubled when doubling the gap distance. Indeed, the reflectivity increases linearly with cell gap,[3] as shown in Figure 6.15(b). Nevertheless, one does need to keep in mind that one cannot simply increase the intensity modulation by increasing the cell gap, because this will lead very quickly to applied voltages which are not practical for driving a device, one of the major drawbacks of reverse mode PSCTs for applications such as reflective paper. Lastly, the reflectivity also depends on the measuring wavelength,[3] as depicted in Figure 6.15(c), with the reflectivity decreasing for increasing wavelength. This behaviour can be attributed to several phenomena. Firstly, the scattering process itself (Mie scattering), which depends on wavelength and size of the "particle", here the polymer network strand. Further, the refractive indices of the liquid crystal depend on the wavelength of the probing light (refractive index dispersion), which also changes the scattering properties, and thus the reflectivity.

So far we have successfully related the polymer network morphology with the electro-optic behaviour on a qualitative basis. There remains the question if a more quantitative relation may be found for this admittedly rather complex problem. Often, complex structures can be described by a simple scaling quantity, the fractal dimension, which is a measure of the scale invariance of a structure.[39–41] The simplest and most common method to determine the fractal dimension, D, is the box counting method, where a grid of variable size is placed over an image, in this case the thresholded black and white image of a polymer network, and the number of network occupied boxes, N, are counted as the box size, d_{box}, is varied. Plotting the relation

$$N \sim \frac{1}{d_{box}^{D}} \tag{6.1}$$

on a log-log scale gives the fractal dimension, D, as the modulus of the slope of a straight line within the scaling regime. This number will be between $1 \leq D \leq 2$. In two-dimensional space, it is $D = 1$ for a line and $D = 2$ for a plane.[40]

The polymer networks do indeed exhibit a fractal structure with scale-invariant morphology over a certain range of scales.[42] This implies that their

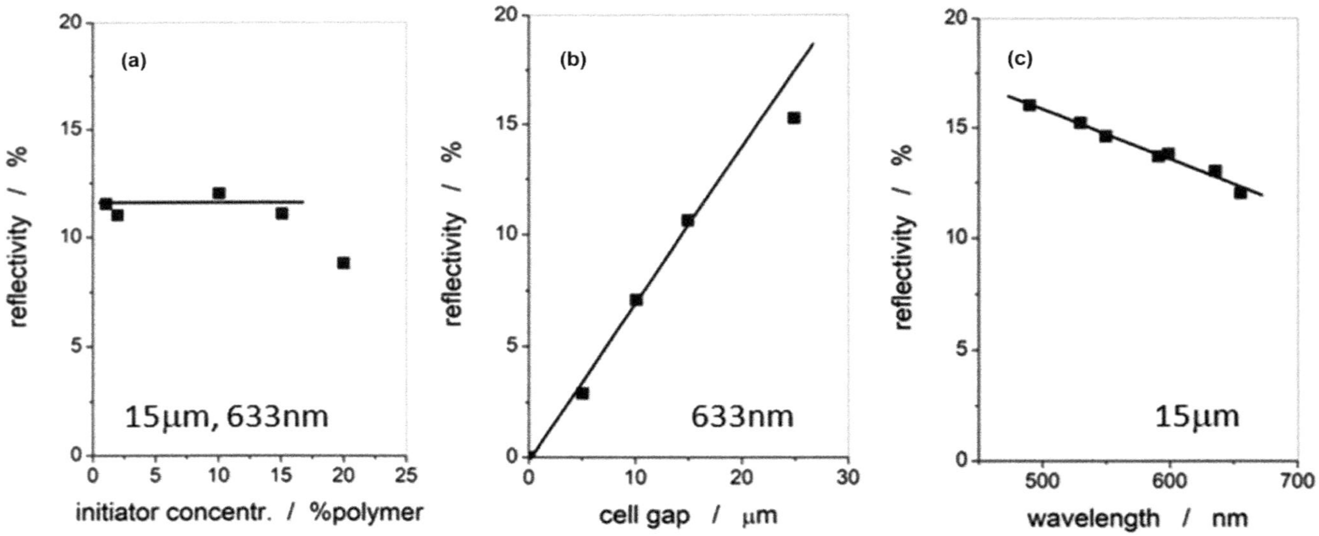

Figure 6.15 Dependence of the diffuse reflectivity on (a) the concentration of photo-initiator as a weight percentage of the monomer concentration, (b) the cell gap, and (c) the measurement wavelength. The photo-initiator concentration does not play any significant role in the determination of network morphology or electro-optic properties, as it is generally chosen to be very small, around 1–2%. The reflectivity increases linearly with cell gap as intuitively expected, and it decreases with increasing wavelength, with the most prominent influence probably being the refractive index dispersion. Solid lines are a guide to the eye.
Graph data reproduced from ref. 3 with permission from John Wiley and Sons, © 2000 WILEY-VCH Verlag GmbH, Weinheim, Fed. Rep. of Germany.

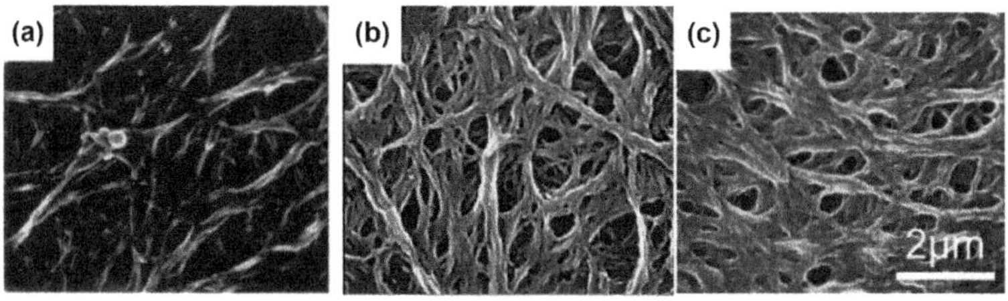

Figure 6.16 Polymer networks prepared under different conditions, can be described by a single numerical value, their fractal dimension, D, which characterizes their morphology. (a) is an open, loose network morphology with fractal dimension $D = 1.75$, (b) is of medium openness with $D = 1.85$ and (c) represents a dense, nearly space-filling network morphology with $D = 1.95$.
Reproduced from ref. 42 with permission from IOP Publishing.

morphology can be characterized by a single number, the fractal dimension, D, which is demonstrated in Figure 6.16 for three different networks of the same polymer prepared under different conditions leading to an open network with large voids and a dimension of $D = 1.75$ (Figure 6.16(a)), a medium morphology with $D = 1.85$ (Figure 6.16(b)) and a dense, nearly space-filling polymer network morphology, with small voids and $D = 1.95$ (Figure 6.16(c)).

One can then plot all reflectivities, R, threshold fields, E_{th}, and decay times, τ_{off}, obtained from different series of varying curing temperature, UV intensity and curing time in terms of the fractal dimension of the observed polymer network morphology. In this case, all data collapses on a single curve for each of the important device performance quantities,[43,44] as shown in Figure 6.17.

The solid curves can be described by a set of simple scaling equations. For the reflectivity this is given by:

$$R_{max} - R = \frac{A}{D - D^*} \tag{6.2}$$

where R is the measured reflectivity, R_{max} the maximum obtainable device reflectivity, D the fractal dimension and D^* the fractal dimension of a network with vanishing influence on the liquid crystal. The constant A is material dependent. For the threshold field the scaling equation is:

$$E_{th} - E_{th}^0 = \frac{B}{2 - D} \tag{6.3}$$

where E_{th} is the measured threshold field, E_{th}^0 is the threshold field of the non-stabilized liquid crystal and B is again a material dependent constant. The decay time obeys the scaling equation:

$$\tau_{off} - \tau = \frac{C}{D - D^*} \tag{6.4}$$

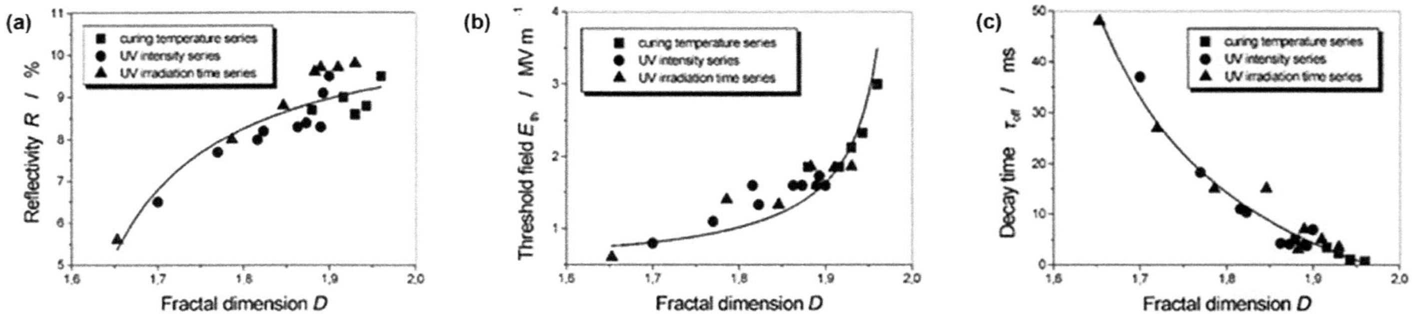

Figure 6.17 Scaling behaviour of (a) the reflectivity, (b) the threshold field and (c) the decay times, in terms of the fractal dimension of the polymer network morphology. Data points are collected from three different preparation series, varying the curing time, UV intensity and the curing temperature for each plot. Note that the solid lines are not a guide to the eye, but a consistent fit to eqn (6.2)–(6.4) for all of the data of the curing temperature series (squares), the UV intensity series (circles) and the UV irradiation time series (triangles).
Reproduced from ref. 43 with permission from John Wiley and Sons, © 2003 WILEY-VCH Verlag GmbH & Co. KGaA, Weinheim.

where $(\tau_{off} - \tau)$ can be interpreted as the decay time for a non-stabilized system, and C is a material constant, which is mainly an indirect measure of the strength of the interaction between network and liquid crystal. Note that the solid lines of Figure 6.17 are not guides to the eye, but a consistent fit to eqn (6.2)–(6.4).[43,44]

This concludes our introduction into the interplay between liquid crystal, polymer network morphology and basic electro-optic properties. More specific behaviour on individual phases like nematics, cholesterics, ferroelectric smectic C* and Blue Phases will be discussed in the chapters to follow.

References

1. D.-K. Yang, L.-C. Chien and J. W. Doane, *Appl. Phys. Lett.*, 1992, **60**, 3102.
2. *Liquid Crystals in Complex Geometries*, ed. G. P. Crawford and S. Zumer, Taylor&Francis, London, 1996.
3. I. Dierking, *Adv. Mater.*, 2000, **12**, 167.
4. D. J. Broer, G. N. Mol and G. Challa, *Makromol. Chem.*, 1989, **190**, 19.
5. D. J. Broer, R. A. M. Hikmet and G. Challa, *Makromol. Chem.*, 1989, **190**, 3201.
6. D. J. Broer, G. N. Mol and G. Challa, *Makromol. Chem.*, 1991, **192**, 59.
7. A. Jakli, L. Bata, K. Fodor-Csorba, L. Rostas and L. Noirez, *Liq. Cryst.*, 1994, **17**, 227.
8. R. A. M. Hikmet, *Mol. Cryst. Liq. Cryst.*, 1992, **213**, 117.
9. R. A. M. Hikmet, *Liq. Cryst.*, 1991, **9**, 405.
10. R. A. M. Hikmet, *Adv. Mater.*, 1992, **4**, 679.
11. Y. K. Fung, D.-K. Yang, S. Ying, L.-C. Chien, S. Zumer and J. W. Doane, *Liq. Cryst.*, 1995, **19**, 797.
12. A. Y.-G. Fuh, M.-S. Tsai and C.-Y. Huang, *Jpn. J. Appl. Phys.*, 1996, **35**, 3960.
13. M. Kleman, *Points, Lines and Walls*, 1983, Wiley, Chichester.
14. H. R. Trebin, *Adv. Phys.*, 1982, **31**, 195.
15. A. Rapini, *J. Phys.*, 1973, **34**, 629.
16. I. Dierking and P. Archer, *RSC Adv.*, 2013, **3**, 26433.
17. H. Kikuchi, M. Yokota, Y. Hisakado, H. Yang and T. Kajiyama, *Nat. Mater.*, 2002, **1**, 64.
18. O. F. Aguilar Gutierrez and A. D. Rey, *Soft Matter*, 2014, **10**, 9446.
19. I. Dierking, L. L. Kosbar, A. Afzali-Ardakani, A. C. Lowe and G. A. Held, *J. Appl. Phys.*, 1997, **81**, 3007.
20. O. D. Lavrentovich and D.-K. Yang, *Phys. Rev. E*, 1998, **57**, R6269.
21. P. Archer and I. Dierking, *Soft Matter*, 2009, **5**, 835.
22. A. S. Achalkumar, D. S. Shankar Rao and C. V. Yelamaggad, *New J. Chem.*, 2014, **38**, 4235.
23. J. Fernsler, L. Hough, R.-F. Shao, J. E. Maclennan, L. Navailles, M. Brunet, N. V. Madhusudana, O. Mondain-Monval, C. Boyer, J. Zasadzinski,

J. A. Rego, D. M. Walba and N. A. Clark, *Proc. Natl. Acad. Sci.*, 2005, **102**, 14191.
24. I. Dierking, L. L. Kosbar, A. Afzali-Ardakani, A. C. Lowe and G. A. Held, *Appl. Phys. Lett.*, 1997, **71**, 2454.
25. C. V. Rajaram, S. D. Hudson and L.-C. Chien, *Chem. Mater.*, 1996, **8**, 2451.
26. C. V. Rajaram, S. D. Hudson and L.-C. Chien, *Polymer*, 1998, **39**, 5315.
27. Y. K. Fung, A. Borstnik, S. Zumer, D.-K. Yang and J. W. Doane, *Phys. Rev. E*, 1997, **55**, 1637.
28. P. Archer and I. Dierking, *J. Phys. D: Appl. Phys.*, 2008, **41**, 155422.
29. P. Archer, I. Dierking and M. Osipov, *Phys. Rev. E*, 2008, **78**, 051703.
30. I. Dierking, *Materials*, 2014, 7, 3568.
31. C. V. Rajaram, S. D. Hudson and L.-C. Chien, *Chem. Mater.*, 1995, 7, 2300.
32. D. S. Muzic, C. V. Rajaram, L.-C. Chien and S. D. Hudson, *Polym. Adv. Technol.*, 1996, 7, 737.
33. I. Dierking, L. L. Kosbar, A. C. Lowe and G. A. Held, *Liq. Cryst.*, 1998, **24**, 397.
34. R. A. M. Hikmet, *J. Appl. Phys.*, 1990, **68**, 4406.
35. R. E. Kraig, P. L. Taylor, R. Ma and D.-K. Yang, *Phys. Rev. E*, 1998, **58**, 4594.
36. I. Dierking, L. L. Kosbar, A. C. Lowe and G. A. Held, *Liq. Cryst.*, 1998, **24**, 387.
37. M. Mitov, A. Boudet, P. Sopena and P. Sixou, *Liq. Cryst.*, 1997, **23**, 903.
38. T. Nakata, T. Gotoh, M. Satoh and E. Hasegawa, *Mol. Cryst. Liq. Cryst.*, 1997, **299**, 389.
39. B. B. Mandelbrot, *The Fractal Geometry of Nature*, (1982), W.H. Freeman, San Francisco.
40. *Fractals in Science*, ed. A. Bunde and S. Havlin, Springer-Verlag, Berlin, 1994.
41. *Fractals and Disordered Systems*, ed. A. Bunde and S. Havlin, Springer-Verlag, Berlin, 2nd edn, 1996.
42. I. Dierking, *J. Phys. D: Appl. Phys.*, 2002, **35**, 2520.
43. I. Dierking, *Adv. Mater.*, 2003, **15**, 152.
44. I. Dierking, *Adv. Funct. Mater.*, 2004, **14**, 883.

CHAPTER 7

Polymer-stabilized Nematics and Their Applications

STEPHEN M. MORRIS

Department of Engineering Science, University of Oxford,
OX1 3PJ, UK
Email: stephen.morris@eng.ox.ac.uk

7.1 Introduction

Arguably, the field of polymer-stabilized nematic liquid crystals (LC) owes a lot to the pioneering work of Broer, Hikmet and co-workers that was carried out in the late 1980s to early 1990s in the Netherlands.[1–6] The innovations in material development that appeared from the Philips Research Labs in Eindhoven combined with other complementary studies carried out during that period[7] began to demonstrate a new class of polymer system that did not readily fit within the classification scheme of LC polymers that existed at that time. These LC–polymer combinations also differed considerably from polymer dispersed LC systems that contained a larger concentration of polymer so that the nematic component formed droplets within a polymer binder, rather than a diffuse interpenetrating network. Remarkably, even at relatively low concentrations by weight (2 to 5wt%), these polymer-stabilized systems can have a dramatic influence on the resulting mechanical and electro-optic properties of the low molecular weight LC host and have, as a result, spawned numerous studies and avenues of research since their realisation in the 1990s.

Soft Matter Series No. 8
Polymer-modified Liquid Crystals
Edited by Ingo Dierking

Published by the Royal Society of Chemistry, www.rsc.org

Prior to the development of polymer-stabilized liquid crystalline materials†, there had been a growing interest in side-chain LC polymers for both display and data storage applications as the polymer backbone provided the much needed rigidity whilst the mesogenic side groups were mobile, albeit within the confines permitted by the polymer.[8,9] However, due to the relatively large viscosities involved, along with the coupled motion of both the polymer backbone and the mesogenic groups, the response times were disappointingly slow. Consequently, researchers began to dedicate their efforts to finding new ways of circumventing the large viscosities through innovative changes to the chemical structure, such as by increasing the length of the flexible spacers as well as the lateral substitution of the mesogenic unit.[10] The studies that focused on chemical alterations of the polymer complemented the fundamental research that had also been carried out on understanding the behaviour of the polymer; for example, how does the backbone influence the mobility of the mesogenic unit?[11,12] In contrast to the LC polymers that went before them, the polymer-stabilized variety provided a means with which to effectively decouple the rigidity of the backbone of the polymer from the mesogenic group. This in turn meant that the impact on mobility could be substantially reduced thereby enabling a much greater freedom of movement in the presence of an electric field. At a fundamental level, and unlike LC polymers, the formation of the network did not involve chemically-grafting the mesogenic unit on to the polymer backbone and this proved to be an important step forward for realising a larger response to external electric fields.

The exact chemical nature of the polyfunctional monomer material used to form the network, *e.g.* whether it is mesogenic, photopolymerizable, monoacrylate or diacrylate, may vary, but for what follows it is sufficient for us to consider that the molecules consist, primarily, of nematogenic units that are terminated with one or more functional groups. When dispersed at low to moderate concentrations by weight into a low molar mass nematic LC host, the doped *reactive mesogen*‡ can be oriented by external forces, such as electric fields and surface alignments, in the same way that dichroic dyes dispersed within a nematic host exhibit a guest–host behaviour in the presence of external stimuli. The presence of the reactive mesogen allows for the macroscopic alignment of the LC molecules to be 'locked-in' following polymerization, which results in the reactive mesogenic units becoming cross-linked to form a distributed and diffuse polymer network. By virtue of the anisotropic shape of the reactive mesogens, the formation of the network occurs without removing the existence of the nematic phase.

†The nomenclature for these systems varies a little throughout the literature with reports using terms such as gels, composites and polymer stabilized networks to refer to ostensibly the same thing. In this chapter, we will endeavour to stick with one terminology to describe these systems, namely polymer-stabilized nematic LCs.

‡Reactive mesogen is a term generally applied to describe the monomers that are dispersed into the nematic LC host that, when combined with a photo-initiator, undergo a cross-linking process.

The formation of the polymer network can be achieved using either thermal or photo-activated polymerization mechanisms. It is, however, the latter technique that is perhaps of greater benefit from a technological point of view as it enables the precise orientational order to be 'frozen-in' at any temperature within the nematic phase. Moreover, by controlling the alignment of the director (the unit vector describing the average orientation of the LC molecules) using external forces such as electric fields, surface treatments *etc.*, it is then possible through the photo-polymerization process of the polymer network to record and memorise the exact molecular orientation at the precise moment of exposure to the ultraviolet (UV) light source. For more discussion on the mechanism by which order is transferred from the host LC to the polymer network for a range of geometries and experimental conditions (*e.g.* UV intensity and wavelength) the reader is referred to the previous chapter of this book.

Studies carried out on the morphology of the polymer networks indicate that they consist, although not exclusively, of an irregular mesh of fibril strands that are arranged into bundles that are anisotropic in nature and that compartmentalize the nematic LC, as evidenced by high resolution imaging techniques such as scanning electron microscopy (SEM) and confocal microscopy.[13] For example, using a combination of SEM imaging as well as birefringent imaging in the isotropic liquid phase, research has revealed that the fibrils in the network are approximately a few polymer chains in diameter and are grouped together to form larger diameter bundles which incarcerate non-reactive nematic LC molecules in the process. These bundles of fibrils are approximately cylindrical in shape and are of the order of a few tenths of a micron in terms of lateral size, with longitudinal lengths spanning hundreds of microns.[14,15] An illustration of the polymer fibril bundles concept is shown in Figure 7.1. From the SEM studies conducted in both the nematic and chiral nematic (cholesteric) phases[14–16] there is clear evidence that the LC alignment is transferred onto the polymer network. Furthermore, for the chiral nematic phase, which is considered separately in the next chapter, SEM images show that the polymer network is also able retain more complex director arrangements such as the splay-bend profile and macroscopic helical structure.[16]

While most reports refer either indirectly or demonstrate directly that the network consists overwhelmingly of a mesh of polymer fibrils, the precise morphology and architecture of the polymer network does, unsurprisingly, depend upon a number of factors including the chemical structure of the monomers; the polymerization conditions; the process by which order is imparted onto the polymer network by the nematic LC; and the anisotropic diffusion process that arises during cross-linking, which can occur at different rates depending upon the nature of the monomers and the conditions under which the network has been formed. As an example of the impact that some of these parameters can have on the resulting morphology of the polymer network, it was noted in an early report by Hikmet and Boots[17] that when using the same non-reactive nematic LC host but two different reactive LC monomers very different morphologies could be obtained even though

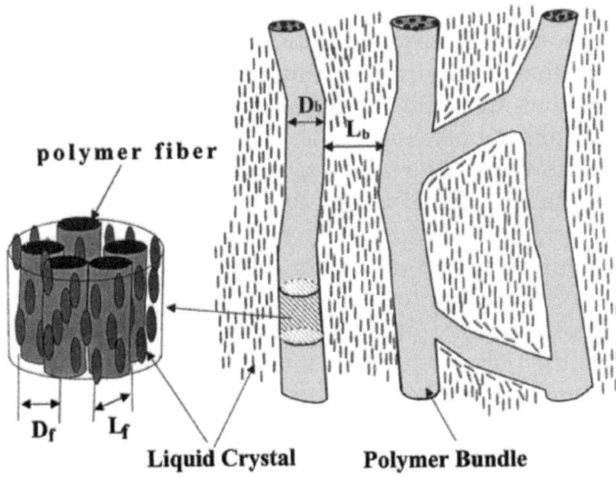

Figure 7.1 An illustration of the polymer bundle model showing the arrangement of polymer fibrils into bundles that trap the non-reactive liquid crystal molecules. D_f and L_f represent the diameter and separation of the polymer fibres, respectively, whereas D_b and L_b represent the diameter and separation of the polymer bundles. The diameters and separation of the bundles are typically of the order of 0.5 to 1 μm and 1 to 4 μm, respectively, as determined by a combination of SEM measurements and theoretical modelling.
Reproduced from ref. 19 with permission from the American Physical Society, Copyright 2000.

the photo-curing conditions were identical. In this case, one type of morphology appeared to consist of the standard polymer fibrils arrangement while the other consisted of thin polymer walls that separated the regions of non-reactive LC host. These rather contrasting morphologies were also reflected in the transmission–voltage characteristics.

The reactive mesogen that has become the *de facto* workhorse for polymer-stabilized LC studies, and has been widely used with nematic LCs, is the compound known as RM257, which exhibits a nematic phase between 70 °C and 126 °C and has a negative dielectric anisotropy of $\Delta\varepsilon = -2$. The chemical structure for this compound is shown in Figure 7.2. This is an example of a diacrylate compound with functional groups at both ends of the molecule. In many ways, the reactive mesogens that are used in polymer-stabilized nematic LCs do not differ substantially from the chemical structures of non-reactive nematic hosts in that they consist of a rigid unit with a flexible chain. To trigger crosslinking, however, requires the addition of a photo-initiator such as Irgacure 819 (or 651), which is also shown in Figure 7.2. There are, of course, a wide range of chemical structures that have been developed over the past 30 years which have been employed to form an anisotropic polymer network. Many of those are bifunctional acrylates or methacrylates.

Given the versatility of the polymer stabilization technique, it is perhaps not surprising that, for the case of nematic LCs, polymer networks have

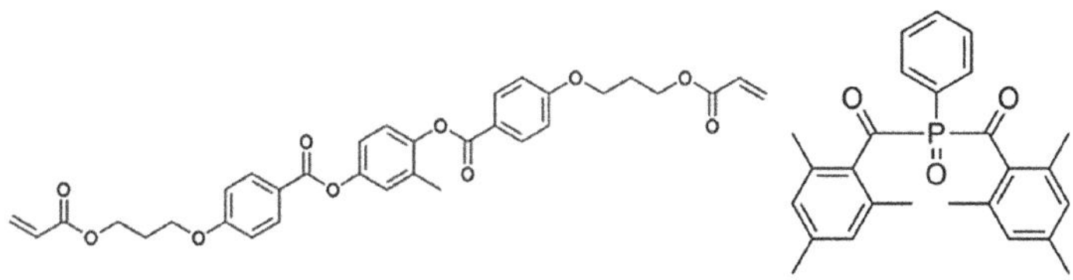

(a) (b)

Figure 7.2 Chemical structures of an example reactive mesogen (a) RM257 and (b) the photo-initiator, Irgacure 819.

been formed in a variety of different geometries ranging from conventional planar-aligned nematic devices that exhibit the traditional Fréedericksz threshold voltage,[18–20] to the incorporation of a photo-polymerized network in twisted nematic, in-plane switching (IPS)[21] and fringe-field switching (FFS) devices.[22–24] For many of the device geometries mentioned previously, improvements in some, but not all, of the electro-optic properties have been observed such as smaller relaxation response times as well as the ability to recover the original alignment upon removal of the external electric field. For IPS and FFS devices, an electric field is applied in the plane of the device rather than along the surface normal by the inclusion of interdigitated electrodes on one of the substrate surfaces. Other geometries that have been considered include the nematic π-cell,[25] which shows a rich variety of transitions and states including the so-called H, V, and bend states and has been the subject of numerous reports as it showed great promise initially as a new LC mode that could switch in under 1 millisecond or less. Nevertheless, irrespective of the geometry, the presence of a polymer network, whether it be in the bulk or within confined regions, can have a profound impact on the resulting electro-optic characteristics of the device.

In this chapter, we will review the research that has been carried out to date on forming polymer networks in nematic LCs and the impact that the network has on the material's response to applied electric fields. This will include a discussion about the work that has been done to understand the origins of the modified electro-optic characteristics. Factors that influence the resulting electro-optic behaviour such as polymer concentration, curing conditions, and device architecture will be considered along with a short summary of the fabrication techniques that have been employed to vary the distribution of the network throughout the nematic device. The role that the chemical structure of the monomer plays in the morphology of the network is, however, considered elsewhere in this textbook. After reviewing different device geometries and the manufacturing processes used to form the polymer network, we will conclude with a brief look at other areas where polymer stabilization of nematic LCs has found application such as in the development of microlenses.

7.2 Influence of Polymer Stabilization on the Electro-optic Characteristics

Over the past twenty-five years, studies have been carried out to determine how the electro-optic characteristics of a nematic LC can be modified in the presence of a polymer network.[6,17–20] Early work revealed that for concentrations of reactive mesogen in the region of 2–10wt%, polymer-stabilized nematics could be used to produce films that scatter light strongly in the presence of an electric field.[6,26] In this work, it was demonstrated that these materials could be switched from a clear, transmissive state to one that was translucent, with the application of a voltage. The opaque (translucent) state arises due to the scattering of light that occurs as a result of a mismatch in the refractive indices between the fixed polymer network and the domains of nematic LC that are free to reorient in the presence of an electric field.

The realisation of the electrically-induced scattering state, along with the benefits seen in polymer-stabilized chiral nematic LCs, provided, amongst other things, practical motivation for studying these systems and it soon became clear that fundamental improvements could also be realised in the electro-optic characteristics of polymer-stabilized nematics when compared with the behaviour observed for their non-polymer-based counterparts. With few exceptions, the device characteristic that undergoes the largest improvement with the addition of polymer stabilization is the relaxation time, which can be reduced quite significantly depending upon the chemical structure and concentration of the reactive mesogen/monomer dispersed in the nematic LC host. Inevitably, this improvement regularly comes at the expense of the driving voltages and the contrast ratio, which we will explore in more detail in the coming sections of this chapter.

In relation to the polymer stabilization process, there exist a range of parameters that can be tweaked that ultimately impact the observed electro-optic behaviour of the nematic LC. The most obvious parameters are the concentration of polymer, the curing conditions, such as the temperature and the intensity of UV light that is used to trigger the polymerization process, and the device architecture. In what follows, we will focus primarily on two of the most important characteristics that define the electro-optic behaviour of a nematic LC device, namely the threshold voltage and the response time, although where possible the impact on the contrast ratio of the device will also be considered. Our discussion will begin by limiting the focus to devices that consist of a planar (homogeneous) alignment of the nematic LC and we will consider how these important device parameters are affected by the presence of the network and the nature of the morphology. We will then extend our discussion to include other device geometries such as twisted and vertically-aligned nematic devices and the nematic π-cell, which exhibit a complex set of transitions not seen in the other geometries.

7.2.1 Planar-aligned Nematic Devices with Transverse Electric Fields

Planar-aligned nematic LC devices consist of anti-parallel rubbed alignment layers that promote a planar (homogeneous) alignment of the director throughout the bulk with some small degree of pre-tilt at the surfaces. An electric field is applied along the surface normal of the device with the aid of transparent electrodes (almost exclusively in the form of indium tin oxide layers) that are coated onto the inner surfaces of both substrates. Although the representation of the polymer morphology in the models that have been applied to date varies, it is generally accepted that the nematic LC molecules that are located within the polymer bundles/domains are free to reorient in the presence of an applied electric field whereas the LC molecules that are trapped by the polymer fibrils are locked in place and are thus unable to respond to the applied stimulus.

When the size of the domains is of the same order as the wavelength of light, then scattering can occur. This was observed originally by Hikmet[6] who showed that for concentrations by weight of monomer/reactive mesogen ranging from 2–10wt% an electrically-induced scattering state could be obtained, as stated previously. Initially, in the absence of an applied voltage, the films appeared transmissive in the planar-aligned configuration. However, with the application of a voltage, and for a nematic host with a positive dielectric anisotropy, the non-reactive LC molecules can be reoriented to align with the electric field direction.[§] This in turn leads to a mismatch in the refractive indices between the fixed, anisotropic polymer network and the mobile LC molecules giving rise to the scattering of light. Scattering can, however, be avoided if longer wavelength incident light is used to illuminate the device, such as wavelengths located in the infrared region of the electromagnetic spectrum (*e.g.* 1550 nm), provided that the domain sizes created during the formation of the polymer network are sufficiently small as to prevent light scattering at this wavelength.[27] Other approaches to reducing scattering involve adjusting the concentration as well as the curing conditions so that smaller domains (below the incident wavelength) are obtained.

7.2.1.1 Influence of the Polymer on the Threshold Voltage

To observe the effect of the polymer network quantitively on the threshold voltage, studies have often concentrated on measuring the capacitance as a function of the applied voltage.[18–20] For the case of a planar-aligned nematic LC, measuring the capacitance in the presence of an applied voltage enables the change in relative dielectric permittivity to be tracked as the director reorients, allowing the threshold (Fréedericksz) voltage to be determined

[§]A similar effect can be seen for a homeotropic alignment with a negative dielectric anisotropy nematic LC.

precisely. In addition to the threshold voltage, measurements of the capacitance–voltage curves also allow for (1) the evaluation of the components of the dielectric permittivity parallel and perpendicular to the director, the former by extrapolating the plots to an infinite voltage, and (2) some combination of the elastic coefficients.

Figure 7.3 shows exemplar plots from two separate studies of the change in the capacitance/relative permittivity as a function of the applied voltage for different concentrations of the reactive mesogen (monomer) in a nematic LC host. In the absence of polymer stabilization and at low voltages ($V < V_{th}$), the LC director configuration remains unaltered from the planar alignment imposed by the anchoring conditions of the substrate surfaces. As the voltage is increased, a threshold is reached above which the director begins to realign with the field direction and continues to align with a further increase in the voltage amplitude. With the addition of the polymer, three features are immediately obvious. Firstly, the capacitance curves appear to shift to higher voltages, leading to higher threshold voltages. Secondly, the gradient becomes smaller as the transition from planar to homeotropic alignment is stretched across a larger range of voltage amplitudes. Thirdly, for the range of voltages shown here (20 V in both cases), a much smaller capacitance/permittivity is recorded at the largest voltage amplitude. These plots clearly indicate that the inclusion of the network restricts the realignment of the LC director when the system is subjected to an external voltage and that the magnitude of this inhibition increases with polymer concentration. It is worth noting that, in general, observations reveal that for concentrations of reactive mesogen below 2wt%, the impact of the network on the threshold voltage is almost negligible and the response that is recorded mimics that of a non-polymer-stabilized nematic LC.

An alternative approach for determining the threshold voltage of a polymer-stabilized planar-aligned nematic LC is to measure the change in the transmitted intensity when the LC device is placed between parallel polarizers with the optic axis aligned at an angle of 45° to their transmission axes.[28] For the case of a nematic LC with a positive dielectric anisotropy, the director rotates to align with the electric field direction, and as a result the transmitted intensity, I, through the LC/polarisers combination varies as

$$I = \frac{1}{2}\cos^2\left(\frac{\Gamma}{2}\right) \tag{7.1}$$

where Γ represents the retardance $(2\pi\Delta nd/\lambda)$,[¶] which is a maximum in the field-off state. The application of the electric field forces the LC director in the 'mobile regions' to realign but, due to the presence of the network, domains form resulting in scattering of the light that reduces the transmission through the LC and the device appears opaque. Accompanying the scattering of light is a change in phase, which arises due to the reorientation of the

¶ Δn is the birefringence ($\Delta n = n_e - n_o$), d is the device thickness and λ is the wavelength of light.

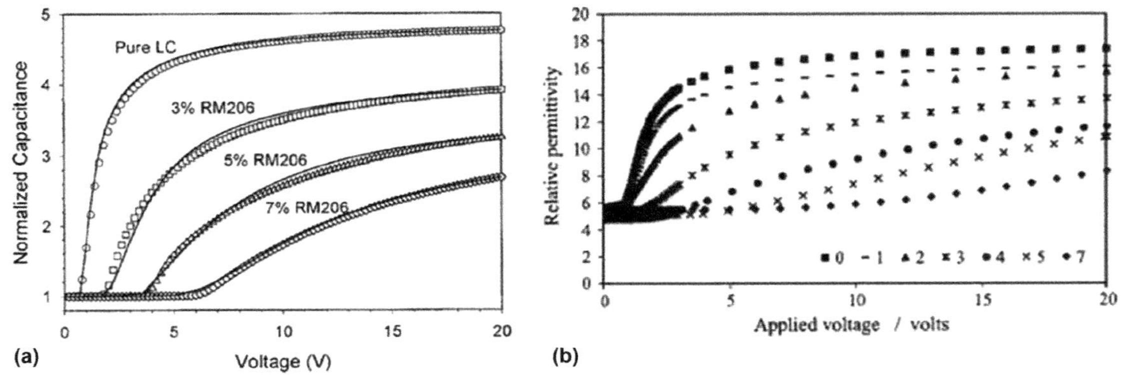

Figure 7.3 (a) Normalized capacitance as a function of voltage for a planar-aligned nematic LC device with different concentrations of the reactive mesogen, RM206, in the nematic host, ZLI4469-100 (device thickness = 10 μm) Reproduced from ref. 19 with permission from the American Physical Society, Copyright 2000. (b) Relative dielectric permittivity as a function of applied voltage for different concentrations of the reactive mesogen, RM257, in the nematic host, E7. The device thickness was 5 μm. Reproduced from ref. 20 with permission from AIP Publishing.

director and is manifested as an oscillation in the transmittance through the polarizer arrangement (due to changes in the retardance as can be seen in eqn (7.1)). By monitoring the change in the transmittance as the bias voltage applied to the cell is increased, it is possible to extract the threshold (critical) voltage.

In accordance with the plots of the capacitance as a function of the voltage, measuring the response using this approach revealed that the addition of the polymer invariably causes these transmission–voltage curves to shift to higher voltages, often introducing measurable hysteresis in the switching behaviour, which could be observed by comparing the plots obtained when either increasing or decreasing the voltage amplitude. A clear presentation of the increase in the threshold voltage with reactive mesogen concentration can be seen in Figure 7.4 from the work of Crawford, Pelcovits, and co-workers.[28] The results show that the dependence of the relative critical electric fields on the polymer network concentration is universal, irrespective of the reactive mesogen that is used – although this may just indicate the morphologies of the networks are very similar. The figure shows that for a relatively small increase in the concentration of reactive mesogen the increase in the threshold voltage can be of the order of a factor of ten larger.

The threshold voltage, derived for the case of non-polymer-stabilized samples in a planar-aligned cell, depends upon the dielectric anisotropy, $\Delta\varepsilon$, and the splay elastic coefficient, K_{11}, and is given by,

$$V_{\mathrm{th}} = \pi\sqrt{\frac{K_{11}}{\varepsilon_0\Delta\varepsilon}} \tag{7.2}$$

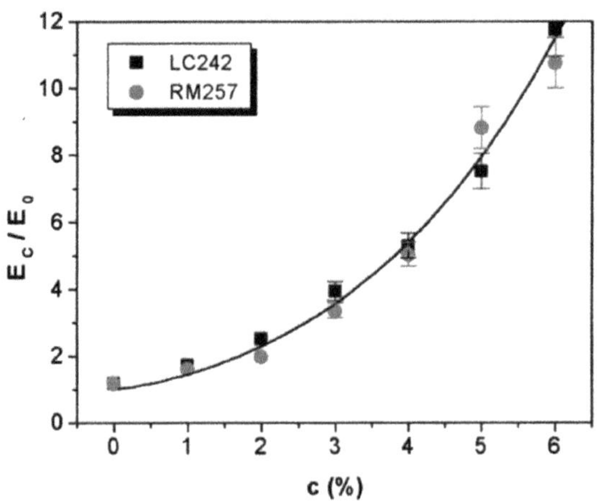

Figure 7.4 Dependence of the relative critical electric field strength as a function of the concentration of reactive mesogen. Results are presented for two different reactive mesogen compounds (LC242 and RM257) when dispersed into the same nematic host.
Reproduced from ref. 28 with permission from AIP Publishing.

Within the context of eqn (7.2), some studies have considered that any changes that have been observed in the threshold voltage must be the result of changes in the material parameters, such as the effective elastic coefficient and/or the dielectric anisotropy. In such cases, it has been shown that the increase in the threshold voltage can be accounted for by changes in both the dielectric anisotropy and the effective elastic coefficient.[29] Ideally, eqn (7.2) needs to be rederived from first principles with the inclusion of a polymer network from the onset. For studies that have avoided following this route and instead have restricted the analysis to the confines of eqn (7.2), then the dielectric anisotropy is observed to decrease and the elastic coefficient increase with polymer concentration, leading, overall, to an increase in the threshold voltage.

Considering the changes in the threshold voltage in terms of modifications to the dielectric anisotropy and elastic constant has its limitations and a more desirable approach is to rederive an expression for the threshold voltage by taking into account the presence of the polymer network. Towards this end, a variety of different models have been proposed to incorporate the role of the polymer in the switching behaviour. As an example, some of these models have involved reducing the complex morphology of the polymer network to either a series of planes or bundles that align parallel to the substrates of the homogeneous nematic LC device.[19,30] The concept of planes that describe the domain structure of a polymer-stabilized system was first introduced by Hikmet and Boots.[17] These models provided reasonable fits to the capacitance–voltage plots (as can be seen in Figure 7.3a), but they do not yield analytical expressions for the threshold voltage directly in terms of the concentration of the polymer.

In an effort to describe the threshold voltage as a function of the concentration of monomer, the team of Crawford developed a phenomenological approach,[28] based upon Frank elastic theory, in which a new expression for the critical threshold was derived. In this case, the influence of the polymer network was accounted for in terms of a characteristic length, ξ, which represents the average distance between neighbouring polymer fibrils and is related to the concentration of reactive mesogen directly through $\xi(c) = \frac{\alpha}{\sqrt{c}} - \beta$, where $\frac{\alpha}{\sqrt{c}}$ is the distance between the centres of neighbouring polymer fibrils and β is the diameter of a typical polymer fibril.[28] By minimizing the free-energy expression, the critical electric field was then found to take the form

$$E_c(c) \approx \left[\frac{\pi^2 K}{\varepsilon_0 \Delta\varepsilon}\left\{\left(\frac{1}{d} + \frac{1}{\xi + 2K/W}\right)^2 + \left(\frac{1}{\xi + 2K/W}\right)^2\right\}\right]^{\frac{1}{2}} \tag{7.3}$$

where d is the device thickness, K is the corresponding elastic coefficient in the one-constant approximation, and W is the anchoring strength, which corresponds to the magnitude of the anchoring imposed on the unreacted LC, provided the surface anchoring energy takes the Rapini–Papoular form.[31] The anchoring strength, W, is concentration dependent and was found to take the form $W = 1.95 \times 10^{-6}\left(\frac{1}{\sqrt{c}} - 0.24\right)^{-2}$ Jm^{-2}. This is smaller than that reported in other studies, where it was found that the anchoring strength at the surface of the polymer bundle was of the same order of magnitude as that of the rubbed polyimide layer on the inner surfaces of the device substrate.[19] This formulation shows that both the distance between the fibrils and the ratio of $K:W$ depend upon the concentration of polymer, as one might expect. Furthermore, in the limit of no concentration of polymer, the expression in eqn (7.3) reduces to the conventional form for the Fréedericksz threshold as given by eqn (7.2).

Other phenomenological models have been developed, such as that by Yang and co-workers[32] who have derived expressions for the threshold voltage in terms of an aligning field imposed by the polymer, $\mathbf{E}_\mathrm{p}$, which in turn is dependent upon the radius of the polymer bundles, R, and the volume fraction, c, of the polymer network. The model adopted in ref. 32 involved the introduction of an additional term, f_p, in the free-energy expression to account for the interaction between the anisotropic polymer network and the nematic LC. This additional term took a familiar form and is analogous to that which describes the coupling between the LC director ($\mathbf{n}$) and an applied electric field and is thus given by,

$$f_\mathrm{p} = -\frac{1}{2}\varepsilon_0\Delta\varepsilon\left(\mathbf{E_p}\cdot\mathbf{n}\right). \tag{7.4}$$

Following the conventional approach of minimizing the free-energy expression using the Euler–Lagrange method, subject to boundary conditions, it was found that the threshold voltage could be expressed as,

$$V_\mathrm{c} \approx V_0\left(1 + \frac{cd^2}{2\pi R^2}\right)^{\frac{1}{2}}, \tag{7.5}$$

where d is the device thickness and V_0 is the conventional Fréedericksz threshold voltage as given by eqn (7.2). The relationship in eqn (7.5) shows that the threshold voltage does indeed increase with the volume fraction of polymer (concentration) and that this increase is more severe for polymer bundles with smaller radii, due to the smaller separation distance between neighbouring bundles. While the general trend of an increasing threshold voltage with polymer concentration fits with experimental observations, this phenomenological model does appear to overestimate the amount by which the threshold voltage increases when the polymer network is formed.

From the collection of studies carried out to date, and as demonstrated in the report by Yang,[32] there is significant evidence to indicate that both the dimensions of the polymer bundles and their separation directly impact the resulting threshold voltage, which is in accordance with the relationships presented in eqn (7.3) and (7.5). As the polymer concentration (typically measured in terms of weight percentage) is increased, the distance between the polymer bundles (described by ξ in the study of the group of Crawford, for example) decreases and the Fréedericksz threshold in turn increases.[28] Using numerical simulations in the framework of the Frank elastic theory, fits to the capacitance data as a function of the voltage have been obtained (Figure 7.3a), which also indicate that the lateral dimensions of the polymer bundles and their separation are inversely proportional to the concentration of reactive mesogen.[19]

Intriguingly, SEM imaging does not entirely support these findings as the images seem to suggest that the lateral size of the polymer bundles does not depend significantly upon the concentration of polymer.[19] An example of a series of SEM images for several different concentrations of reactive mesogen is shown in Figure 7.5. However, SEM imaging is not without its difficulties as the polymer-stabilized LC devices usually need to be delaminated before the non-reactive LC host is removed and the polymer network can be imaged. The process of delamination and removal of the LC can cause the network to collapse and it is thus difficult to establish the precise dimensions of the network when in combination with the nematic LC. The discrepancy observed between the SEM images and the model could also be due to the non-uniform formation of the network across the sample. According to the Beer–Lambert relation, absorption will vary across the depth of the LC device as[‖] $I = I_0 \exp(-\alpha x)$ and may therefore lead to an uneven distribution in the polymer network dimensions.

Even though most reports find that the threshold voltage increases with the addition of the polymer, it turns out that this is not necessarily the case if an electric field is applied during the photo-polymerization process to cause a reorientation of the LC director. For example, Mora and co-workers demonstrated that for low concentrations of the reactive mesogen, the threshold appears to decrease in conjunction with an increase in the dielectric constant in the field-off condition.[18] As the polymer concentration increases so too does the dielectric constant in the field-off state leading to the case where almost no transition is observed, as shown in Figure 7.6. This behaviour results from the fact that more of the LC molecules are frozen-in by the polymer network when aligned by the applied voltage and the surface anchoring on the substrates is insufficient to force the director to return to a homogeneous state in the absence of an applied field. Based upon these results, it is clear that the actual electro-optic behaviour that is observed also depends upon whether a voltage is applied during the curing procedure as well as the polymer concentration that is used.

It is well known that for non-polymer-stabilized nematic LCs, the threshold voltage is found to be independent of the device thickness, which is evident from eqn (7.2). Reports have considered whether the threshold

---

[‖] Here I_0 is the incident intensity, I is the intensity at depth x, and α is the spectral absorption coefficient.

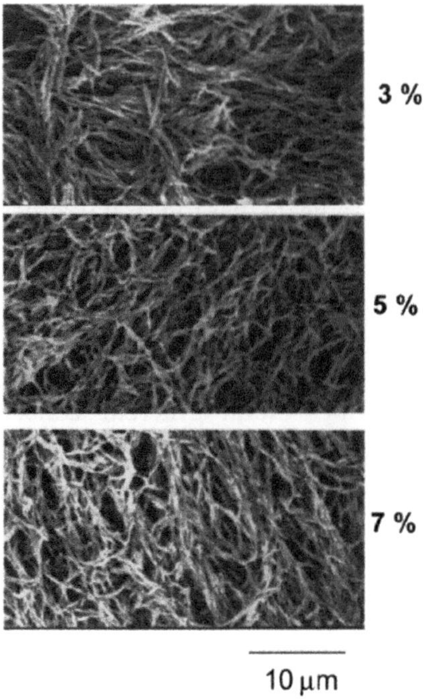

Figure 7.5 Images from a Scanning Electron Microscope of a polymer-stabilized nematic LC in a planar-aligned configuration for three different concentrations of the reactive mesogen. In this case the LC mixture is comprised of the nematic host, ZLI4469-100, the diacrylate monomer, RM206 (Merck) and the photoinitiator benzoin methyl ether (Polyscience). These mixtures were filled into devices that were 10 μm thick.
Reproduced with permission from ref. 19 with permission from the American Physical Society, Copyright 2000.

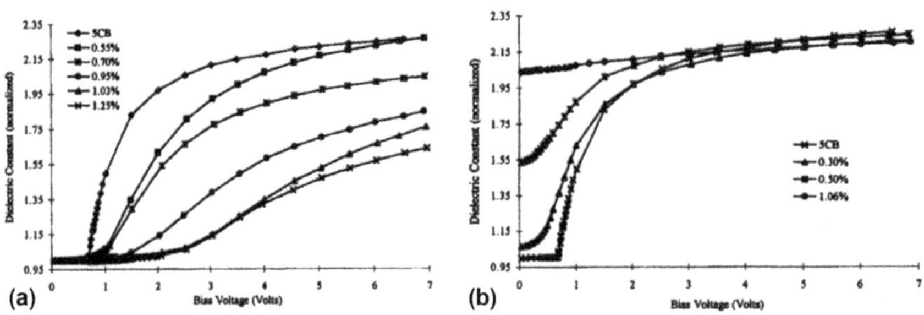

Figure 7.6 Plots of the normalized dielectric constant as a function of the bias voltage for polymer-stabilized nematic liquid crystal devices that are polymerized without (a) and with (b) a bias voltage applied during the photopolymerization process. Plots are shown for different concentrations of the reactive mesogen 4,4′-bis(acryloyloxy)biphenyl in the nematic host 5CB.
Reproduced from ref. 18 with permission from Taylor & Francis Ltd (www.tandfonline.com).

voltage of polymer-stabilized nematic LC devices varies with the cell (device) thickness and it is clear from eqn (7.3) and (7.5) that these can have different dependencies. One report, for example, has shown that, irrespective of the concentration of reactive mesogen that is used, the threshold voltage has an almost linear dependence on the layer thickness.[19] Regardless of the exact form of the dependence, which varies to some degree across reports in the literature, the clear dependence upon device thickness contrasts with non-polymer-stabilized nematic devices, which show no dependence on thickness. In the latter case, the alignment of the LC in the absence of an applied voltage is controlled by the anchoring at the substrate surfaces. However, for polymer-stabilized nematic LCs, this is not the case, as the polymer network also imposes an alignment condition on the non-reactive LC components and the spacing of the polymer bundles is much smaller than that of the cell gap. Results also appear to indicate that the dimensions of the polymer bundles do not vary with cell thickness as elucidated from both fits to experimental data using theoretical modelling and SEM imaging.

7.2.1.2 Influence of the Polymer Network on the Response Time

A vitally important parameter in LC devices is, of the course, the response time. However, there are far less reports on the behaviour of this parameter in planar-aligned nematic LC devices in the presence of a polymer network when compared to studies of the threshold voltage. This reason for this is two-fold: 1. the apparatus that is typically used for measuring the capacitance (which is employed to measure threshold voltages) is unable to record the response times and 2. measuring the transmittance characteristics, which is an alternative way of measuring voltage as described in the previous section, is sometimes complicated by oscillations that arise due to retardance values of the LC device that far exceed $\pi/2$. Nonetheless, researchers have attempted to consider quantitatively the impact of the network on the response time, an example being the report of Crawford *et al.* who considered the relaxation time as a function of the concentration,[28] which is shown in Figure 7.7a. It can be seen from the results that, unlike the threshold voltage, the response time decreases as the concentration of the reactive mesogen increases, demonstrating an obvious benefit from the point of view of device performance. Theoretical results presented in ref. 32 also corroborate these findings and are shown in Figure 7.7b, where it can be seen that the turn-off (relaxation) time decreases with increasing polymer concentration. Both sets of results are illustrative examples of the typical behaviour that is observed when adding 1–10wt% reactive mesogen. Figure 7.7b also shows that based upon theoretical predictions it is expected that a decrease in the radius of the polymer bundles leads to a concomitant decrease in the response time as the bundles become more tightly packed together.

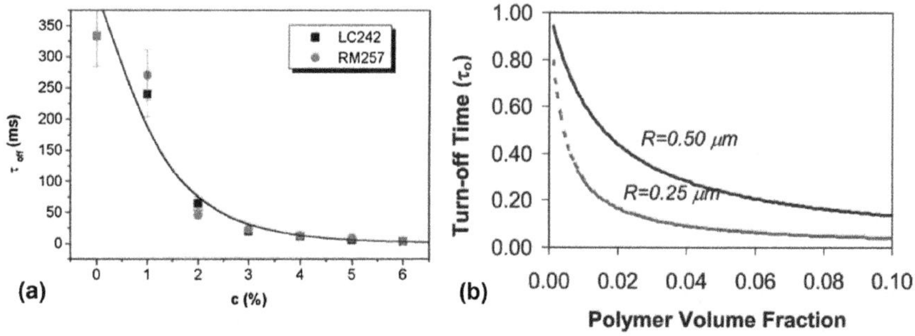

Figure 7.7 Relaxation times of a polymer-stabilized nematic LC device in a planar-aligned configuration as a function of the concentration of polymer. (a) Relaxation time as a function of reactive mesogen concentration for two different reactive mesogens in the nematic host, BL038 (Merck). The device thickness was 5 μm. The solid line represents the fit using eqn (7.6). Reprinted from ref. 28, with the permission of AIP Publishing. (b) Calculated relaxation times as a function of polymer concentration using eqn (7.7) for polymer networks with different radii of the polymer bundles. In this case the device thickness was also 5 μm. Reproduced from ref. 32 with permission from AIP Publishing.

In terms of analytical expressions describing the response time, it was shown in ref. 28 that the relaxation time depends upon the concentration of reactive mesogen through the critical electric field, E_c, as,

$$\tau_{\text{off}} = \frac{\pi^2 \gamma}{\varepsilon_0 \Delta\varepsilon E_c^2} \tag{7.6}$$

where γ is the rotational viscosity and E_c takes the form of that given in eqn (7.3). The solid line in Figure 7.7a represents the fit to the experimental data using eqn (7.6) and the comparable magnitudes of the response times for the two different reactive mesogens infer that the morphologies of the two polymer networks are actually very similar. A slightly different form was obtained by Yang *et al.* who also used Frank elastic theory to derive analytical expressions for the response time by considering the polymer network as an effective field. In their report,[32] they found that

$$\tau_{\text{off}} \approx \left(\frac{\gamma d^2}{K_{11}\pi^2}\right)\left(\frac{1}{1 + c(d/R)^2/2\pi}\right) \tag{7.7}$$

where the first term represents the relaxation of a nematic LC without polymer stabilization and the expression collapses to this form when the volume fraction of the polymer is zero (*i.e.* no polymer network). Both eqn (7.6) and (7.7) are in accord with experimental observations in so much as the response time is found to decrease with an increase in the concentration of polymer. However, eqn (7.7) typically underestimates the response times when compared to experimental results, although determining volume fractions and polymer bundle dimensions are non-trivial, which leads to some discrepancy. Verification of the role of the dimensions of the polymer bundles is a little less clear cut and, while the theoretical predictions suggest that the turn-off time can be further improved by reducing the size of the polymer bundles (at the expense of the threshold voltage), it is difficult to marry this up with experimental results, as precise control of the dimensions of the polymer network is not straightforward. Potential routes for controlling the radii of the polymer bundles and their separation is considered in the next sub-section. Despite the shortcomings of the models developed thus far in terms of accurate predictions of the threshold voltage and relaxation time, they do serve as an excellent guide in terms of predicting the behaviour of polymer-stabilized planar-aligned nematic LC devices and provide some form of design motif for the mixture formulations in order to obtain a combination of fast response times combined with acceptable threshold voltages.

7.2.1.3 Curing Temperature and UV Intensity

Several studies have considered the role of curing temperature on the resultant electro-optic characteristics of polymer-stabilized nematic LC devices.[15,19,33–35] By changing the temperature at which the polymer network forms, it is possible to modify the electro-optic properties, namely the threshold voltage and the response time. This modification in the electro-optic behaviour arises due to a number of factors. Firstly, the diffusion process of the monomer is temperature dependent and this occurs much more rapidly at high temperatures, leading to a greater dispersion and, as a result, a relatively coarse network. In this case, both the diameter and separation of the polymer fibrils increase with the curing temperature. It has also been revealed, using a technique known as gel permeation chromatography, that the molecular weight of the polymer decreases with increasing temperature.[34] With the formation of a coarser network, the Fréedericksz threshold voltage is observed to decrease with increasing temperature along with a simultaneous reduction in the hysteresis in the transmission–voltage curves recorded for repeated sweeps of the voltage amplitude.

While the approach of altering the photo-curing temperature shows promise in terms of achieving lower threshold voltages, the downside is that the improvements in the response time that are ordinarily obtained are then sacrificed. This is because the confinement imposed upon the non-reactive nematic LC is relaxed as the polymer network becomes more distributed. Nonetheless, adjusting the curing temperature at which the network is formed provides a mechanism by which a trade-off can be achieved such that a lower response time can be obtained with only a marginal increase in the threshold voltage. There also appears to be an optimum trade-off point in terms of temperature, as the contrast ratio is found to exhibit a maximum value at a certain temperature.[34] The phase in which the polymer is formed (*e.g.* nematic and isotropic phase) and the intensity of the UV radiation have an influence on the resulting switching behaviour and this is due to changes in the morphology and the cross-linking process.[13,19] For low curing intensities, the polymer bundles grow much larger than they do at high UV intensities because a lower proportion of the photo-initiator molecules are converted into free radicals, which then result in fewer activation sites for cross-linking with the reactive mesogen units.

When preparing mixture formulations for the fabrication of polymer-stabilized nematic LC devices, it is also important to consider whether the nematic LC host exhibits an absorption band in the UV and at wavelengths that overlap with the photopolymerization light source, as this can interfere with the cross-linking process. Absorption by the nematic host during curing has been considered previously by Dessaud and co-workers,[20] where subtle variations in the capacitance–voltage behaviour were recorded for two different polymer-stabilized compositions. These polymer compositions contained the same reactive mesogen but different nematic hosts: one that absorbed UV light (E7) and one that did not (ML1001). The differences in the capacitance data were attributed to the formation of different polymer architectures, although in this case no SEM images were reported to substantiate this claim.

7.2.2 In-plane and Fringe-field Switching Nematic Devices

In-plane switching nematic devices[36,37] have their roots firmly entrenched in the 1970s,[38,39] but their commercial success was very slow in coming due in part to several initial technological stumbling blocks. Despite the clear advantages with regards to viewing angle, there were a number of potential showstoppers when compared with other modes, such as the twisted nematic mode. Drawbacks with this technology included the reduced optical throughput due to the presence of interdigitated electrodes (see Figure 7.8 for an illustration of the typical electrode architecture), large driving voltages (due to the relatively large electrode spacings), and slow response times. While the first two parameters could be tolerated using desktop and large flat panel technology, the slow response times was its Achilles' heel. As a result, research groups explored alternative routes to improve the relaxation time such as the addition of viscosity reducing agents.[40,41]

Escuti and co-workers were among the first to consider the use of polymer stabilization to combat the problem of slow response times in IPS devices.[21,42] To achieve this they biased the alignment in the voltage-off configuration using a polymer network. Subsequent studies have confirmed the decrease in the turn-off (relaxation) response time while attempting to minimise the adverse impact on the contrast ratio.[43] They showed conclusively that both the relaxation time and the voltage-on time could be improved (the former dramatically so) using a relatively low-density polymer network. In accordance with the observations made for the more conventional, planar-aligned devices with transverse electric fields, the

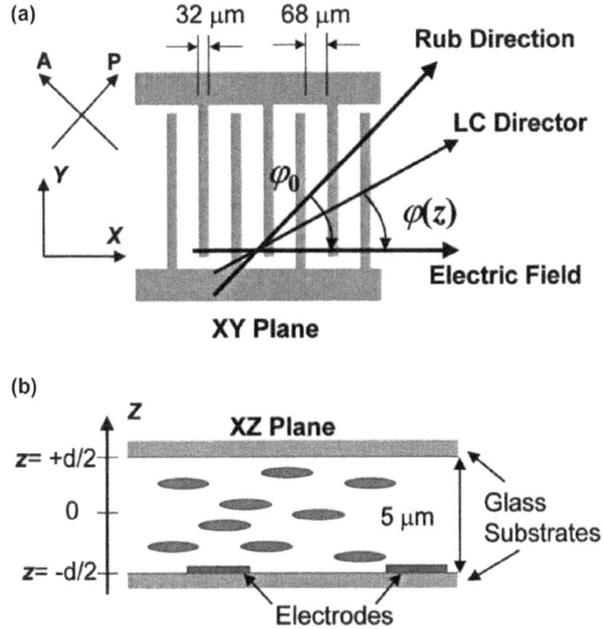

Figure 7.8 Example of the in-plane switching geometry when viewed from above (a) and from the side (b).
Reproduced from ref. 21 with permission from AIP Publishing.

improvement in the response times occurred at the expense of an increase in the driving voltage. In these studies, the authors used the reactive mesogen RM257 at concentrations ranging from 0.5 to 2.0wt% and found that the fall times decreased with an increase in the polymer concentration irrespective of the magnitude of the electric field.

An improvement in the rise time (voltage-on transition) was also noted for low electric field amplitudes but was independent of the concentration of polymer at larger field amplitudes. The reduction in the response time arises due to the presence of the network which, in combination with the anchoring imposed by the substrate surfaces, encourages the LC director to return to the zero-voltage alignment. The polymer network also serves to limit the amount by which the LC director reorients when the device is subjected to an external voltage. As the LC director does not rotate by as large an amount when a voltage is applied, the rise time is thus observed to be shorter for low electric field amplitude. The restriction in the reorientation of the LC director can also be seen in the transmission voltage curves. These show a much more staggered response as the equivalent electric field amplitude is unable to cause the same degree of reorientation of the director as that observed for the non-polymer-stabilized device. This in turn leads to a decrease in the transmission as the polymer concentration is increased.

To account for the observed experimental behaviour, the authors of ref. 21 introduced an effective field term into the free-energy expression based upon continuum theory, in a similar way to that of Yang *et al.* for planar-aligned nematic devices with transverse electric fields.[32] The effective polymer field term, E_p, which is minimized when the LC is aligned to the preferential direction of the network (governed by the rubbing direction at the substrate surfaces), is analogous in this case to a surface anchoring term. By considering a homogeneous electric field in the plane of the device and starting with an expression for the free-energy, with the effective polymer field term included, Escuti and colleagues showed that the 'on' and 'off' response times could be expressed as,[21]

$$t_{\mathrm{on}} \cong \frac{\gamma}{\varepsilon_0|\Delta\varepsilon|\sqrt{E^4 + E_{\mathrm{p}}^4}} \tag{7.8}$$

$$t_{\mathrm{off}} \cong \frac{\gamma}{\varepsilon_0|\Delta\varepsilon|E_{\mathrm{p}}^2} \tag{7.9}$$

where both response times are found to depend upon the effective polymer field in accordance with experimental data. Note the similarities here to those of the planar-aligned nematic devices with transverse electric fields described in Section 7.2.1.2.

Polymer stabilization clearly has the potential to play an important role in IPS devices, but the approach of sacrificing the contrast ratio in favour of the response time is not desirable for high definition display panels. Trade-offs can be made to find a compromise between a short response time and an acceptable contrast ratio, although the preferable solution would be to form the network in such a way that did not deteriorate the contrast ratio. The reduction in contrast ratio with the formation of the polymer network results, primarily, from light scattering that arises from the mismatch in the refractive indices between the polymer network and the non-reactive LC. Scattering, in turn, causes a depolarization of the light, which can then 'leak' through the polarizers in the voltage-off (dark state).

To address the issue of the decrease in the contrast ratio, Zhou *et al.* compared the switching characteristics of two IPS devices consisting of different reactive mesogens, both of which included functional groups at either end of the molecule with a central rigid core.[43] In accordance with the earlier work of Escuti,[21,42] a measurable reduction in the response time was observed for both devices, but the impact on the contrast ratio was quite different and was attributed to differences in the morphology of the polymer network. The first reactive mesogen tested was the ubiquitous RM257 and the findings were very similar to those reported in ref. 21. A second mixture consisting of the reactive mesogen, referred to as HCM-009, dispersed into a lower birefringence nematic host, was tested in an IPS device and it was found that there was a noticeable reduction in the relaxation time (above a certain concentration of reactive mesogen), yet the contrast ratio appeared relatively unaffected, which is in conflict with the observations made for the RM257 compound. It was speculated, based upon the images obtained from SEM, that the morphology was the key factor responsible for the differences in performance, most notably because the network formed using RM257 consisted of more tightly packed fibrils that had smaller lateral dimensions and therefore imposed stronger anchoring on the nematic LC.[32] This hypothesis would indeed account for the higher threshold voltages, but the impact on the contrast ratio is less clear. A further complication arises from the fact that two different nematic hosts were used in the study, the second of which had a lower birefringence, which would invariably lead to less scattering of light and therefore a higher contrast ratio. Nevertheless, the morphology of the network and the combination of reactive mesogen and nematic LC needs to be selected judiciously.

Fringe-field switching[22–24] (FFS) was developed as a successor to IPS, providing the much-needed improvement in the optical throughput while maintaining the wide viewing angle characteristics. However, even with the improvement in light efficiency, the technology still suffered from slow response times due to relatively large electrode spacings, which are limited by the fabrication process and the associated manufacturing costs. Unsurprisingly, polymer stabilization has also been applied to FFS devices in an attempt to improve device performance such as the response times.[44,45] In this case, reports have demonstrated new transflective display modes through the polymer stabilization of FFS devices.[44]

7.2.3 Polymer-stabilized Twisted Nematic and Vertically-aligned Nematic Devices

Polymer stabilization of twisted nematic (TN) devices has also been considered,[46] although it has not received quite the same level of attention as the homogeneously (planar) aligned nematic devices. The presence of the twisted director profile results in a more complex morphology of the polymer network than is observed in the devices described in the previous sections. As before, an improvement in the response time can be achieved, but the inclusion of the network can also introduce additional benefits, such as the suppression of backflow, which is an unwanted feature that is commonly associated with TN devices.[46] Moreover, despite the presence of an interpenetrating network, the twisted nematic configuration still maintains its waveguiding characteristics.

For the case of a TN device, the turn-off (τ_{off}) and turn-on (τ_{on}) response times are given by

$$\tau_{\text{off}} = \frac{\gamma_1 d^2}{K_{22}\pi^2} \tag{7.10}$$

$$\tau_{\text{on}} = \frac{\tau_{\text{off}}}{\left(\frac{V}{V_{\text{th}}}\right)^2 - 1} \tag{7.11}$$

where γ_1 is the rotational viscosity, V_{th} is the threshold voltage, and K_{22} is the twist elastic constant. For most monomesogenic compounds, K_{22} is the smallest of the elastic constants, and therefore, assuming all other factors are equal, the relaxation (turn-off) time for the TN device is slower than that observed for the planar-aligned nematic LC devices described in the previous sections. With the addition of 3wt% reactive mesogen (monomer), it is found that both response times reduce, but that the threshold voltage is increased by a factor of four. A substantive theoretical model has not been developed that describes the changes observed for a TN cell, but in their study, Lu and co-workers attributed the faster relaxation time to a larger effective twist elastic constant.[46]

Vertically-aligned (VA) nematic technologies were first proposed by Fujitsu in the mid-1990s as a mode that has a high optical contrast at normal incidence and does not require rubbing of the substrate surfaces. There are two main categories of VA technology namely the multidomain (MVA) and patterned VA (PVA) modes.[47–51] A VA mode exhibits a high contrast ratio, particularly at normal incidence, because of the high optical extinction obtained in the dark state. A downside, however, is that this mode requires negative dielectric anisotropy materials when a transverse electric field is employed, and such materials are typically characterised by a rotational viscosity that is inherently large compared with that observed for their positive dielectric anisotropy counterparts. The consequence, therefore, is that the response times are longer than those obtained in the devices described in this chapter thus far. However, polymer stabilization can be used to improve the response times of the VA mode[52,53] and studies have been conducted that show that an anisotropic polymer network morphology is preferable for reducing both the turn-on and turn-off times.[53] Using an isotropic polymer structure, it is possible to reduce the turn-on time, but the relaxation time is more or less unaffected.[53] As before, the improvements in the response times come at the expense of the threshold voltage and the maximum transmittance.

The VA mode, whilst offering much promise in terms of optical contrast, suffered from a very poor viewing angle. To overcome this, Fujitsu introduced the MVA mode. This involved dividing the domains in each pixel into separate domains in the form of chevrons and using protrusions to force the necessary alignment so as to improve the viewing angle characteristics. Since the development of the MVA, multiple incarnations of this mode have been introduced, often assigned superlative prefixes in the form of Super-MVA, Premium-MVA, and Advanced-MVA, with further improvements in the off-axis viewing angle and response time being offered.

On the other hand, the PVA mode, which was developed by Samsung as an alternative to the MVA mode, wherein the pixel and common electrodes are patterned alternately leading to an electric field that has both horizontal and vertical components, also provided improvements in terms of a faster decay time as well as a rubbing-free fabrication process. As the PVA technology has advanced, new generations of this mode have been introduced, in a similar manner to that described for the MVA modes (*e.g.* S-PVA and A-PVA).

Unfortunately, in their native states both the MVA and PVA suffer from a relatively slow rising (turn-on) time, which is in contrast to most other LC modes where it is the relaxation (turn-off) time that is the limiting factor. The issue of slow rise time in these LC modes was alleviated, in part, by the introduction of the so-called over-drive** technology, although this brings with it its own issues in terms of complexity of the drive electronics and the manufacturing costs. In an effort to reduce the rise time, and improve other device characteristics, researchers have considered applying the technique of polymer stabilization to the MVA and PVA modes.

Structuring the pixel into separate domains enables the viewing angle to be increased, but the downside is that the inter-domain regions, as well as the edges of the pixels, encourage the formation of disclinations, which result from the inhomogeneous electric field profile that exists in these regions. This is shown in the optical micrograph of a PVA pixel in Figure 7.9a and an illustration of the corresponding alignment of the nematic director is presented in Figure 7.9b. The presence of these disclinations results in a degradation of two principal device characteristics: the response time and the transmittance. To combat the formation of these unwanted disclinations, researchers have considered redesigning the pixel configuration.[54] Alternatively, as an example, Kim and co-workers[55] have applied the technique of polymer stabilization to the PVA mode in an attempt to improve the transmittance and suppress the formation of the disclinations. In their study, a combination of reactive mesogen and photoinitiator were dispersed into a super-fluorinated nematic LC mixture that had a negative dielectric anisotropy of $\Delta\varepsilon = -4$. Formation of the polymer network was then carried out with the application of an electric field so as to cause a reorientation of the LC director away from the layer normal thereby reducing the pre-tilt of the molecules in the absence of a voltage (see Figure 7.9c). Polymerization at larger voltage amplitudes resulted in a further decrease in the pre-tilt angle when the electric field was removed.

Subsequent transmission–voltage (T–V) studies revealed that, contrary to previous observations for the planar-aligned devices described in the preceding sections, the T–V curves were found to shift towards lower voltages resulting in lower threshold voltages (Figure 7.10a) when the polymer network was formed in the presence of a large amplitude electric field. As the results show, the larger the voltage amplitude applied during polymerization, the lower the threshold voltage. Simulations carried out using the software package LCD Master showed that the experimental results were indeed consistent with a decrease in the pre-tilt angle in the field-off state and that an improved transmittance could be achieved (Figure 7.10b). The introduction of polymer stabilization was therefore found to minimize the formation of disclinations and defects by imposing a lower pre-tilt in the field-off state, which in turn was manifested as an improved transmittance

**Over-drive technologies often appear under the alternative pseudonym of *dynamic capacitance compensation*, for which there have been multiple variants.

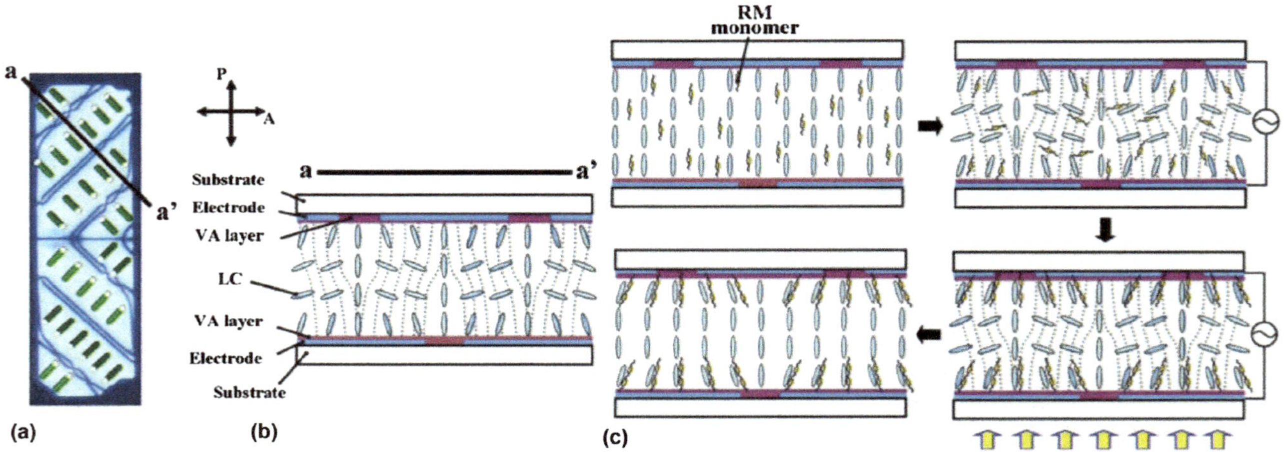

Figure 7.9 Polymer stabilization of a patterned vertically-aligned (PVA) nematic LC device. (a) Optical micro-photograph of a pixel in the white state of the PVA cell with an illustration of the LC molecules. (b) A cross-section of the device. (c) Illustration of how polymer stabilization can be used to force a pre-tilt of the director within the device.
Reproduced from ref. 55 with permission from AIP Publishing.

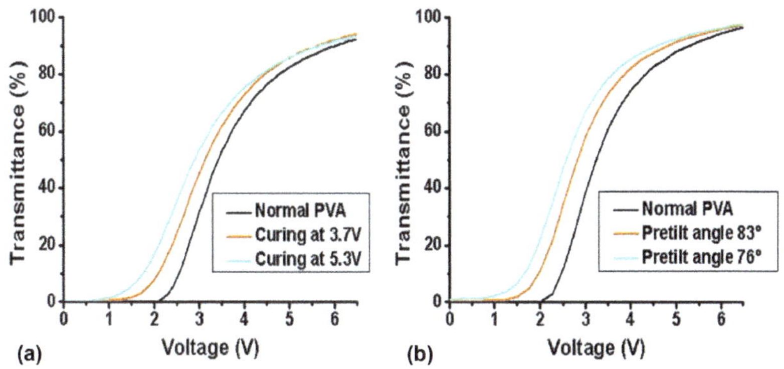

Figure 7.10 The transmittance–voltage curves for a polymer-stabilized PVA nematic device when a polymer network is formed in the presence of an applied voltage of different amplitudes. (a) Experimental results of the T–V curves for a 3.8 μm-thick device. (b) Simulations of the transmittance–voltage curves for a conventional PVA device with different pre-tilts in the absence of an applied voltage.
Reproduced from ref. 55 with permission from AIP Publishing.

and lower rising time. The larger the voltage amplitude applied during photo-polymerization, the smaller the pre-tilt in the field-off (black) state and the better the device performance characteristics. It was noted that the polymer network was confined to the surfaces although no additional data, such as SEM images, were provided to support this conclusion.

A study has shown that the relaxation time of a vertically-aligned nematic LC can be reduced if polymer stabilization is carried out at sub-zero temperatures.[56] In contrast to the study by Kim *et al.*,[55] an electric field was not applied during the photo-curing process and it was found that, in accordance with the results obtained for planar-aligned nematic devices, the T–V curves were shifted to higher voltages with the introduction of the polymer network along with a decrease in the maximum transmittance. A further increase in the polymer concentration results in a greater shift in the threshold voltage. By curing at different temperatures, it was found that the threshold voltage increases with decreasing temperature, as noted for the planar-aligned devices, and that the relaxation time is decreased. In their report, Park and co-workers[56] showed that both the turn-on and turn-off response times are reduced by increasing the concentration of the reactive mesogen. By forming the network at low temperatures, the optimum performance, as far as response times are concerned, occurs when the cross-linking takes place at a temperature of −20 °C. This behaviour is most likely due to an increase in the density of the network and a decrease in the domain size (distance between polymer bundles/domains); however, SEM images carried out on these polymer structures were not very clear and somewhat inconclusive, perhaps due to a degradation to the network during the preparation procedure. These notable improvements in the response time, however, come at considerable expense to the threshold voltage and the contrast ratio of the device. One approach in the future may be to consider adopting the pre-tilt biasing demonstrated in ref. 55, but with the formation of the network taking place at the lower end of the temperature range of the nematic phase.

7.2.4 Polymer Stabilization in π-cells

The nematic π-cell, which was first proposed in 1984 as a new fast-switching LC mode,[25] has continued to attract interest for use in field sequential display devices. The novelty of the approach, compared with existing technology at the time, was that the rubbed surface alignment layers were aligned parallel rather than anti-parallel so as to produce no "backwards" torque on the director near the centre of the cell upon relaxation after an electric field had been removed. In a conventional planar-aligned cell with pre-tilts in opposite directions on the inner surfaces of the opposing substrates, such as those described in Section 7.2.1, the backward torque at the centre of the cell causes a 'slowing' of the relaxation. For the optically compensated bend (OCB) mode, which involves using a π-cell combined with an optical compensation film, the desired 'on' and 'off' states of the device are the vertically aligned (V) state and the bend state, respectively (Figure 7.11a). This mode is of interest as it offers fast response times and potentially wide viewing angles. In practice, however, the bend state is not stable and instead relaxes through a twisted (π) state to eventually finish in a splay configuration, which has the lowest free-energy in the absence of an applied electric field (Figure 7.11b). This set of elaborate transitions[††] complicates the device behaviour as the transition from the splay to the bend state is by no means straightforward since the two states are not topologically continuous with one another.

To circumvent some of these unwanted transitions and ensure that the initial state is either a bend or π-twisted configuration, polymer stabilization has been employed to lock-in these states so as to avoid the need for complex driving schemes required to trigger a transition from the splay state, which typically occurs above a critical voltage and requires a long time to stabilize.[57–62] One approach that has been adopted is to stabilize the bend state through the formation of polymer walls.[57] First, an electric field is applied to the device to induce the bend alignment. Following this, photo-polymerization is carried out in selected regions, as defined by a mask, to form polymer walls within the device (Figure 7.12). After fabrication, the splay state is suppressed, and the initial zero-voltage state is now the bend-state, thereby ensuring that the 'on' to 'off' transition with an electric field consists of only the V and bend states, respectively.

[††]In practice, the scheme shown in Figure 7.11 over simplifies the nature of the transitions as the device also passes through the so-called symmetric and asymmetric H states.

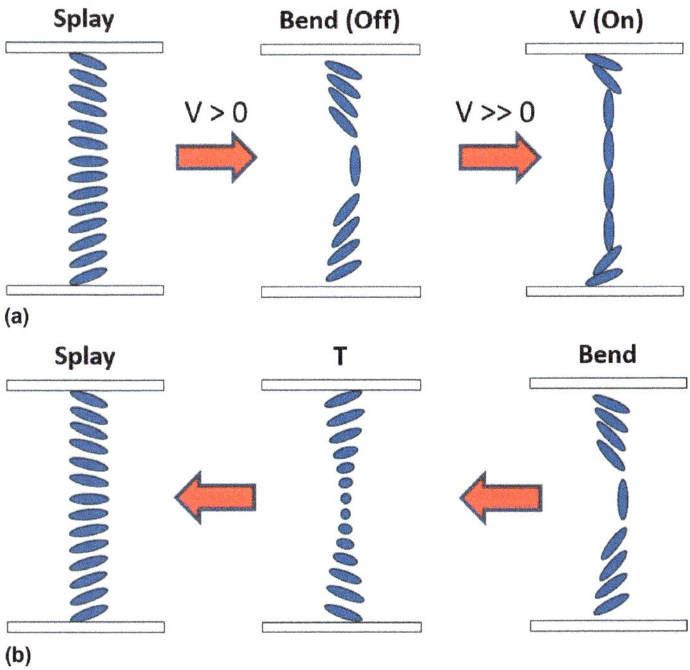

Figure 7.11 The transitions in the director profile for a nematic π-cell. (a) The transition from the initial splay (H) state at zero voltage to the bend and vertical (V) states above certain critical voltage amplitudes. (b) The relaxation of the bend state to the splay state through the π-twisted (T) state.

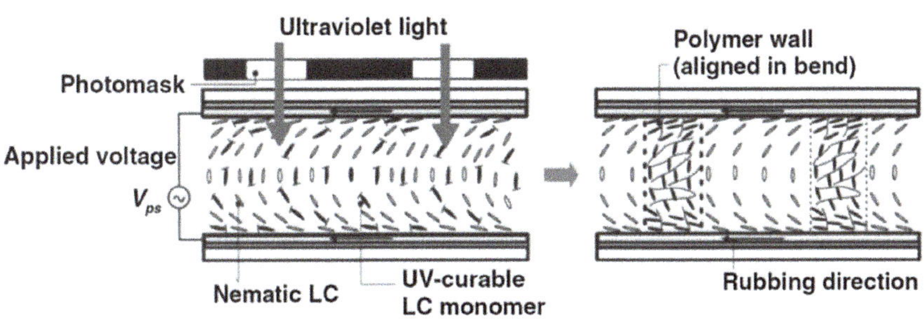

Figure 7.12 Illustration of the use of polymer stabilization to form polymer walls to stabilize the bend alignment in a nematic π-cell.
Reproduced from ref. 57 with permission from the Japan Society of Applied Physics.

Polymer stabilization in the bulk has also been shown to be a viable way of achieving a spontaneous and voltage-free alignment of the bend state.[58,59] In this case, a voltage is applied to form the bend state before it is then locked in place using photopolymerization. The result is a polymer scaffold that consists of fibrils that mimic the bend configuration, which was confirmed using SEM imaging. Such devices were shown to switch rapidly between the 'on' and 'off' states. However, the polymer concentration was found to be a critical parameter in this approach as too little polymer is insufficient to serve as a template for the bend-alignment, while too much results in very poor optical contrast and transmittance–voltage characteristics. Therefore, an optimum concentration is required to ensure that there is sufficient polymer to form a template for the bend alignment, but not too much that the optical properties are compromised.

In addition, experimental and theoretical research has shown that the rise and decay times of the transient response are adversely affected by the presence of the polymer network, which has been attributed to a polymer network-induced alteration in the flow properties of the device.[60] Ultimately, careful control of the polymerization conditions,[61,63] the dimensions of the grid-like mask[64] (if one is used), and the mixture formulation are of importance when considering the impact of polymer stabilization on the resulting electro-optic properties of OCB devices. The formation of thick polymer walls can decrease the optical throughput considerably whereas polymer/LC domains formed in the bulk can also reduce the transmission properties as a result of unwanted light scattering. There is also a trade-off to be made in terms of stabilizing the bend state in the voltage-off configuration and the resulting response times.

7.3 Advanced Fabrication Techniques for Polymer-stabilized Nematic Devices

With the advent of advanced photolithographic and direct laser writing techniques, ever more sophisticated morphologies are being explored and reported. Through new manufacturing processes such as two-photon polymerization,[65] it is now possible to form volume elements (often referred to as voxels) of polymer networks on the micron scale, leading to some rather exotic features and behaviour. The small, localized dimensions of the polymer network result from the nonlinear nature of the two-photon absorption since only a small region around the focal spot of the laser is of a sufficiently high intensity to trigger the cross-linking process. Using such an approach provides the experimental tools with which to stabilize transient alignments and improve device performance.

When combined with adaptive optics components, it is possible to use two-photon polymerization to write directly into fully assembled devices allowing intricate networks to be formed in very small regions that memorize the LC alignment at the moment of exposure to the laser writing beam. An example of using this technique to lock-in different alignments in a nematic π-cell is shown in Figure 7.13. Here, different regions of V (vertical), H (splay), and π-twisted (T) states co-exist leading to interesting new electro-optic behaviour and the formation of disclinations/defects that separate

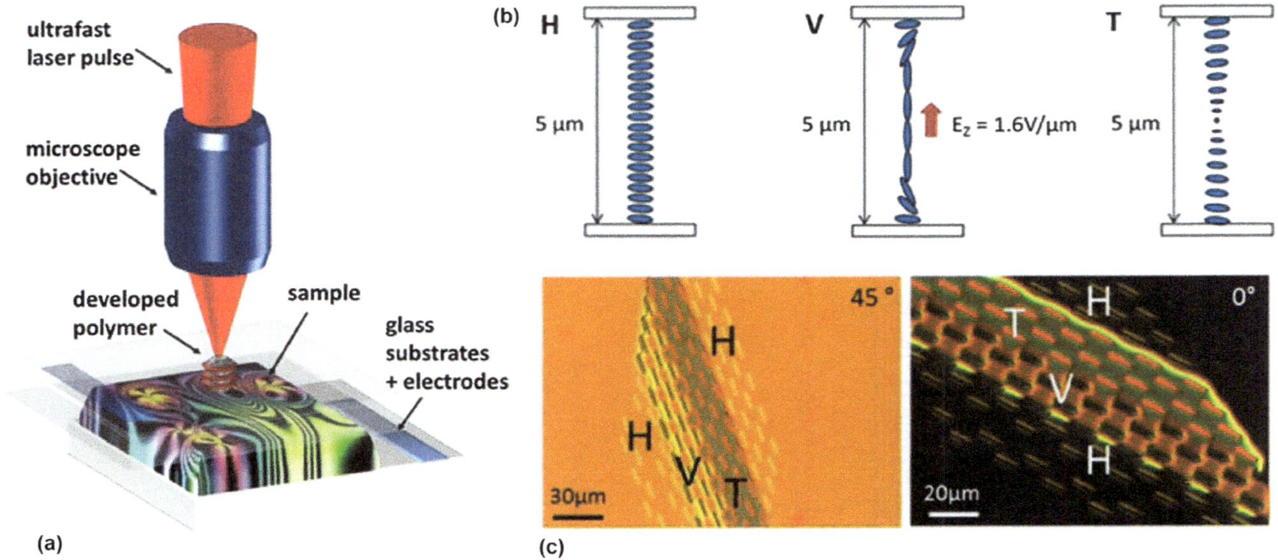

Figure 7.13 Direct laser writing to stabilize different director alignments in a nematic π-cell. (a) Illustration of direct laser writing in a fully assembled device. (b) Director profiles for the H, V, and T states (c) polarizing optical microscope images of the laser-written H, T, and V states.
Reproduced from ref. 65 with permission from the Royal Society of Chemistry.

the topologically-distinct states. These findings show that in addition to the formation of polymer-networks in the bulk or confined to walls/surfaces, it is possible to form intricate networks extending over small volumes thereby stabilizing only a localized region of the LC alignment.

7.4 Polymer-stabilized Nematic Liquid Crystal Microlenses

Polymer stabilization can also be used to form nematic microlenses[66] (as shown in Figure 7.14). One of the earliest approaches to forming polymer-stabilized nematic microlenses involved the use of a hole-patterned electrode, which provides a parabolic refractive index distribution when an electric field is applied. This gradient index profile is then frozen-in by photocuring the polymerizable LC mixture, resulting in the formation of a microlens or an array of microlenses as shown in Figure 7.14. With the application of an electric field (voltage) after the fabrication process, it is possible to tune the refractive index profile and thus the focussing

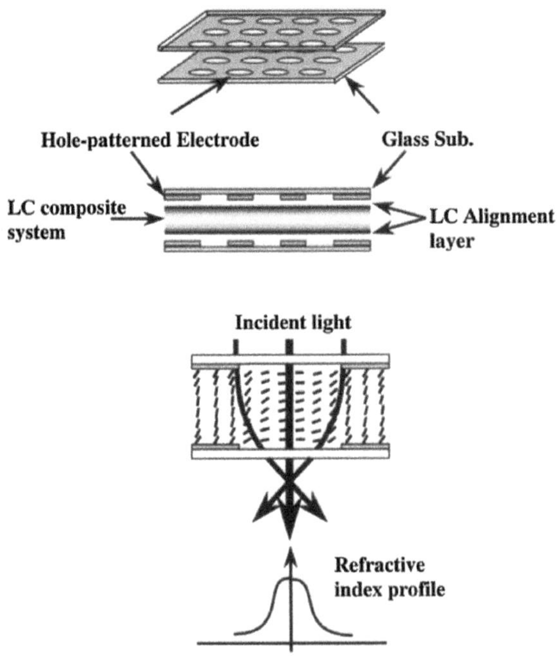

Figure 7.14 The formation of a microlens array using hole-patterned electrodes and a polymer network to stabilize the alignment of a nematic liquid crystal to form a parabolic refractive index profile.
Reproduced from ref. 67 with permission from the Japan Society of Applied Physics.

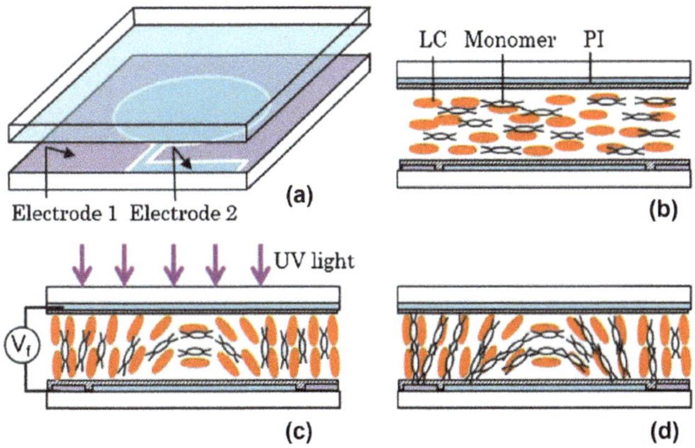

Figure 7.15 Fabricating tunable nematic liquid crystal microlenses using fringe-fields and polymer stabilization.
Reproduced from ref. 72 with permission from The Optical Society.

properties of the microlens. As we have seen in previous sections, altering the voltage amplitude allows for different director profiles to be frozen-in by the polymer network resulting in lenses with different optical characteristics.

Over the years, researchers have considered alternative ways with which to form the polymer network. For example, Presnyakov and co-workers[68,69] and have shown that a non-uniform network can be obtained using an incident beam that has a Gaussian or non-uniform spatial intensity distribution, which is then recorded into the polymer network. This in turn creates the desired refractive index profile thereby negating the need for the patterned electrodes. Subsequently, the application of a voltage to the device leads to tuning of the focussing properties in the same manner as that observed for the patterned electrode devices.

There have been a number of studies on the development of microlenses that are based, fundamentally, on polymer-stabilized nematic LCs.[70–74] Other routes that have been explored involve replacing the hole-patterned electrodes with polymer moulds in the form of plano-concave polymer microlenses, the recesses of which are then filled with a polymerizable LC mixture.[70] After filling, a polymer network is formed to produce the resulting tunable microlens device, which increases in focal length as the voltage amplitude is increased. Other techniques that have been successfully adopted to form microlens arrays include the use of photomasks to form centro-symmetric polymer networks that, in the presence of a homogeneous electric field, result in a gradient refractive index profile,[71] and the use of fringe-fields to create the necessary refractive index profile that is then locked-in by a polymer network[72] (Figure 7.15). A key challenge in forming polymer-stabilized LC microlenses is achieving sufficient stability to maintain the necessary refractive index profile without drastically limiting the tuning range.

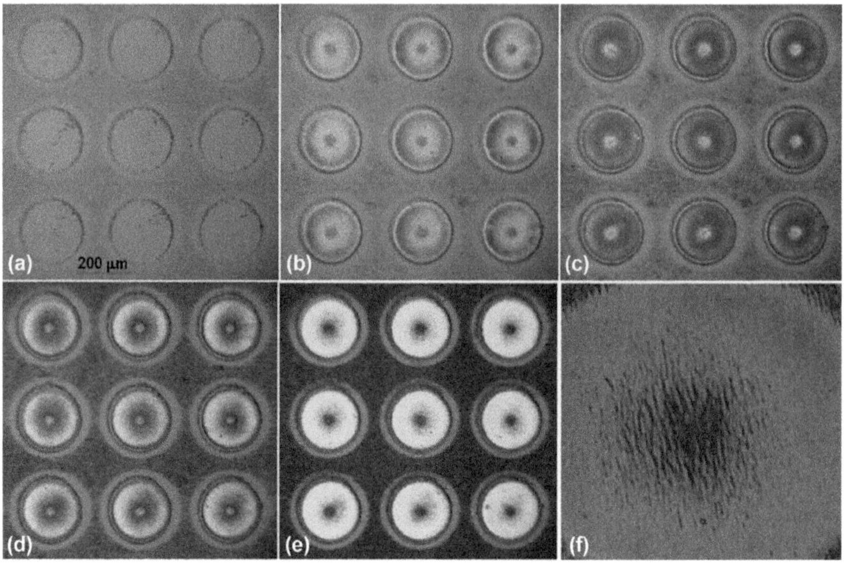

Figure 7.16 Example of polymer-stabilized nematic liquid crystal microlens viewed on an optical polarizing microscope at different applied voltage amplitudes. (a) V = 0, (b) V = 1.5 Vrms, (c) V = 3 Vrms, (d) V = 4 Vrms, (e) V = 6 Vrms, and (f) the magnified morphology at V = 10 Vrms. Reproduced from ref. 74 with permission from AIP Publishing.

An example of a polymer-stabilized nematic liquid crystal microlens viewed by optical polarized microscopy at different applied voltage amplitudes is shown in Figure 7.16.

7.5 Summary

In this chapter, we have considered how polymer stabilization can be used to alter the electro-optic characteristics of a range of nematic devices, including the conventional planar-aligned geometries that exhibit the ubiquitous Fréedericksz threshold voltage, in-plane switching, twisted nematic, PVA and MVA devices, and the OCB mode. We have seen that polymer stabilization can be applied to a rich variety of nematic modes, providing important improvements in a number of the performance characteristics albeit at the expense of other device parameters that are no less important. In general terms, the most common observation is that polymer stabilization improves the response time (relaxation time), but at the cost of the driving voltage, transmission–voltage curves, and in some cases the contrast ratio. However, careful control of the curing conditions, mixture formulation, and device architecture, means that trade-offs can be identified that provide the necessary improvement in one parameter, such as the response time, whilst at the same time there is only a tolerable degradation in another device property (such as the voltage). We have also seen how polymer stabilization in nematic LCs can be used effectively to form microlenses.

The ability to form the polymer network using photo-exposure unlocks a broad range of possibilities in terms of the network morphology and the resulting electro-optic characteristics. As we have seen, the network can be created at different temperatures and when subjected to electric fields so as to alter the alignment of the director in the voltage-off state (as we have seen for the MVA, PVA, and OCB devices). Even though it may seem that the field of polymer-stabilized nematic LCs is already suitably well-trodden and mature, the emergence of new materials and cutting-edge fabrication techniques ensures that there is still a rich and diverse landscape to explore.

References

1. D. J. Broer, J. Boven, G. N. Mol and G. Challa, *Makromol. Chem.*, 1989, **190**, 2255.
2. D. J. Broer, R. G. Gossink and R. A. M. Hikmet, *Angew. Makromol. Chem.*, 1990, **183**, 45.
3. D. J. Broer and I. Heynderickz, *Macromolecules*, 1990, **23**, 2474.
4. R. A. M. Hikmet, *Liq. Cryst.*, 1991, **9**(3), 405.
5. R. A. M. Hikmet, *Liq. Cryst.*, 1992, **12**(5), 831.
6. R. A. M. Hikmet, *J. Appl. Phys.*, 1990, **68**(9), 4406.
7. P. Mariani, B. Samorig, A. S. Angeloni and P. Ferruti, *Liq. Cryst.*, 1986, **1**(4), 327.
8. H. W. Schmidt, *Adv. Mater.*, 1989, **101**, 964.
9. T. Nakamura, T. Uneo and C. Tani, *Mol. Cryst. Liq. Cryst.*, 1989, **169**, 167.
10. P. A. Gemmell, G. W. Gray and D. Lacey, *Mol. Cryst. Liq. Cryst.*, 1985, **122**, 205.
11. P. Fabre, C. Casagrande, M. Veyssié and H. Finkelmann, *Phys. Rev. Lett.*, 1984, **53**(10), 993.
12. J. P. Casquilho and F. Volino, *Mol. Cryst. Liq. Cryst.*, 1990, **180**, 357.
13. C. V. Rajaram, S. D. Hudson and L. C. Chien, *Chem. Mater.*, 1995, 7, 2300.
14. Y. K. Fung, D.-K. Yang, S. Ying, L.-C. Chien, S. Zumer and J. W. Doane, *Liq. Cryst.*, 1995, **19**(6), 797.
15. C. V. Rajaram, S. D. Hudson and L. C. Chien, *Chem. Mater.*, 1996, **8**(10), 2451.
16. I. Dierking, L. L. Kosbar, A. Afzali-Ardakani, A. C. Lowe and G. A. Held, *J. Appl. Phys.*, 1997, **81**, 3007.
17. R. A. M. Hikmet and H. M. J. Boots, *Phys. Rev. E*, 1995, **51**, 5824.
18. S. Mora, A. M. Jamieson and L. C. Chien, *Mol. Cryst. Liq. Cryst.*, 1997, **292**, 323.
19. R.-Q. Ma and D.-K. Yang, *Phys. Rev. E*, 2000, **61**, 1567.
20. N. Dessaud, P. Raynes and P. Bonnett, *J. Appl. Phys.*, 2004, **96**, 4366.
21. M. J. Escuti, C. C. Bowley, G. P. Crawford and S. Žumer, *Appl. Phys. Lett.*, 1999, **75**, 3264.

22. S. H. Lee, S. L. Lee and H. Y. Kim, *Appl. Phys. Lett.*, 1998, **73**, 2881.
23. M. Lee, *et al.*, *J. SID*, 2002, **10**, 117.
24. Y. Chen, Z. Luo, F. Peng and S. T. Wu, *J. Disp. Technol.*, 2013, **9**(2), 74–77.
25. P. J. Bos and K. Rickey Koehler/Beran, *Mol. Cryst. Liq. Cryst.*, 1984, **113**, 329.
26. A. Jákli, D. R. Kim, L. C. Chien and A. Saupe, *J. Appl. Phys.*, 1992, **72**, 3161.
27. Y.-H. Fan, Y.-H. Lin, H. Ren, S. Gauza and S.-T. Wu, *Appl. Phys. Lett.*, 2004, **84**, 1233.
28. P. A. Kossyrev, J. Qi, N. V. Priezjev, R. A. Pelcovits and G. P. Crawford, *Appl. Phys. Lett.*, 2002, **81**(16), 2986.
29. R. A. M. Hikmet and J. A. Higgins, *Liq. Cryst.*, 1992, **12**(5), 831.
30. C.-C. Chang, L.-C. Chien and R. B. Meyer, *Phys. Rev. E*, 1997, **56**(1), 595.
31. A. Rapini and M. J. Papoular, *J. Phys., Colloq.*, 1969, **30**, C4.
32. D.-K. Yang, Y. Cui, H. Nemati, X. Zhou and A. Moheghi, *J. Appl. Phys.*, 2013, **114**, 243515.
33. F. Du, S. Gauza and S.-T. Wu, *Opt. Express*, 2003, **11**, 2891.
34. F. Du and S.-T. Wu, *Appl. Phys. Lett.*, 2003, **83**, 1310.
35. T. Murashige, H. Fujikakei, S. Ikehata and F. Sato, *Jpn. J. Appl. Phys.*, 2002, **41**, L1152.
36. M. Oh-e and K. Kondo, *Appl. Phys. Lett.*, 1995, **67**(26), 3895.
37. Z. Ge, X. Zhu, T. X. Wu and S. T. Wu, *J. Disp. Technol.*, 2006, **2**(2), 114.
38. T. Shimojo, *Proc. Soc. Inf. Disp.*, 1973, **14**, 110.
39. R. A. Soref, *J. Appl. Phys.*, 1974, **45**, 5466.
40. D. Klement and K. Tarumi, *SID Dig.*, 1998, **29**, 393.
41. K. Kondo, *SID Dig.*, 1998, **29**, 398.
42. M. J. Escuti, C. C. Bowley, S. Zumer and G. P. Crawford, *SID Dig.*, 1999, **30**, 32.
43. X. Zhou, G. Qin, Y. Dong and D.-K. Yang, *J. SID*, 2015, **23**(7), 333.
44. X. Zhou, G. Qin and D.-K. Yang, *Opt. Lett.*, 2016, **41**(2), 257.
45. C.-Y. Chien, L.-Y. Wu, C.-R. Sheu, C.-J. Hsu, C. Huang, C.-H. Chen, L.-Y. Huang, S.-L. Lee and W.-C. Tsai, *Opt. Mater. Express*, 2017, 7(8), 2740.
46. Y.-Q. Lu, F. Du, Y.-H. Lin and S.-T. Wu, *Opt. Lett.*, 2004, **12**, 1221.
47. H. Yoshida, *Vertically Aligned Nematic (VAN) LCD Technology Handbook of Visual Display Technology*, 2012, pp. 2133–2155.
48. A. Takeda, S. Kataoka, T. Sasaki, H. Chida, H. Tsuda, K. Ohmuro, T. Sasabayashi, Y. Koike and K. Okamoto, *Dig. Tech. Pap. - Soc. Inf. Disp. Int. Symp.*, 1998, 1077.
49. K. Ohmuro, S. Kataoka, T. Sasaki and Y. Koike, *Dig. Tech. Pap. - Soc. Inf. Disp. Int. Symp.*, 1997, 845.
50. K. H. Kim, J. J. Ryu, S. B. Park, J. K. Song, B. W. Lee, J. S. Byun and J. H. Souk, *Dig. Tech. Pap. - Soc. Inf. Disp. Int. Symp.*, 2001, 58.
51. P. Kumara, C. Jaggi, V. Sharma and K. K. Raina, *Micron*, 2016, **81**, 34.
52. C.-Y. Huang, W.-Y. Jhuang and C.-T. Hsieh, *Opt. Express*, 2008, **16**, 3859.

53. J.-I. Baek, K.-H. Kim, J. C. Kim, T.-H. Yoon, H. S. Woo, S. T. Shin and J. Hyung Souk, *Jpn. J. Appl. Phys.*, 2009, **48**, 056507.
54. G.-D. Lee, J. H. Son, Y. H. Choi and S. H. Lee, *Appl. Phys. Lett.*, 2007, **90**, 033509.
55. S. G. Kim, S. M. Kim, Y. S. Kim, H. K. Lee, S. H. Lee, G.-D. Lee, J.-J. Lyu and K. H. Kim, *Appl. Phys. Lett.*, 2007, **90**, 261910.
56. B. W. Park, S.-W. Oh, J.-W. Kim and T.-H. Yoon, *J. Opt. Soc. Korea*, 2014, **18**(4), 395.
57. H. Kikuchi, H. Yamamoto, H. Sato, M. Kawakita, K. Takizawa and H. Fujikake, *Jpn. J. Appl. Phys.*, 2005, **44**, 981.
58. S. H. Kim and L.-C. Chien, *Jpn. J. Appl. Phys.*, 2004, **43**, 7643.
59. T.-J. Chen, Y.-P. Kuo, J.-J. Wu and C.-H. Sun, *Jpn. J. Appl. Phys.*, 2006, **45**, 2702.
60. Y. Asakawa, T. Takahashi and S. Saito, *Jpn. J. Appl. Phys.*, 2007, **46**, 7774.
61. C. Y. Huang, R. -X. Fung and Y.-G. Lin, *Jpn. J. Appl. Phys.*, 2007, **46**, 5230.
62. Y. Asakawa, K. Yokota, M. Nanaumi, N. Takatuka, T. Takahashi and S. Saito, *Jpn. J. Appl. Phys.*, 2006, **45**, 5878.
63. S. Y. Jeong, D. W. Kim, K. W. Rhie, M. P. Hong, P. Bos, W. G. Jang, K. H. Kim and S. T. Shin, *Jpn. J. Appl. Phys.*, 2008, **47**, 1647.
64. S. R. Lee, J.-H. Shin, J.-I. Baek, M.-C. Oh, T.-H. Yoon and J. C. Kim, *Appl. Phys. Lett.*, 2007, **90**, 163513.
65. C. C. Tartan, P. S. Salter, T. D. Wilkinson, M. J. Booth, S. M. Morris and S. J. Elston, *RSC Adv.*, 2017, 7, 507.
66. T. Nose, S. Masuda, S. Sato, J. Li, L.-C. Chien and P. J. Bos, *Opt. Lett.*, 1997, **22**, 351.
67. S. Masuda, T. Nose and S. Sato, *Jpn. J. Appl. Phys.*, 1998, **37**, L1251.
68. V. V. Presnyakov, K. E. Asatryan, T. V. Galstian and A. Tork, *Opt. Express*, 2002, **10**, 865.
69. V. V. Presnyakov and T. V. Galstian, *J. Appl. Phys.*, 2005, **97**, 103101.
70. H. Ren, Y.-H. Fan and S.-T. Wu, *Opt. Lett.*, 2004, **29**, 1608.
71. H. Ren, Y.-H. Fan, S. Gauza and S.-T. Wu, *Opt. Commun.*, 2004, **230**, 267.
72. H. Ren, S. Xu and S.-T. Wu, *Opt. Lett.*, 2013, **38**, 3144.
73. H. Ren, Y.-H. Fan and S.-T. Wu, *J. Phys. D: Appl. Phys.*, 2004, **37**, 400.
74. M. Xu, Z. Zhou, H. Ren, S. H. Lee and Q. Wang, *J. Appl. Phys.*, 2013, **113**, 053105.

CHAPTER 8

Electrical Color Tuning in Polymer-stabilized Cholesteric Liquid Crystals

DENG-KE YANG

Liquid Crystal Institute and Chemical Physics Interdisciplinary Program, Kent State University, Kent, OH, 44242, USA
Email: dyang@kent.edu

8.1 Introduction

Cholesteric liquid crystals (also called chiral nematic liquid crystals) consist of chiral rod-like or disk-like molecules.[1–4] In this chapter we only discuss rod-like molecules. A chiral molecule does not have reflection symmetry. In a cholesteric (abbreviated as N* or Ch) liquid crystal, the rod-like molecules self-assemble into a helical structure as shown in Figure 8.1(a). The molecules rotate periodically along an orthogonal helical axis. In a plane perpendicular to the helical axis, the molecules are oriented parallel to one another. The averaged direction of the long molecular axis is the liquid crystal director denoted by $\vec{n}$. The distance for the liquid crystal director to twist 360° is the helical pitch denoted by P. Because the physical properties of the liquid crystal are the same along $\vec{n}$ and $-\vec{n}$, the periodicity is $P/2$. Cholesteric liquid crystals are 1-dimensional photonic crystals. Because of their periodic helical structure, they Bragg-reflect light. Their reflection band centers at the wavelength $\lambda=[(n_e+n_o)/2]P$ and has the bandwidth $\Delta\lambda=(n_e-n_o)P$, where n_e and n_o are the extraordinary and ordinary refractive indices of the liquid crystal, respectively. Many cholesteric liquid crystals

Soft Matter Series No. 8
Polymer-modified Liquid Crystals
Edited by Ingo Dierking

Published by the Royal Society of Chemistry, www.rsc.org

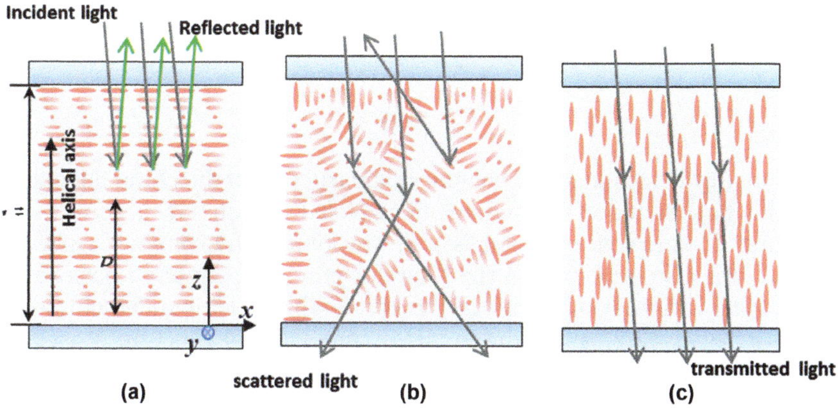

Figure 8.1 Schematic diagrams of the structure of cholesteric liquid crystals. (a) Planar state, (b) focal conic state, and (c) homeotropic state.

have pitches comparable to the wavelength of visible light, and thus reflect visible light and have a colorful appearance.

Because of their unique optical properties, cholesteric liquid crystals have been exploited in many applications. They are used to make reflective displays.[5–9] Differing from transmissive liquid crystal displays (such as those used in smartphone screens and TV), where there is backlight (or edgelight) and the liquid crystal modulates the light intensity with the help of polarizers, cholesteric reflective displays reflect ambient light. They have good readability under sunlight and are very energy-saving. They do not need polarizers and are compatible with plastic substrates (which may have non-uniform birefringence). They are very suitable for electronic paper and flexible display applications.[10] Cholesterics are also used to make tunable color filters.[11–13] In this application, the cholesteric liquid crystal reflects light with certain preselected wavelength in one state, and transmits the light with nearly 100% transmittance in another state. They can be used to protect human eyes, in detectors and in communication equipment. When cholesteric liquid crystals are doped with fluorescent dyes, they can be used to make mirrorless lasers.[14–16] Near the edges of the reflection band, the effective refractive index varies sharply with the wavelength, and therefore the density of states is high and stimulated emission is promoted. In some cholesterics, the helical pitch is very sensitive to temperature.[17–20] When the ambient temperature varies, their colorful appearance changes. They can be used to make thermometers and detect temperature variation on the surfaces of human skin and electronic circuit boards.

8.2 Construction of Cholesteric Liquid Crystal

Cholesteric (Ch) liquid crystals used in practical applications are usually mixtures of nematic liquid crystals and chiral dopants. The pitch P depends on the concentration x and helical twisting power *HTP* of the chiral dopant according to the equation

$$P = \frac{1}{HTP \cdot x} \tag{8.1}$$

HTP is mainly determined by the molecular structure of the chiral dopant and is slightly dependent on the nematic host. Note that eqn (8.1) is correct when the chiral dopant concentration is low. The unit of *HTP* is μm^{-1}. The sign of the *HTP* is also important. A right-handed chiral dopant has a positive *HTP* and induces a right-handed helical structure, while a left-handed chiral dopant has a negative *HTP* and induces a left-handed helical structure. Most chiral dopants have *HTP* around 10. As an example, chiral dopant R811 (Merck) has $HTP = 10\ \mu m^{-1}$. If the cholesteric liquid crystal is required to reflect green light with the central wavelength of 0.55 μm the concentration of the chiral dopant should be $x = 1/[(\lambda/\bar{n}) \cdot HTP] = 1/[(0.55\ \mu m/1.6) \times 10\ \mu m^{-1}] = 29\%$. The concentration is relatively high and attention should be paid to the solubility of the chiral dopant in the nematic host. The highest *HTP* of commercially available chiral dopants is about 150 μm^{-1} [R5011 (Merck)].

In identifying nematic hosts for chiral nematic liquid crystals, the physical parameters to be considered are the birefringence Δn, dielectric anisotropy $\Delta\varepsilon$, elastic constants and phase transition temperatures. Most chiral dopants are in the solid state at room temperature, therefore nematic hosts with lower melting points are desirable. In applications where the color of the reflected light is required to not change with temperature, a nematic host having smectic phases below the nematic phase should be avoided.

8.3 States of Cholesteric Liquid Crystal and their Optical Properties

In applications, a cholesteric liquid crystal is usually sandwiched between two parallel substrates with transparent electrodes. The optical properties of the liquid crystal depend on the orientation of the helical axis. When the helical axis is perpendicular to the cell substrate as shown in Figure 8.1(a), it reflects light. This state is known as the planar state (sometimes it is called the Grandjean state, or planar texture). The liquid crystal director is given by

$$n_x = \cos(qz), \quad n_y = \sin(qz), \quad n_z = 0 \tag{8.2}$$

where $q = 2\pi/P$ is the helical wave number of the liquid crystal. A microphotograph, which is taken under a polarizing optical microscope under reflection mode, of the planar state of a cholesteric liquid crystal is shown in Figure 8.2(a). The cholesteric reflects green light. The web-like lines shown in the photograph are known as oily streaks. In order to explain the structure of the oily streaks, we can consider the cholesteric liquid crystal as a lamellar structure with the layer thickness equal to half pitch when the helical pitch is much smaller than the cell thickness. Similar to smectic liquid crystals,

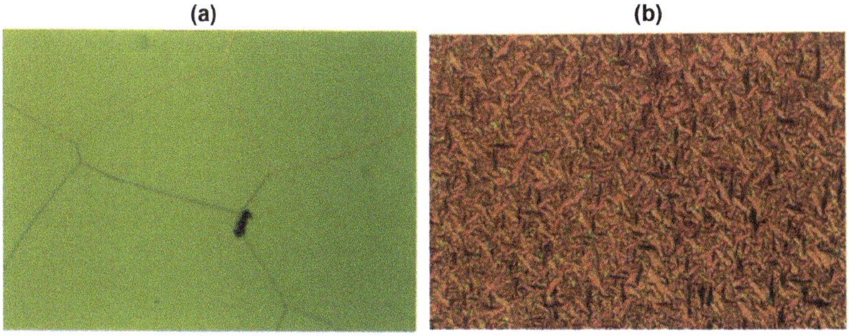

Figure 8.2 Microphotographs of cholesteric liquid crystal. (a) Planar state, (b) focal conic state.

changing the layer thickness costs a lot of elastic energy while bending the layer does not. The oily streaks are bent Ch layers and usually exist in the planar state, because this does not cost much elastic energy. A good planar state with less or no defects can be achieved by coating a homogeneous alignment layer on the inner surfaces of the substrates. The planar state is compatible with the anchoring of the homogeneous alignment layer and the surface energy is minimized. In the absence of an external electric field, the planar state has the minimum free energy and is stable.

When an intermediate electric field is applied across the cholesteric liquid crystal (with positive dielectric anisotropy; details will be discussed in next section), the helical axis will be tilted away from the cell normal direction to reduce the electrical energy and the liquid crystal is switched to the focal conic state whose structure is shown in Figure 8.1(b). It is a poly-domain structure with the helical axis oriented more or less randomly through the cell. The domains have a fan shape. In each domain, the Ch layer is curved and the helical axis is along the radial direction of the fan. When observed under a polarizing optical microscope, the texture of the focal conic state looks like that of a smectic-A liquid crystal, as shown in Figure 8.2(b). Because the helical axis in this state is no longer perpendicular to the cell substrate, the liquid crystal does not reflect light incident around the cell normal direction. If the domain size is comparable to the visible light, it will scatter light. When the applied electric field is removed, the focal conic state is stable under weak homogeneous anchoring or homeotropic anchoring conditions, because the helical structure is preserved in this state. Under a strong homogeneous anchoring condition, the focal conic state is metastable and the liquid crystal will relax slowly (in the order of a minute) back to the planar state.

When a sufficiently high electric field is applied across the Ch liquid crystal, the helical structure is unwound, and the liquid crystal is switched to the homeotropic state where the liquid crystal director is perpendicular to the cell substrate, as shown in Figure 8.1(c). In this state the liquid crystal does not reflect light and is transparent. When the applied electric field is removed, the homeotropic state is unstable, and the liquid crystal will transform to the planar state or focal conic state (please see the details in Section 8.4).

Now we quantitatively consider the reflection of cholesteric liquid crystals. When the pitch of a cholesteric is a constant and the incident angle of the light is 0°, the reflection spectrum can be calculated analytically. In this chapter, we will consider the cases where the pitch varies spatially and the incident angle is not 0°, and the reflection spectrum can be calculated numerically by using the Berremann 4×4 method.[4,21–23] In the simulation, the cholesteric cell consists of two parallel glass substrates and the liquid crystal layer is sandwiched between them. Outside the cell, the medium is air with refractive index 1.0. The refractive index of the glass is 1.5. The refractive indices of the cholesteric liquid crystal are $n_o = 1.5$ and $n_e = 1.7$ and the cell thickness is 10 μm.

We first consider how the incident light angle affects the reflection. The incident angle specified, θ, is the incident angle outside the cholesteric cell. The internal incident angle, θ_{internal}, is different from θ, as shown in Figure 8.3, because of the refraction at the interface. The internal incident angle can be calculated from $\bar{n}\sin\theta_{\text{internal}} = \sin\theta$. For example, when the incident angle is 60°, the internal incident angle is about 33°. The reflection spectrum at various incident angles is shown in Figure 8.4. The pitch is 400 nm and the incident light is unpolarized. When the incident angle is 0°, the central wavelength of the reflection band is located at the wavelength $\bar{n}P = [(1.5 + 1.7)/2] \times 400 = 640$ nm. The bandwidth is $\Delta nP = (1.7 - 1.5) \times 400 = 80$ nm. The peak

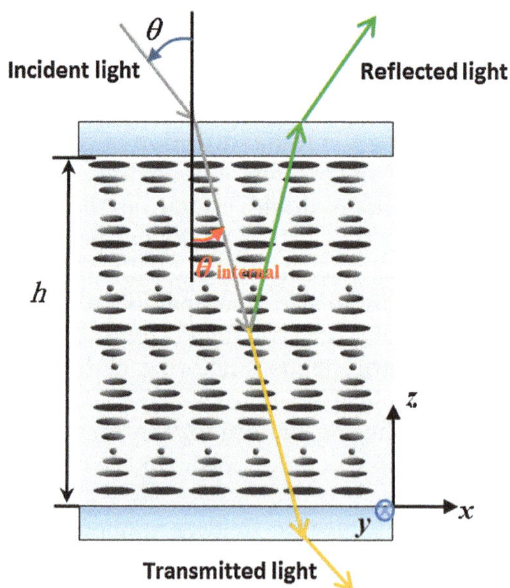

Figure 8.3 Schematic diagram showing the propagation of light in a cholesteric liquid crystal.

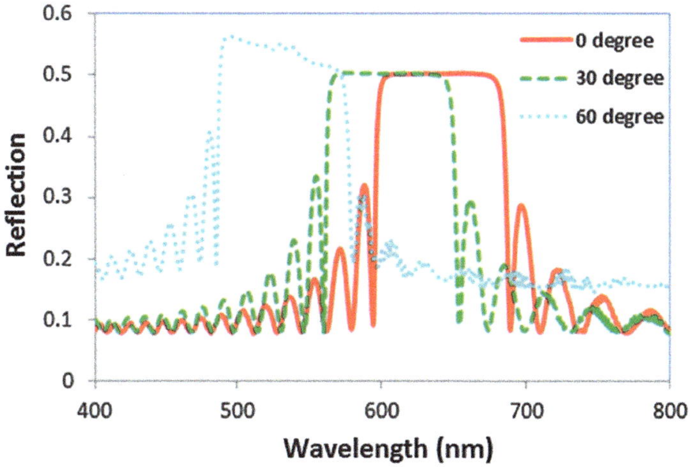

Figure 8.4 Reflection spectrum of the cholesteric liquid crystal for various incident light angles.

reflectance is 0.5, because only circularly polarized light with the same handedness as the liquid crystal helical structure is reflected. When the incident angle is increased, the reflection band is shifted to a shorter wavelength according to the Bragg law, $\bar{n}P\cos\theta_{\text{internal}}$. Furthermore, the reflection bandwidth and peak reflectance increase.

In polymer-stabilized cholesteric liquid crystals, it is possible to have a pitch gradient such that the pitch varies spatially in the cell. For a linear variation, the pitch as a function of the position is described by

$$P(z) = P_o + \alpha(z - h/2), \tag{8.3}$$

where P_o is the pitch at the middle of the cholesteric layer, $z = h/2$ and α is the pitch variation rate. The twist angle of the liquid crystal director at position z is given by

$$\phi(z) = \int_0^z \frac{2\pi}{P(\xi)}\mathrm{d}\xi = \int_0^z \frac{2\pi}{P_o + \alpha(\xi - h/2)}\mathrm{d}\xi \tag{8.4}$$

The simulated reflection spectra of cholesteric liquid crystals with different pitch gradients is shown in Figure 8.5. The cell thickness is 10 micron. The refractive indices of the liquid crystal are $n_e = 1.7$ and $n_o = 1.5$ and $P_o = 350$ nm. When the pitch is a constant, namely at $\alpha = 0.0$, the reflection band width is $\Delta\lambda = \Delta nP = 70$ nm. When the variation rate is $\alpha = 0.1$, the shortest pitch is 300 nm and the longest pitch is 400 nm. The reflection band width is $\Delta\lambda = n_e(P_o + \alpha h/2) - n_o(P_o - \alpha h/2) = 230$ nm. When the variation rate is $\alpha = 0.2$, the shortest pitch is 250 nm and the longest pitch is 450 nm. The reflection bandwidth is $\Delta\lambda = n_e(P_o + \alpha h/2) - n_o(P_o - \alpha h/2) = 390$ nm. As the pitch variation rate is increased, the reflection bandwidth increases and the edges of the band become less steep.

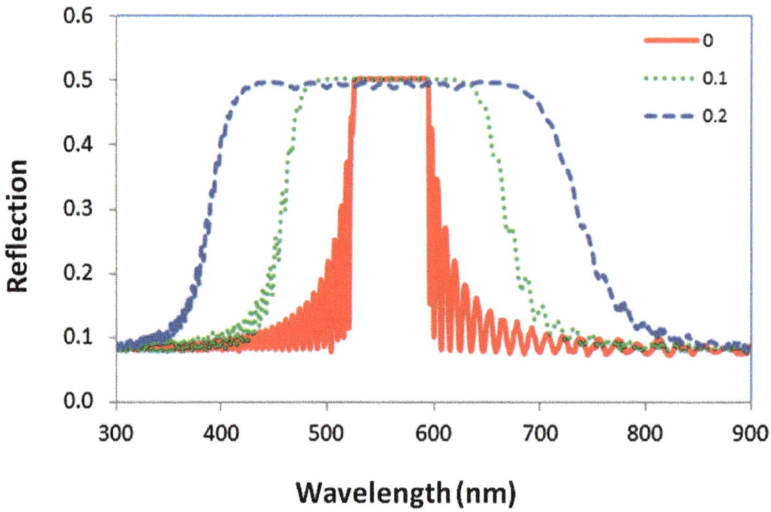

Figure 8.5 Reflection spectrum of the Ch liquid crystals with three pitch variation rates: 0.0, 0.1 and 0.2.

8.4 Transitions Between Cholesteric States

Cholesteric liquid crystals are anisotropic dielectric materials and their molecular orientation can be changed by applying electric fields. Here we consider the case where transparent electrodes, such as indium tin oxide (ITO), are coated on the top and bottom inner surfaces of the substrates. When a voltage is applied across the cell, the generated electric field is parallel to the normal of the cell substrate. For a cholesteric liquid crystal with a negative dielectric anisotropy ($\Delta\varepsilon = \varepsilon_{//} - \varepsilon_{\perp} < 0$), the molecules tend to align perpendicular to the electric field to minimize the electrical energy. If the liquid crystal is initially in the planar state, when an electric field is applied, it will remain in the planar state, because the molecules are already perpendicular to the electric field. If the liquid crystal is initially in the focal conic state, when an electric field is applied, the state is unstable and it will transform to the planar state to reduce the electrical energy.

For a cholesteric liquid crystal with positive dielectric anisotropy ($\Delta\varepsilon > 0$), the molecules tend to align parallel to the electric field to minimize the electrical energy. The transitions driven by electric fields are schematically shown in Figure 8.6. If the liquid crystal is initially in the planar state, when an intermediate voltage $V_{H \to FC}$ is applied, it will transform to the focal conic state where the helical structure is preserved but the electrical energy is decreased.[4,24] The transition is a nucleation process and is slow. When the applied voltage is removed, it will remain in the focal conic state (stable under weak homogeneous anchoring and homeotropic anchoring conditions and metastable under strong homogeneous anchoring conditions). In order to switch the liquid crystal back to the planar state, a high voltage

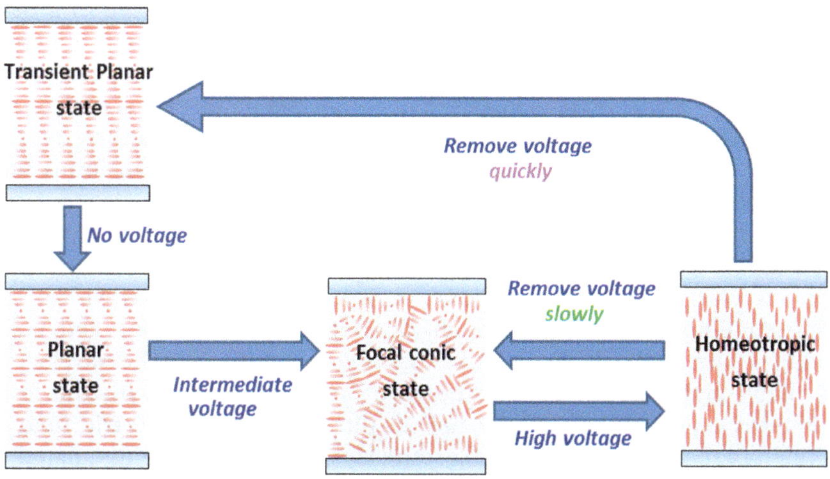

Figure 8.6 Transitions in a Ch liquid crystal with a positive dielectric anisotropy.

$V_{FC\rightarrow H}$ must be applied, which will switch the liquid crystal to the homeotropic state where the helical structure is unwound. The critical voltage needed for the switching is given by[1,4,25]

$$V_c = \frac{\pi^2 h}{P}\sqrt{\frac{K_{22}}{\varepsilon_o \Delta\varepsilon}}. \tag{8.5}$$

When the applied voltage is removed, the homeotropic state is unstable and the helical structure will be restored. There are two possible ways to restore the helical structure. If the voltage is removed slowly or reduced to the intermediate voltage, the liquid crystal will relax to the focal conic state.[4,8] This relaxation is a nucleation process and is slow. The transition time $\tau_{H\rightarrow FC}$ is of the order of a hundred ms. If the voltage is removed quickly, the liquid crystal will first transform to a transient planar state with a pitch $P_t = (K_{33}/K_{22})P$.[4,26–28] This transition is a homogenous transition and fast, in the order of one ms (for P comparable to visible light wavelength). The transition time is given by

$$\tau_{H\rightarrow TP} = \frac{\gamma P^2}{K_{22}}, \tag{8.6}$$

where γ is the rotational viscosity coefficient of the liquid crystal. The transient planar state is unstable, because its pitch is different from the intrinsic pitch, and the liquid crystal will further transform to the stable planar state with the intrinsic pitch.[4,28,29] This transition is also a nucleation and slow. The transition time $\tau_{TP\rightarrow P}$ is of the order of a hundred ms. The resulting state of the liquid crystal after the removal of the high applied voltage depends on how the voltage is removed. If the high voltage is reduced to 0 in a time interval less than $\tau_{H\rightarrow TP}$, the liquid crystal transforms to the transient planar state and then further relaxes to the planar state; if the high voltage is reduced to 0 in a time interval longer than $\tau_{H\rightarrow FC}$, the liquid crystal transforms to the focal conic state. If the high voltage is reduced to 0 in a time interval longer than $\tau_{H\rightarrow TP}$ but less than $\tau_{H\rightarrow FC}$, the liquid crystal transforms to a poly-domain structure where some domains are in the planar state and the other domains are in the focal conic state.

In cholesteric liquid crystals with positive dielectric anisotropies, an applied electric field may produce a layer undulation, known as the Helfrich deformation,[9,30–35] which can result in reflectance and color change. When the helical pitch of a CLC is comparable with the wavelength of visible light, it can be considered a lamellar system (with the layer thickness equal to half of the helical pitch). When an electric field is applied perpendicular to the cholesteric layers, the layers are undulated in a sinusoidal or zigzag manner to reduce the electrical energy as shown in Figure 8.7(a). In some regions of the deformation, the liquid crystal is tilted toward the electric field direction and thus the electrical energy is reduced. The tilt of the liquid crystal as well as expansion and contraction of the helical pitch, however, increase the elastic energy. The electric field must be sufficiently high so that the decrease of the electrical energy can compensate the increase the elastic energy. In the deformation, both twist and bend elastic energy are involved. When the wavelength of the undulation is[1,2,36]

$$L = \left(\frac{K_{22} + 3K_{33}}{8K_{22}}\right)^{1/4} \sqrt{hP}, \tag{8.7}$$

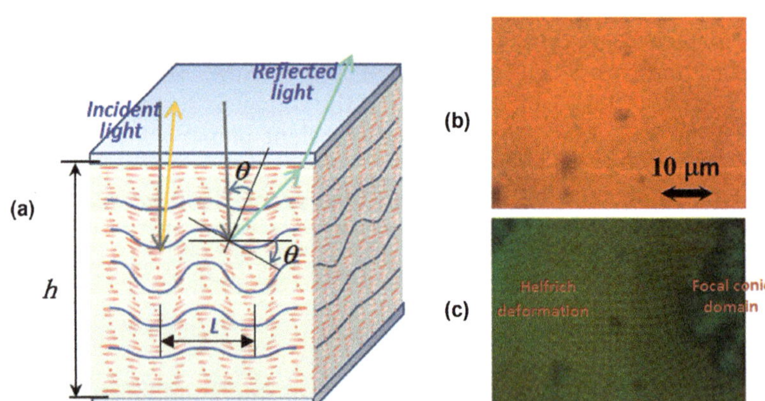

Figure 8.7 (a) Schematic diagram of the structure of the Helfrich deformation; (b) reflection microphotograph of the planar state; (c) reflection microphotograph of the Helfrich deformation.
Reproduced from ref. 36 with permission from the Royal Society of Chemistry.

the total elastic energy is minimized. The critical electric field for the Helfrich deformation is given by

$$E_H = \sqrt{\frac{2\pi^2}{hP} \frac{\left[2K_{22}(K_{22} + 3K_{33})\right]^{1/2}}{\Delta\varepsilon\varepsilon_o}}, \tag{8.8}$$

which depends on the twist elastic constant, K_{22}, and the bend elastic constant, K_{33}. The transition from the planar state to the Helfrich deformation is a homogeneous transition. It does not need nucleation seeds and is fast.

When no electric field is applied, the liquid crystal is in the planar state and reflects light at the wavelength $\lambda = \bar{n}P\cos\theta$. An optical microphotograph of the planar state of a cholesteric phase is shown in Figure 8.7(b). The liquid crystal reflects red light. When an electric field above the threshold field is applied, the liquid crystal transforms to the Helfrich deformation. A microphotograph of the Helfrich deformation is shown in Figure 8.7(c). The two-dimensional grid reflects the two-dimensional undulation of the Helfrich deformation. When the cholesteric layer is tilted by the angle, θ, as shown in Figure 8.7(a), for incident light in the cell normal direction, the incident angle of the light with respect to the normal of the cholesteric layer is also θ, the reflection band is now shifted to a shorter wavelength given by $\bar{n}P\cos\theta$, and therefore the color of the reflected light changes to green as shown in Figure 8.7(c).

In reality there is, however, a problem with the Helfrich deformation. As mentioned in the previous paragraphs, the liquid crystal can also be switched to the focal conic state by electric fields. The critical field needed for the transition from the planar state to the focal conic state is lower than that for the Helfrich deformation. Although the transition to the focal conic state is slower than the transition to the Helfrich deformation, the liquid crystal can only stay in the Helfrich deformation for a short time period and eventually the focal conic state will take over, as shown in Figure 8.7(c).

In summary, for pure cholesteric liquid crystals in cells with top and bottom electrodes, the reflectance can be changed electrically, but the color cannot be tuned electrically. In order to electrically tune the color, polymer stabilization is needed.

8.5 Polymer-stabilized Ch Liquid Crystals

Polymer-stabilized cholesteric liquid crystals (PSCLCs) are composites of cholesteric or chiral nematic liquid crystals and polymers.[37–43] In the construction of a polymer-stabilized liquid crystals, first a cholesteric is mixed with a monomer with functionality larger than 1 (usually 2). The concentration of the monomer is usually less than 10%. The mixture is in the cholesteric phase. A small amount of photo-initiator is also added and the monomer is photo-polymerized under UV light to form an anisotrpic polymer network. In order to get a stable polymer network that has a strong aligning effect on the liquid crystal, the monomer should be mesogenic (having a molecular structure similar to liquid crystal molecules) and bifunctional. The molecular structure of a commonly used monomer, RM257 (Merck), is shown in Figure 8.8(a). The formed polymer network mimics the structure of the state of the liquid crystal during the polymerization.[44] The polymer network is dispersed in the liquid crystal, as shown in Figure 8.9. It has a strong aligning effect and can stabilize the state of the liquid crystal.[45–52]

The morphology of the polymer network can be studied by using SEM.[44,45] The polymer-stabilized Ch liquid crystal cell is immersed in a solvent, such as hexane. The solvent dissolves the liquid crystal but not the polymer network. The cell is then split and put in a vacuum oven to let the solvent evaporate. Finally, the polymer network attached to the cell substrate is studied under SEM. A typical polymer network formed in a cholesteric liquid crystal is shown in Figure 8.8(b). The photograph plane is parallel to the cell substrate. The polymer network consists of interconnected anisotropic fibrils. The longitudinal length of the fibril is longer than 10 microns and the lateral size is sub-micron. The anisotropic property of the polymer network is caused by the aligning effect of the liquid crystal on the monomer and the anisotropic diffusional property of the monomer molecules in the liquid crystal.

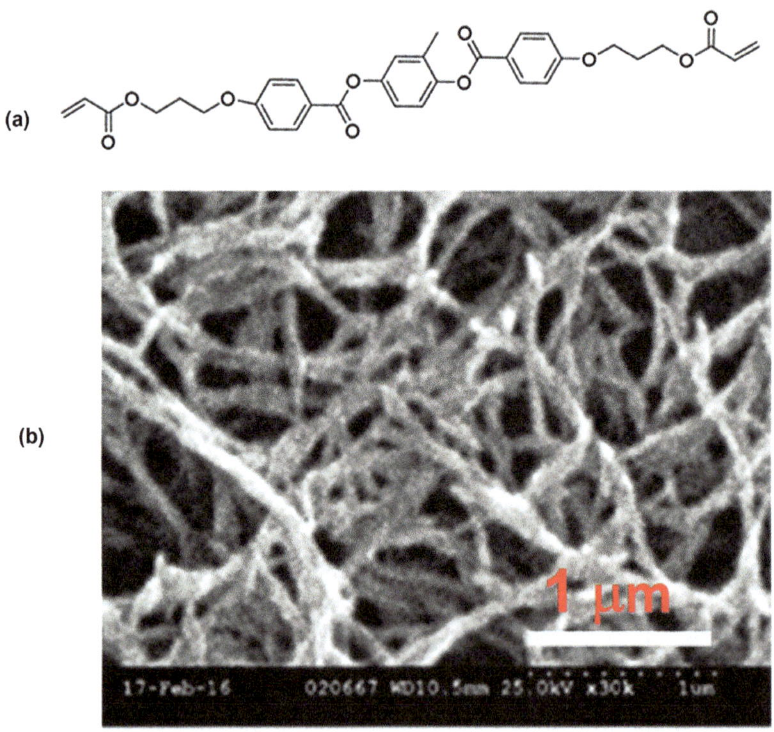

Figure 8.8 (a) Molecular structure of monomer, (b) SEM photograph of polymer network.

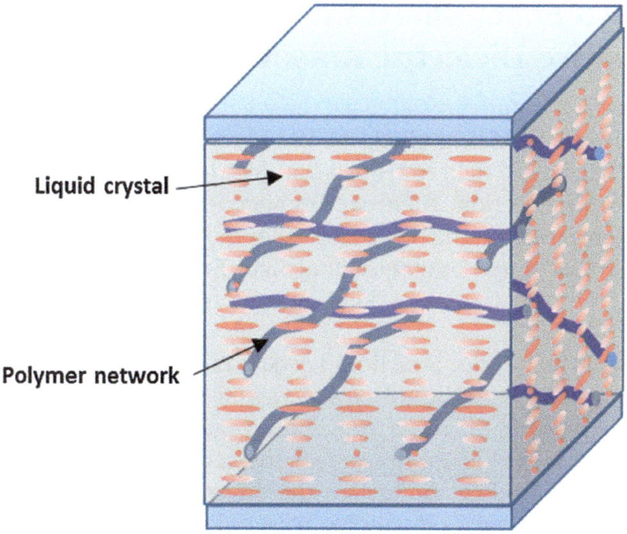

Figure 8.9 Schematic diagram of a polymer-stabilized cholesteric liquid crystal.

The morphology of the polymer network is determined by the following factors.[45] The first factor is the molecular structure of the monomer. Both the rigid core and two side flexible tails are necessary to obtain a stable anisotropic polymer network. The second factor is the structure of the double bond at two ends. If it is acrylic, the polymerization rate is faster and fibrils with smaller lateral size are obtained. If it is methyl acrylic, the polymerization rate is slower and fibrils with larger lateral size is obtained. The third factor is the conecentration of the monomer. If the monomer concentration is high, the formed fibril density is high. The fourth factor is the state of the mixture of the liquid crystal and the monomer during the polymerization. If the mixture is in the N* phase, the anisotropic fibrils are parallel to the local liquid crystal director. If the mixture is in the isotropic phase, a polymer with bead shape is formed. The fifth factor is the photo-initiator concentration. If the photo-initiator concentration is high, the formed fibril density is high and the lateral size of the fibrils is small. The sixth factor is the UV light intensity. If the UV intensity is high, the formed fibril density is high and the lateral size is small. The seventh factor is the temperature. If the polymerization takes place at a high temperature, the lateral size of the fibrils is bigger.

After the polymerization, the anisotropic polymer network has a strong aligning effect on the liquid crystal and can stabilize the cholesteric liquid crystal in a desired state. The aligning effect of the polymer network depends on its density.[45–47] The higher the polymer network concentration, the stronger the aligning is. For a given concentration, the aligning effect depends on the lateral size of the fibrils. When the lateral size is small, the density of the fibrils is high and the distance between the fibrils is short, and therefore the aligning effect is strong.

8.6 Polymer-stabilized Ch Liquid Crystals With Positive Dielectric Anisotropy

The cholesteric liquid crystal used has a positive dielectric anisotropy. When an electric field is applied, the liquid crystal molecules tend to orient parallel to the electric field to reduce the electrical energy. In the discussions presented in this section, there are homogeneous alignment layers on the two inner surfaces of the cells and the monomers are polymerized in the planar state, unless otherwise specified. Furthermore, the frequency of the applied electric field is 1 kHz, at which the polymer network does not move.

After the polymerization, the liquid crystal is stabilized in the planar state. The reorientation of the liquid crystal in the PSCLCs under externally applied electric fields is different from that in pure cholesteric liquid crystals. When an electric field is applied across the cell, the generated electric field tends to tilt the liquid crystal toward the direction perpendicular to the cell substrate, while the polymer network tries to keep the liquid crystal parallel to the cell substrate. The polymer network prevents the liquid crystal from transforming to the focal conic state when an electric field is applied, and therefore it becomes possible to electrically tune the reflection.[9,35,36,53–56] Furthermore, the polymer network can help the liquid crystal relax back to the planar state quickly when the applied electric field is removed.

There are two types of PSCLC catagorized by their reflection bandwidth. The first one is narrow band PSCLC[9,35,37] which is fabricated by using cells with relatively thin thickness, low UV intensity and achiral monomers. The reflection bandwidth is about 50 nm. The second one is broad band PSCLC which is fabricated by using relatively thick thickness, high UV intensity and chiral monomers. The reflection bandwidth can be as large as 300 nm, covering the entire visible light region.

8.6.1 Narrow Reflection Band PSCLC

The narrow band PSCLC is made from a cholesteric liquid crystal and an achiral monomer. The mixture consisting of 96% Ch liquid crystal and 4% monomer is filled into cells with a thickness of 10 micron. When the monomer concentration is relatively low, the formed polymer network is not strong. Although the planar state is stabilized, only the reflectance can be tuned by applying electric fields. The reflection spectra of the PSCLC under various applied electric fields are shown in Figure 8.10(a).[35] As the electric field is increased, the reflectance decreases, but the central wavelength of the reflection band remains at the same wavelength. When no electric field is applied, the peak reflectance is about 40%. When the electric field is increased to 5 V μm^{-1}, the peak reflectance decreases to about 3%. The photographs of the cell are shown in Figure 8.10(b). When no voltage is applied, the sample reflects orange colored light. When the applied electric field is 5 V μm^{-1}, the sample becomes black.

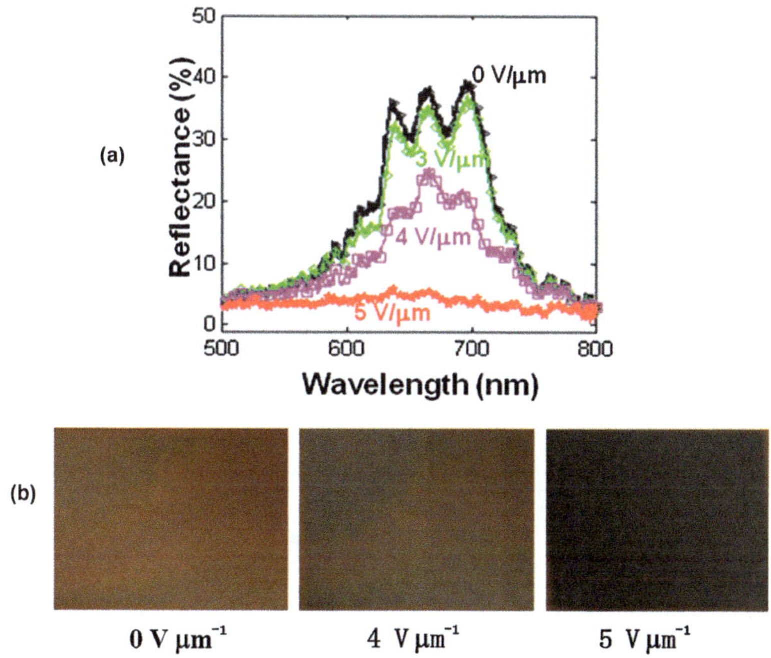

Figure 8.10 Reflection of the narrowband PSCLC with weak polymer network under various electric fields. (a) Reflection spectra, (b) photographs.

The reason for the reflectance tuning but not color tuning is probably that the polymer network is not strong enough. The periodic cholesteric layer undulation collapes before a sufficiently large tilt angle is created to produce a color change. Nevertheless the polymer network can help the liquid crystal relax back to the planar state when the applied electric field is removed. Figure 8.11(a) shows the dynamic response of the PSCLC. After the removal of the electric field, the liquid crystal relaxes back to the planar state in less than 5 ms. For comparision, the dynamic response of a Ch liquid crystal cell without polymer stabilization is shown in Figure 8.11(b). The transition time from the homeotorpic state to the planar state is about 1000 ms, much longer than the transition time of the PSCLC.

When the polymer network in a PSCLC is strong enough, it becomes possible to electrically tune the color. The reflection spectra of the PSCLC with a strong polymer network under various applied electric fields are shown in Figure 8.12(a).[35] As the applied electric field is increased, the reflection band is shifted to shorter wavelengths. The photographs of the sample are shown in Figure 8.12(b). When no voltage is applied, the sample reflects orange colored light. When the electric field is 4 $\mathrm{V\,\mu m^{-1}}$, the sample reflects yellow colored light. When the applied electric field is 5 $\mathrm{V\,\mu m^{-1}}$, the sample reflects green colored light.

As shown by eqn (8.8), the critical electric field to switch a cholesteric liquid crystal from the planar state to the Helfrich deformation depends on the twist

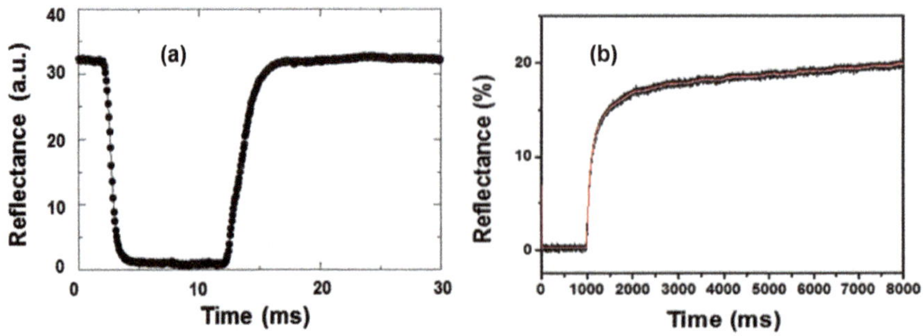

Figure 8.11 (a) Reflectance *vs.* time after the removal of the applied electric field. (a) Polymer-stabilized cholesteric liquid crystal. (b) Cholesteric liquid crystal without polymer network.

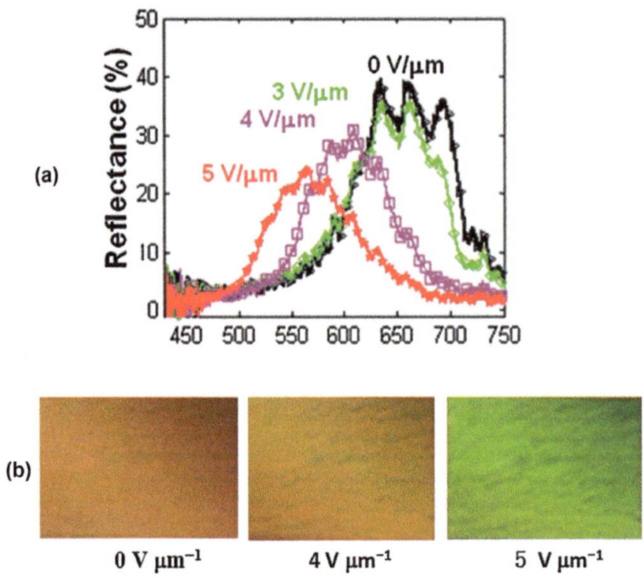

Figure 8.12 (a) Reflection of the narrow band PSCLC with strong polymer network under various electric fields; (a) reflection spectra, (b) photographs.

and bend elastic constants. Recently it was discovered that liquid crystal dimers, having a bent shape, with a flexible linkage in the middle, can significantly reduce the bend elastic constant. Therefore, doping cholesteric liquid crystals with the dimer can help stabilize the Helfrich deformation. By using both the dimer and polymer network, it becomes much easier to stabilize the Helfrich deformation and electrically tune the reflection color of cholesterics. The reflection spectra of a liquid crystal stabilized by dimer and polymer network at various applied electric fields are shown in Figure 8.13(a). As the applied electric field is increased, the reflection band is shifted to shorter wavelengths. The photographs of the sample are shown in Figure 8.13(b).

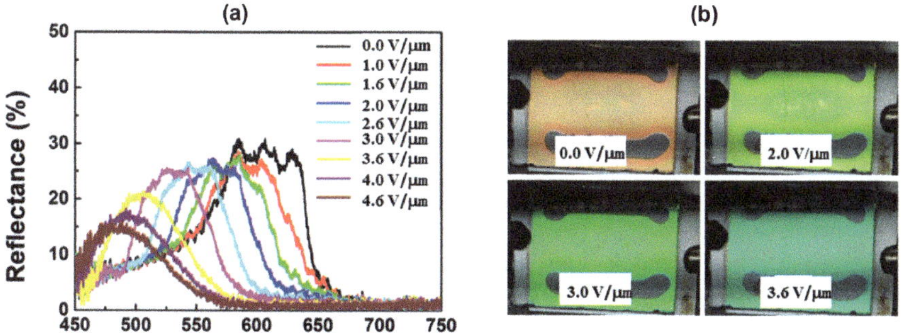

Figure 8.13 (a) Reflection of the narrow band Ch liquid crystal stabilized by dimer and polymer network under various applied electric fields; (a) reflection spectra, (b) photographs.
Reproduced from ref. 36 with permission from the Royal Society of Chemistry.

When no voltage is applied, the sample reflects orange colored light. When the applied electric field is 2.0 V μm^{-1}, the sample reflects green colored light. When the applied electric field is 3.5 V μm^{-1}, the sample reflects cyan colored light. The driving voltage is lower and the color tuning range is larger than those of the cholesteric liquid crystal stabilized only by polymer.

8.6.2 Broad Reflection Band PSCLC

In some applications, it is desired that the reflection band covers the entire visible light region. For a cholesteric with a uniform pitch, no matter whether it is polymer-stabilized or not, the reflection band is governed by the pitch and the birefringence of the liquid crystal, which is usually around 50 nm. Visible light covers the wavelength region from 400 nm to 700 nm. As discussed in Section 8.3, a gradient of pitch is needed to cover the visible light region. The required pitch gradient can be achieved by using polymer stabilization. The initial material is a mixture of a low molecular weight ChLC and bifunctional (achiral) monomer (as used in the narrow band PSCLC) as well as a monofunctional chiral monomer.[57–61] Of course, a small amount of photo-initiator must be added as usual. The material is filled into a cell coated with a homogeneous alignment layer, and is in the planar state. The cell is then irradiated by UV light to polymerize the bifunctional and monofunctional monomers.

Now the main issue is how to create a stable pitch gradient. Inside the cell, the local pitch is determined by the local chiral dopant concentration according to the formula:

$$P = \frac{1}{x_1 (HTP)_1 + x_2 (HTP)_2}, \tag{8.9}$$

where x_1 and $(HTP)_1$ are the concentration and helical twisting power of the low molecular weight chiral dopant, respectively; x_2 and $(HTP)_2$ are the concentration and helical power of the chiral monomer, respectively.

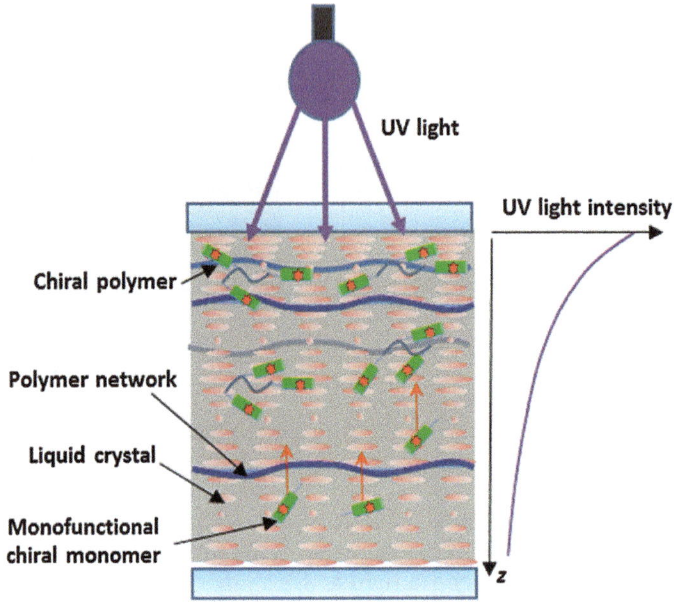

Figure 8.14 Schematic diagram of polymer-stabilized broadband cholesteric liquid crystal.

The concentration, x_1, of the low molecular weight chiral dopant is more or less uniform throughout the cell. The chiral monomer is the one whose concentration can be spatially varied. In order to achieve this, a dye, which absorbs UV light, is added to the initial mixture. When the cell is irradiated by UV light during the polymerization, the UV light intensity is higher in the top side of the cell, which faces the UV light, than the bottom side as shown in Figure 8.14. In the top side, more free radicals are produced, and therefore the chiral monomers (and the bi-functional monomers) are polymerized at a higher rate. The density of the unpolymerized chiral monomer at the top side becomes lower than the bottom side, which results in a lower chemical potential at the top side. Therefore, some chiral monomers diffuse from the bottom side to the top side of the cell. The higher concentration of the chiral monomer generates a shorter pitch at the top side. When the chiral monomers are polymerized, they become side groups of the polymer network, and still have helical twisting power (which may be different from the helical twisting power of the chiral monomer).

Li *et al.*, at Kent Optronics Inc., have developed a switchable mirror based on a broad reflection band PSCLC. It consists of two layers of PSCLC: one with left-handed helix and the other with right-handed helix. It exhibits high reflectance for unpolarized light. The reflection spectra of the switchable mirror under various applied voltages are shown in Figure 8.15(a). At 0 V, the material is in the planar state exhibiting a high reflectance close to 100%. The reflection bandwidth is about 250 nm, covering the entire visible light

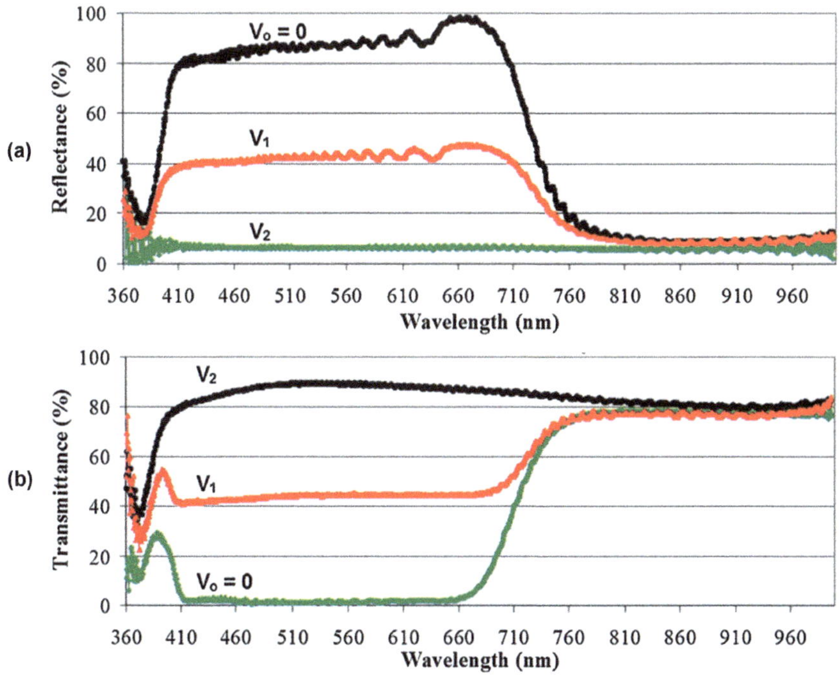

Figure 8.15 (a) Reflection and (b) transmission spectra of the broad reflection band polymer-stabilized cholesteric liquid crystal.
Photo courtesy of Kent Optronics Inc.

region. Its reflectance can be continously adjusted by applying voltages. When an intermediate voltage V_1 (100 V) is applied, it becomes semi-reflecting with a transmittance near 40%. When the applied voltage is increased to V_2 (260 V) , the material is switched to the homeotropic state, and the reflectance is decreased to 10% which is mainly due to the reflection from the two glass-air interfaces. The transmission spectra of the switchable mirror at various applied voltages are shown in Figure 8.15(b). At 0 V, the transmittance is low because of the high reflection of the planar state. As the applied voltage is increased, the transmittance increases. When the applied voltage is increased to V_2, the transmittance becomes close to 90%. This indicates that there is no light scattering in the high voltage state. Photographs of the switchable mirror are shown in Figure 8.16. A vase with flowers is placed in front of the mirror and a paper with print is placed behind the mirror. At 0 V, the mirror is highly reflecting and shows the mirror image of the vase, and the paper behind the mirror cannot be seen, as shown in Figure 8.16(a). When the high voltage V_2 is applied, the mirror is switched to the transparent state; the paper behind the mirror now can be seen, as shown in Figure 8.16(b).

Besides making use of chiral monomers and UV intesity gradients, Mitov *et al.* have also used thermochromic cholesteric liquid crystals and

Figure 8.16 Photographs of switchable mirror based on the broad reflection band polymer-stabilized cholesteric liquid crystal; (a) in reflection state, (b) in transmissive state.
Photo courtesy of Kent Optronics Inc.

temperature gradients to produce broad band polymer-stabilized cholesteric liquid crystals.[62] For a thermochromic cholesteric liquid crytal, the pitch and thus the wavelength of the reflection band change dramatically with temperature. During polymerization, if there is temperature gradient across the cell, the liquid crystal has different pitches at different positions. The *in situ* formed polymer network stabilizes the cholesteric layers with the various pitches. After polymerization, the liquid crystal has a pitch gradient and thus exhibits broad band reflection.

8.7 PSCLCs with Negative Dielectric Anisotropy

The cholesteric liquid crystal used has a negative dielectric anisotropy. When an electric field is applied, the liquid crystal molecules tend to orient perpendicular to the electric field to reduce the electrical energy. As an example, a mixture, consisting of 66.5% nematic liquid crystal MLC-2079 (with dielectric anisotropy $\Delta\varepsilon = -6.7$ and birefringence $\Delta n = 0.15$), 28% chiral dopant R811, 5% monomer RM257 and 0.5% photo-initiator benzoin methyl ether, is used to fill a 10 micron cell coated with homogeneous alignment layers on the two inner surfaces. The material is in the planar state with the liquid crystal parallel to the cell substrate. Before polymerization, when an (AC or DC) electric field is applied across the cell, the liquid crystal remains in the planar state, because it is already perpendicular to the electric field. The reflection spectrum of the cell remains the same, independent of the applied electric field, as shown in Figure 8.17(a).[63,64] The reflection bandwidth (FWHM) is about 60 nm. After polymerization of the monomer under UV irradiation at 0 V, the reflection band is same as that before polymerization. However, when an DC electric field is applied across the cell, surprisingly, the reflection changes; usually the reflection band is broadened, as shown in Figure 8.17(b). When 50 V is applied, the bandwidth is increased to 150 nm. This type of reflection bandwidth broadening was first reported by

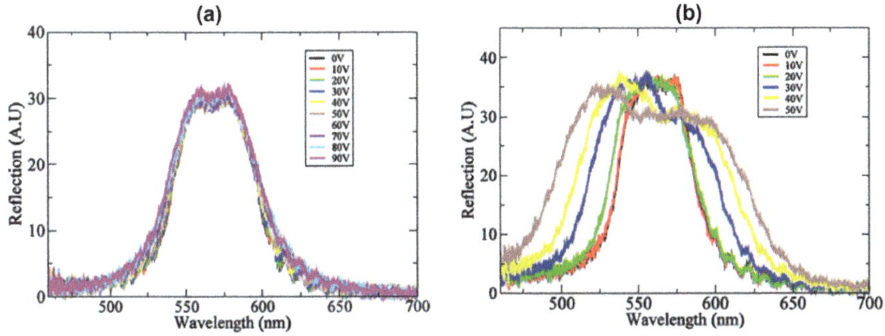

Figure 8.17 Reflection spectrum of the polymer stabilized cholesteric liquid crystal with negative dielectric anisotropy; (a) before polymerization, (b) after polymerization.

Tondiglia *et al.*[65–68] The response time of the bandwidth broadening is about 1 second. Therefore, the material has no response to AC electric fields with frequency higher than 10 Hz.

It was discovered that the reflection bandwidth broadening of the polymer-stabilized cholesteric liquid crystal with a negative dielectric anisotropy is caused by the motion of the dispersed polymer network. There are ions (both positive and negative), due to dissociation of impurities, in the liquid crystal. Before polymerization, when an electric field is applied, the ions move, but they have little effect on the orientation of the liquid crystal, except for creating turbulence when the applied electric field is very high. After polymerization, the ions are trapped on the surface of the formed polymer network due to electrostatic interaction. More ions with one polarity are trapped on the surface of the polymer fibrils than ions with the other polarity, because of their different molecular structures, as shown in Figure 8.18. When no electric field is applied, the liquid crystal is in the planar state with a uniform pitch P as shown in Figure 8.18(a). Therefore the reflection spectrum is the same as that of the cell before polymerization. Because the monomers are *in situ* polymerized, the formed polymer fibrils also have a helical structure, known as the structural helix, which matches the helical structure of the liquid crystal. The helical pitch of the polymer fibrils is the same as that of the liquid crystal. When an electric field is applied, the polymer fibrils move to the top side of the cell under the electrostatic force as shown in Figure 8.18(b). At the bottom side of the cell, the polymer fibrils are expanded and the pitch of the structural helix becomes longer, while at the top side of the cell, the polymer fibrils are contracted and the pitch of the structural helix becomes shorter. The polymer fibrils have a strong aligning effect and force the liquid crystal near their surface to align parallel to the fibrils. The liquid crystal is forced to adapt to the helical structure of the polymer fibrils. Therefore at the bottom side of the cell, the pitch of the liquid crystal becomes longer than the original pitch P, while at the top side of the cell, the pitch of the liquid crystal becomes shorter than P. Thus the reflection bandwidth is broadened.

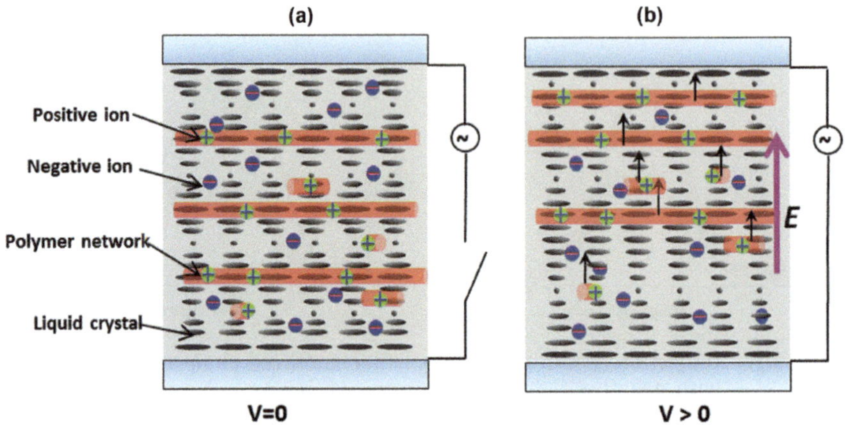

Figure 8.18 Schematic diagrams of the polymer-stabilized N* liquid crystal with negative dielectric anisotropy; (a) no applied voltage, (b) when a DC voltage is applied.

It is difficult to directly observe the motion of the polymer fibrils in the PSCLC cell under DC electric fields using a polarizing optical microscope, because the motion is in the vertical direction perpendicular to the viewing plane of the microscope. The difficulty can be circumvented by using an IPS cell with interdigitated electrodes on one surface of the cell substrate and using a nematic liquid crystal, as shown in Figure 8.19(a).[64] The polymer-stabilized nematic liquid crystal cell is prepared in the same way as the polymer-stabilized Ch liquid crystal. The cell is then studied by polarizing optical microscopy. When a voltage is applied to the interdigitated electrodes, an electric field parallel to the cell substrate is generated. Now the polymer fibrils move in the horizontal direction parallel to the viewing plane. Some microphotographs of the polymer fibrils under various applied voltages are shown in Figure 8.19. Because both the liquid crystal and the polymer fibrils exhibit birefringence at room temperature, it is difficult to distinguish the polymer fibrils from the liquid crystal under the optical microscope. This problem can be solved by heating the cell to an elevated temperature such that the liquid crystal transforms to the isotropic phase and no longer exhibits birefringence. At 0 V, the polymer fibrils are uniformly distributed, as shown in Figure 8.19(b). When a voltage is applied across the electrode, the polymer fibrils start to move toward the negative electrode, as shown in Figure 8.19(c) and (d). The region from which the polymer fibrils move away becomes darker. The region near the edge of the negative electrode, where the polymer fibrils accumulate, becomes brighter.

Now we discuss a phenomenological theory that can be used to describe the reflection bandwidth broadening in the PSCLC with a negative dielectric anisotropy.[64] The polymer network has an anchoring effect which causes the liquid crystal on the surface of the polymer fibril to align parallel to the fibril. The polymer fibrils are formed in the planar state where the liquid crystal

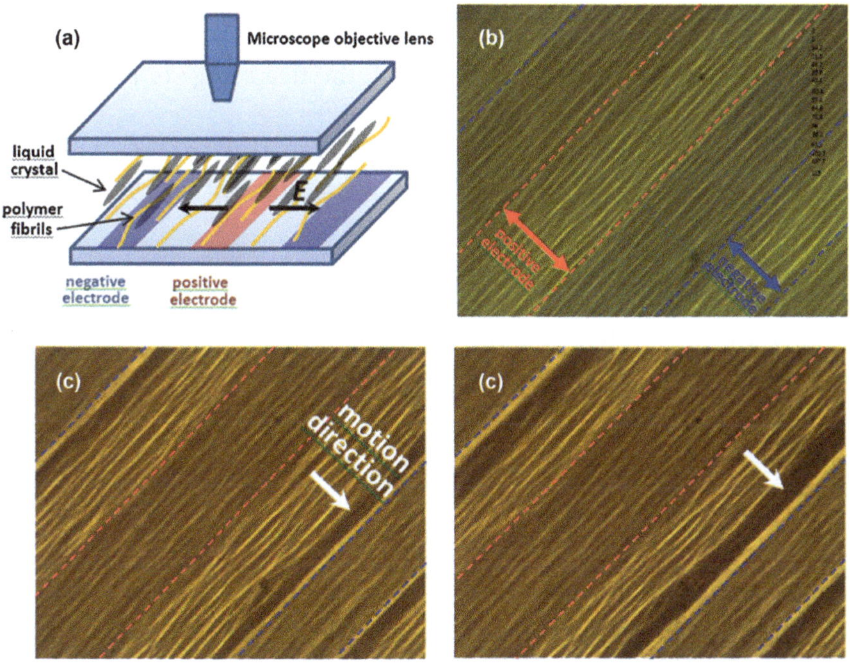

Figure 8.19 (a) Schematic diagram of polymer-stabilized nematic liquid crystal in IPS cell. (b)–(d) Microphotographs of the polymer fibrils in the polymer-stabilized nematic liquid crystal cell in the isotropic phase; (b) 0 V, (c) 40 V, (d) 80 V.
Reproduced from ref. 64 with permission from the Royal Society of Chemistry.

has the uniform intrinsic helical structure. After polymerization, when no electric field is applied, the liquid crystal orientation matches the orientation of the polymer fibrils. When a DC voltage is applied, under the electrostatic force, the fibrils move to new positions at which the orientation of the fibrils does not match the orientation of the liquid crystal. The fibrils force the liquid crystal to reorient such that the orientation of liquid crystal matches that of the polymer fibrils. The displacement of the polymer fibrils is described by $u(z)$ and the twist angle of the liquid crystal director is described by $\phi(z)$, where z is the coordinate perpendicular to the cell surface. The LC director is in the xy plane and is given by

$$n_x = \cos\phi, \quad n_y = \sin\phi \quad \text{and} \quad n_z = 0. \tag{8.10}$$

When no voltage is applied,

$$u = 0, \quad \phi = q_o z, \tag{8.11}$$

where q_o is the intrinsic helical wavenumber and is related to the intrinsic pitch by $q_o = 2\pi/P_o$. When a DC electric field, E, is applied, the polymer fibril originally located at $(z-u)$, with the twist angle $q_o(z-u)$, moves to the position z at which the twist angle of the liquid crystal is $\phi(z)$. The difference between the twist angles of the polymer fibril and the liquid crystal is $\phi(z) - q_o(z-u)$. For a small displacement, the free energy of the system is given by

$$f = \frac{1}{2}K_{22}\left(\frac{d\phi}{dz} - \frac{2\pi}{P_o}\right)^2 + \frac{1}{2}B\left(\frac{du}{dz}\right)^2 + \frac{1}{2}A[\alpha - q_o(z-u)]^2 - \frac{1}{2}CEu \qquad (8.12)$$

Now we discuss the terms on the right side of eqn (8.12). The first term is the twist elastic energy of the liquid crystal. A typical value of the twist elastic constant, K_{22}, is 8×10^{-12} *N*. The second term is the elastic energy of the polymer network with B as the Young's modulus. By fitting the experimental data, it is found that B has a value around 10^3 J m^{-3}. The third term is the surface anchoring energy of the liquid crystal on the surface of the fibrils. The coefficient A depends on the anchoring strength, W, of the fibrils and the surface area, S, of the fibrils per unit volume. This term is analogous to the surface anchoring energy of a liquid crystal at a polymer surface introduced by Rapini and Papoular. If the fibrils are cylinders with the radius R and occupy a volume fraction ν, the average distance, L, between the neighboring fibrils is given by $\pi R^2/L^2 = \nu$. The surface area of the fibrils per unit volume is given by

$$S = \frac{2\pi R}{L^2} = \frac{2\pi R}{(\pi R^2/\nu)} = \frac{2\nu}{R}. \qquad (8.13)$$

The coefficient A is

$$A = WS = \frac{2W\nu}{R} \qquad (8.14)$$

For example, if $W = 10^{-4}$ J m^{-2}, $\nu = 8\%$ and $R = 100$ nm, then $A = 1.6\times 10^2$ J m^{-3}. The last term is the electrical energy of the fibrils because of the trapped charges on them. It is assumed that the ions trapped on the fibrils are independent of the applied electric field. The coefficient C depends on the amount of net charge trapped on the fibrils. By fitting the experimental data, C is found to be around 5×10 C m^{-3}.

By minimizing the free energy with respect to ϕ and u, we get

$$\frac{\delta f}{\delta\phi} = \frac{\partial f}{\partial\phi} - \frac{d}{dz}\left(\frac{\partial f}{\partial(\partial\phi/\partial z)}\right) = A[\phi - q_o(z-u)] - K_{22}\frac{d^2\phi}{dz^2} = 0 \qquad (8.15)$$

$$\frac{\delta f}{\delta u} = \frac{\partial f}{\partial u} - \frac{d}{dz}\left(\frac{\partial f}{\partial(\partial u/\partial z)}\right) = -q_oA[\phi - q_o(z-u)] - CE - B\frac{d^2u}{dz^2} = 0 \quad (8.16)$$

Because the polymer network is bound to the cell surface, the boundary conditions for u are $u(z=0)=0$ at the bottom surface of the cell and

$u(z=h)=0$ at the top surface. Also, because the liquid crystal is anchored by the alignment layer on the two surfaces of the cell, the boundary conditions for ϕ are therefore $\phi(z=0)=0$ and $\phi(z=h)=2\pi h/P_o$. Eqn (8.15) and (8.16) can be analytically solved. For example, when the following parameters are used: $P_o=360$ nm, $K_{22}=8\times10^{-12}$ N, $A=1.6\times10^2$ J m^{-3}, $C=45$ C m^{-3}, $B=10^3$ J m^{-3} and cell thickness $h=10$ μm, the theoretically calculated pitch values as a function of position z under various applied voltages are shown in Figure 8.20. If the refractive indices of the liquid crystal are $n_o=1.5$ and $n_e=1.7$ and the applied voltage is 40 V, the reflection bandwidth is $\Delta\lambda=1.7\times410-1.5\times320=217$ nm, which is similar to the experimentally observed value.

There a few factors which affect the bandwidth broadening.[69] The first is the polymer network. A polymer network is necessary for the electric field induced bandwidth broadening. There is an optimal concentration, which is about 5% of the monomer. If the monomer concentration is too low, the formed polymer network does not have an aligning effect on the liquid crystal. Even if the polymer network moves under electric fields, the pitch of the liquid crystal does not change. If the monomer concentration is too high, the formed polymer network becomes too rigid and then does not move under electric fields. The rigidity of the polymer network can be controlled not only by the monomer concentration but also by the functionality of the monomer. When a bi-functional monomer (such as RM257) and a mono-functional monomer (such as RM23) are mixed together, the average

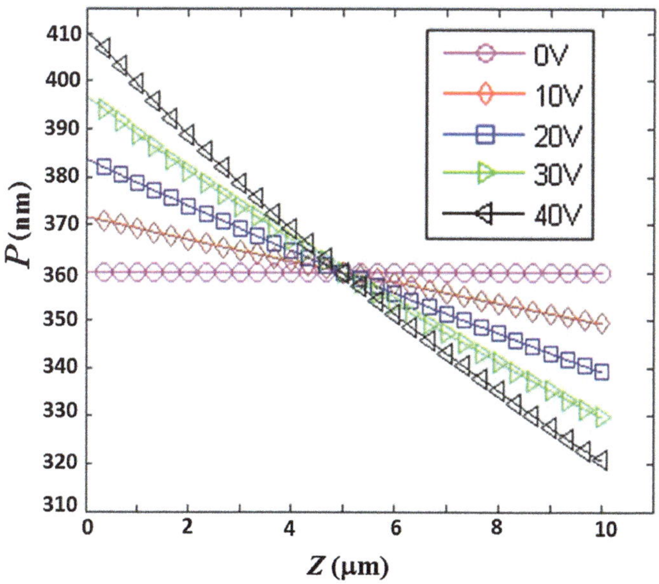

Figure 8.20 Theoretically calculated pitch as a function of position under various applied voltages.

Figure 8.21 Photographs of the polymer-stabilized cholesteric liquid crystal under various applied voltages.

functionality can be adjusted by the relative concentration ratio of the two monomers. When the total concentration of the two monomers is fixed, as the concentration of the mono-functional monomer is increased, the rigidity of the formed polymer network decreases, and the broadening increases. But if the concentration of the mono-functional monomer is too high, the formed polymer network becomes too soft, and its aligning effect becomes too weak to change the pitch. Furthermore, when high voltages are applied, the polymer network becomes unstable and the planar orientation of the fibrils is destroyed.

The second factor is the ion density. Generally speaking, the higher the ion density, the larger the broadening under a given applied voltage. There are a few ways to increase the ion density. Experiments show that when an organic salt is added to the liquid crystal, it will dissociate into ions, and thus increase the ion density. Experiments also show that when the concentration of the photo-initiator is increased, there are more ions after polymerization. An insulation layer between the ITO electrode and the liquid crystal can also affect the ion density. A polymer alignment layer is usually coated on top of the ITO. If the alignment layer is thin or the ITO is not coated, more ions will be injected from the electrode into the liquid crystal, and thus increase the ion density.

Photographs of a 10 micron polymer-stabilized cholesteric liquid crystal with a negative dielectric anisotropy under various applied voltages are shown in Figure 8.21. When the applied voltage is 0 V, the reflection bandwidth is 60 nm, and the cell has a green appearance. When the applied voltage is 10 V, the bandwidth is increased to 200 nm, and the cell has a white-yellow appearance. When 20 V is applied, the bandwidth becomes more than 300 nm, and the cell has a white appearance.

8.8 Conclusion

Cholesteric liquid crystals exhibit selective reflection because of their periodic helical structure. The wavelength of their reflection band is controlled by the helical pitch P and the bandwidth is controlled by both the pitch and the birefringence Δn. Because of this unique optical property, they are used in many applications such as reflective displays, E-books, writing tablets, tunable color filters and mirrorless lasing. It is, however, very difficult to electrically tune the color of the reflection of low molecular weight N* liquid crystals in cells with top and bottom electrode geometry. By introducing polymer networks to cholesteric liquid crystals, known as polymer stabilization, it becomes possible to electrically tune the color and intensity of the reflection. In one case, the reflection intensity can be adjusted by applying voltages: the reflectance is high at 0 V and decreases continuously to 0 as the applied voltage is increased. When the applied voltage is turned off, the dispersed polymer network can help the liquid crystal relax back quickly to the state with high reflectance. In another case, the wavelength of the reflection band can be tuned. It is also possible to obtain stable states with pitch gradients, whose reflection band can cover the entire visible light region and reflectance can be adjusted by applying voltages. For PSCLCs with negative dielectric anisotropies, the reflection bandwidth can be dramatically broadened by applying voltages. The applications of polymer-stabilized cholesteric liquid crystals are enormous and many new applications should be explored.

References

1. P. G. de Gennes, *The Physics of Liquid Crystals*, Clarendon, Oxford, 1974.
2. S. Chandrasekhar, *Liquid Crystals*, Cambridge University Press, New York, 2nd edn, 1997.
3. L. M. Blinov and V. G. Chigrinov, *Electrooptical Effects in Liquid Crystal Materials*, Springer-Verlag, New York, 1994.
4. D.-K. Yang and S. T. Wu, *Fundamental of Liquid Crystal Devices*, Wiley, 2nd edn, 2015.
5. S.-T. Wu and D.-K. Yang, *Reflective Liquid Crystal Displays*, John Wiley & Sons, Ltd., 2001.
6. D.-K. Yang, J. W. Doane, Z. Yaniv and J. Glasser, *Appl. Phys. Lett.*, 1994, **65**, 1905.
7. D.-K. Yang, Reflective Cholesteric Liquid Crystal Displays, in *Mobile Displays*, ed. A. Bhowmik, Z. Li and P. J. Bos, John Wiley & Sons, Ltd., ch. 16, 2008.
8. D.-K. Yang, J. L. West, L. C. Chien and J. W. Doane, *J. Appl. Phys.*, 1994, **76**, 1331.
9. H. Xianyu, T. H. Lin and S. T. Wu, *Appl. Phys. Lett.*, 2006, **89**, 091124.
10. J. W. Doane and A. Khan, Cholesteric Liquid Crystals for Flexible Displays, in *Flexible Flat Panel Displays*, ed. G. P. Crawford, John Wiley & Sons, Ltd., ch. 17, 2005.
11. U. A. Hrozhyk, S. V. Serak, N. V. Tabiryan, T. J. White and T. J. Bunning, *Opt. Express.*, 2010, **18**, 9651.
12. U. A. Hrozhyk, S. V. Serak, N. V. Tabiryan, T. J. White and T. J. Bunning, *Opt. Mater. Express*, 2011, **1**, 943.

13. T. J. White, R. L. Bricker, L. V. Natarajan, V. P. Tondiglia, L. Green, Q. Li and T. J. Bunning, *Opt. Express*, 2010, **18**, 173.
14. N. V. Kukhtarev, *Quantum Electron.*, 1978, **8**, 774.
15. V. I. Kopp, B. Fan, H. K. M. Vithana and A. Z. Genak, *Opt. Lett.*, 1998, **23**, 1707.
16. H. Finkelmann, S. T. Kim, A. Munoz, P. Palffy-Muhoray and B. Taheri, *Adv. Mater.*, 2001, **13**, 1069.
17. I. Sage, Thermochromic Liquid Crystal Devices, in *Liquid Crystals-applications and Uses*, ed. B. Bahadur, World Scientific, New Jersey, 1990, vol. 3.
18. R. S. Pindak, C. C. Huang and J. T. Ho, *Phys. Rev. Lett.*, 1974, **32**, 43.
19. P. N. Keating, *Mol. Cryst. Liq. Cryst.*, 1969, **8**, 315.
20. F. Zhang and D.-K. Yang, *Liq. Cryst.*, 2002, **29**, 1497.
21. D. W. Berreman, *J. Opt. Soc. Am.*, 1972, **62**, 502.
22. D. W. Berreman and T. J. Scheffer, *Mol. Cryst. Liq. Cryst.*, 1970, **11**, 395.
23. M. Xu, F. D. Xu and D.-K. Yang, *J. Appl. Phys.*, 1998, **83**, 1938.
24. O. D. Lavrentovich and D.-K. Yang, *Phys. Rev. E, Rapid Commun.*, 1998, **57**, R6269.
25. R. B. Meyer, *Appl. Phys. Lett.*, 1969, **14**, 208.
26. W. Greubel, U. Wolf and H. Kruger, *Mol. Cryst. Liq. Cryst.*, 1973, **24**, 103.
27. M. Kawachi, O. Kogure, S. Yosji and Y. Kato, *Jpn. J. Appl. Phys.*, 1975, **14**, 1063.
28. D.-K. Yang and Z.-J. Lu, *SID Intl. Symp. Digest Tech. Papers*, 1995, **26**, 351.
29. P. Watson, J. E. Anderson, V. Sergan and P. J. Bos, *Liq. Cryst.*, 1999, **26**, 1307.
30. W. Helfrich, *Appl. Phys. Lett.*, 1970, **17**, 531.
31. W. Helfrich, *J. Chem. Phys.*, 1971, **55**, 839.
32. C. J. Gerritsma and P. Van Zanten, *Phys. Lett. A*, 1971, **37**(1), 47.
33. T. J. Scheffer, *Phys. Rev. Lett.*, 1972, **28**, 593.
34. J. P. Hurault, *J. Chem. Phys.*, 1973, **59**, 2068.
35. R. S. Zola, H. Nemati, Y.-C. Yang, D.-K. Yang, K.-L. Cheng, C.-C. Liang, F. Shiu and C.-C. Tsai, *SID Symp. Dig. of Tech. Papers*, 2012, **40**, 551.
36. M. Yu, H. Yang and D.-K. Yang, *Soft Matter*, 2017, **13**, 8728.
37. D. J. Broer, Networks Formed by Photoinitiated Chain Cross-linking, in *Liquid Crystals in Complex Geometries*, ed. G. P. Crawford and S. Zumer, Taylor & Francis, London, 1996.
38. R. A. M. Hikmet, Anisotropic Gels Obtained by Photopolymerization in the Liquid Crystal State in *Liquid Crystals in Complex Geometries*, ed. G. P. Crawford and S. Zumer Taylor & Francis, London, 1996.
39. D. J. Broer, R. G. Gossink and R. A. M. Hikmet, *Die Angew. Makromol. Chem.*, 1990, **183**, 45.
40. D.-K. Yang, L.-C. Chien and Y. K. Fung, Polymer Stabilized Cholesteric Textures: Materials and Applications in *Liquid Crystals in Complex Geometries*, ed. G. P. Crawford and S. Zumer, Taylor & Francis, London, 1996.

41. I. Dierking, *Polym. Chem.*, 2010, **1**, 1153.
42. D.-K. Yang, Polymer Stabilized Liquid Crystal Displays, in *Progress in Liquid Crystal Science and Technology*, ed. H.-S. Kwok, S. Naemura and H. L. Ong, World Scientific, 2013.
43. G. A. Held, L. L. Kosbar, I. Dierking, A. C. Lowe, G. Grinstein, V. Lee and R. D. Miller, *Phys. Rev. Lett.*, 1997, **79**, 3443.
44. Y. K. Fung, D.-K. Yang, Y. Sun, L. C. Chien, S. Žumer and J. W. Doane, *Liq. Cryst.*, 1995, **19**, 797.
45. R. Q. Ma and D.-K. Yang, *Phys. Rev. E.*, 2000, **61**, 1576.
46. D.-K. Yang, Y. Cui, H. Nemati, X. Zhou and A. Moheghi, *J. Appl. Phys.*, 2013, **114**, 243515.
47. M. J. Escuti, C. C. Bowley, S. Zumer and G. P. Crawford, *SID Intl Symp. Digest Tech. Papers*, 1999, **30**, 32.
48. Z. Ge, S. Gauza, H. Xianyu and S. T. Wu, *Appl. Phys. Lett.*, 2009, **94**, 101104.
49. L. Rao, Z. Ge, S. T. Wu and S. H. Lee, *Appl. Phys. Lett.*, 2009, **95**, 231101.
50. S. G. Kim, S. M. Kim, Y. S. Kim, H. K. Lee, S. H. Lee, G.-D. Lee, J.-J. Lyu and K. H. Kim, *Appl. Phys. Lett.*, 2007, **90**, 261910.
51. Y. R. Kwon, Y. E. Choi, P. Wen, B. H. Lee, J. C. Kim, M.-H. Lee, K.-U. Jeong and S. H. Lee, *J. Phys. D: Appl. Phys.*, 2016, **49**, 165501.
52. S. E. Hicks, S. P. Hurley, R. S. Zola and D.-K. Yang, *J. Disp. Technol.*, 2011, **7**, 619.
53. J. Guo, H. Cao, J. Wei, D. Zhang, F. Liu, G. Pan, D. Zhao, W. He and H. Yang, *Appl. Phys. Lett.*, 2008, **93**, 201901.
54. J. Sun, H. Wang, L. Wang, H. Cao, H. Xie, X. Luo, J. Xiao, H. Ding, Z. Yang and H. Yang, *Smart Mater. Struct.*, 2014, **23**, 125038.
55. M. E. McConney, V. P. Tondiglia, J. M. Hurtubise, L. V. Natarajan, T. J. White and T. J. Bunning, *Adv. Mater.*, 2011, **23**, 1452.
56. M. E. McConney, T. J. White, V. P. Tondiglia, L. V. Natarajan, D.-K. Yang and T. J. Bunning, *Soft Matter*, 2012, **8**, 318.
57. H. Guillard, P. Sixou, L. Reboul and A. Perichaud, *Polymer*, 2001, **42**, 9753.
58. D. J. Broer, J. Lub and G. N. Mol, *Nature*, 1995, **378**, 467.
59. L. Li and S. M. Faris, *SID Tech. Digest*, 1996, **27**, 111.
60. M. Mitov, *Adv. Mater.*, 2012, **24**, 6260.
61. C. Binet, M. Mitov and M. Mauzac, *J. Appl. Phys.*, 2001, **90**, 1730.
62. M. Mitov, E. Nouvet and N. Dessaud, *Eur. Phys. J. E*, 2004, **15**, 413.
63. H. Nemati, *Color Tuning in Polymer Stabilized Cholesteric Liquid Crystals*, dissertation, Kent State University, 2015.
64. H. Nemati, S. Liu, R. S. Zola, V. P. Tondiglia, K. M. Lee, T. White, T. Bunning and D.-K. Yang, *Soft Matter*, 2015, **11**, 1208.
65. V. T. Tondiglia, L. V. Natarajan, C. A. Bailey, M. M. Duning, R. L. Sutherland, D.-K. Yang, A. Voevodin, T. J. White and T. J. Bunning, *J. Appl. Phys.*, 2011, **110**, 053109.
66. V. P. Tondiglia, L. V. Natarajan, C. A. Bailey, M. E. McConney, K. M. Lee, T. J. Bunning, R. Zola, H. Nemati, D.-K. Yang and T. J. White, *Opt. Mater. Express*, 2014, **4**, 1465.

67. C. A. Bailey, V. P. Tondiglia, L. V. Natarajan, M. M. Duning, R. L. Bricker, R. L. Sutherland, T. J. White, M. F. Durstock and T. J. Bunning, *J. Appl. Phys.*, 2010, **107**, 013105.
68. K. M. Lee, V. P. Tondiglia, M. E. McConney, L. V. Natarajan, T. J. Bunning and T. J. White, *ACS Photonics*, 2014, **1**, 1033.
69. M. Yu, L. Wang, H. Nemati, H. Yang, T. Bunning and D.-K. Yang, *J. Polym. Sci.*, 2017, **55**, 835–846.

CHAPTER 9

Polymer Stabilized Ferroelectric Liquid Crystals and their Applications

HIROKAZU FURUE

Tokyo University of Science, Department of Materials Science and Technology, 6-3-1 Niijuku, Katsushika, Tokyo 125-8585, Japan
Email: hfurue@rs.noda.tus.ac.jp

9.1 Introduction

A Ferroelectric Liquid Crystal (FLC) is most often a chiral smectic C (SmC*) phase, as illustrated in Figure 9.1.[1] For reasons of symmetry, these phases exhibit a spontaneous polarization by uniform orientation of the permanent dipole moment of molecules. Since an applied electric field directly couples to the spontaneous polarization, it is possible to achieve an order of magnitude faster response times than observed for the paraelectric nematic liquid crystal, tens of microseconds in contrast to tens of milliseconds. Also, since the direction of the spontaneous polarization is perpendicular to the molecular long axis, when the electric field, E, is applied in the direction perpendicular to the sheet of paper as shown in Figure 9.1, the molecular orientation change is approximately within the paper plane (in-plane switching, IPS), which gives rise to enhanced viewing properties and contrast.

Therefore, it is relatively easy to produce displays with wide viewing angles. Furthermore, the possibility of optical bistability and thus the property of optical memory, in which the orientation direction is maintained

Soft Matter Series No. 8
Polymer-modified Liquid Crystals
Edited by Ingo Dierking

Published by the Royal Society of Chemistry, www.rsc.org

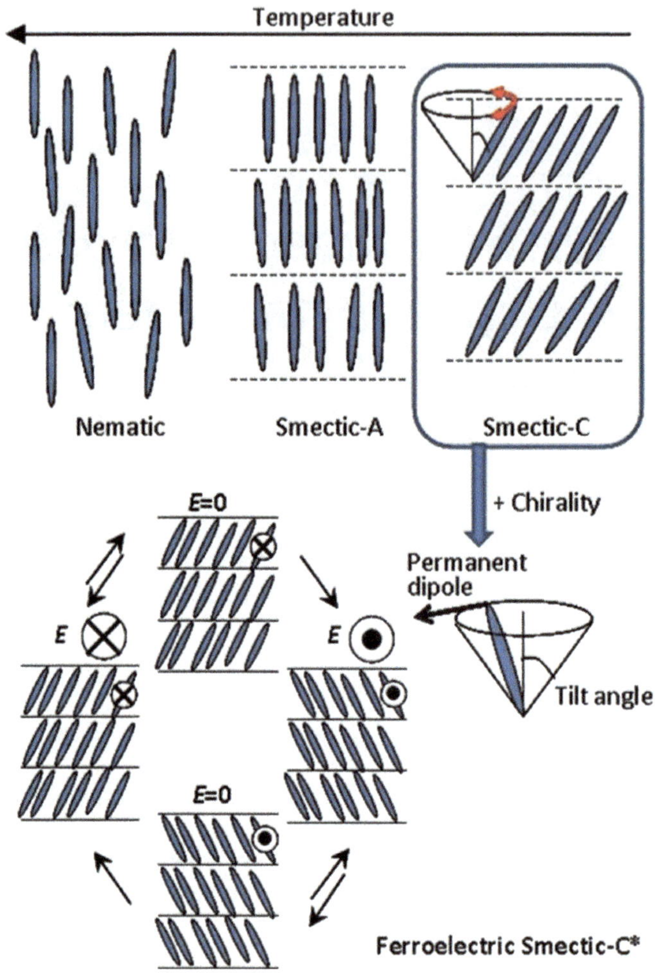

Figure 9.1 Molecular alignment structure of a ferroelectric liquid crystal.

even after electric field removal ($E = 0$), is also possible.[2] Based on the above characteristics, FLCs are expected to provide a next-generation liquid crystal display material, for example, by enabling beautiful moving video images and low power consumption by the field sequential color method.[3–8] However, among these positive characteristics, the bistability between only two stable states that can be adopted, depending on the polarity of the applied electric field, makes it difficult to display grayscale images, albeit making it very useful for monochrome displays. Therefore, for advanced display applications, it is necessary to eliminate the bistability of FLCs, that is, to have a monostable state when no electric field is applied and to be able to continuously modulate the molecular orientation direction by the electric field amplitude.

For the monostabilization of FLCs, the technique of polymer stabilization has been proposed.[9–11] Polymer stabilization is a technique for stabilizing a weak liquid crystal molecule alignment structure with strong polymer networks generated by adding a photopolymerizable monomer to a liquid crystal medium and irradiating with ultraviolet (UV) irradiation. Since the orientation structure at the time of polymerization is stabilized, this can be controlled by the right choice of temperature, thus LC phase, electric field application or other external parameters. It is therefore possible to variously improve the characteristics of the liquid crystal medium and develop novel optic and electrooptic characteristics. As an example of polymer stabilization, its application to the Blue Phase is currently widely investigated, as also pointed out on several occasions in this book.[12] The Blue Phase originally appears in a narrow temperature range between the isotropic liquid phase and the chiral nematic phase, but it can be expanded to a very wide temperature range by polymer stabilization. Furthermore, optically isotropic liquid crystals can also be prepared by performing polymerization in an isotropic liquid phase.[13,14] In addition, it is also possible to accelerate the electric field response of the liquid crystal by polymer stabilization.[15,16] In the following, we will discuss polymer-stabilized ferroelectric liquid crystals, but polymer stabilization technology in general is an attractive technology for various liquid crystal research and development areas as it can improve the properties of the liquid crystals and facilitate expression of novel characteristics.

Monostabilizing FLCs (which originally were bistable) through polymer stabilization can be said to represent a novel property. The bistability of an FLC consists of two oriented structures with different molecular orientation directions and it is possible to switch the orientation direction by the polarity of the applied electric field. However, even if the electric field is removed, it does not return to the original initial orientation. Therefore, if FLC molecules can be returned to the initial orientation due to anchoring of an introduced polymer, monostability can be obtained. This is the basic principle of FLC monostabilization by polymer stabilization. There have been three methods of polymer stabilization (polymerization conditions) proposed so far. Each will be described below.

9.2 Polymerization in SmC* Phase Under DC Electric Field

If one of the bistable states in an FLC could be strongly stabilized by the introduction of a polymer network, then monostability might be possible. As shown in Figure 9.2, to achieve this a photopolymerizable liquid crystal monomer is added to an FLC sample, a DC electric field is applied to align the molecular orientation into one of the bistable states, and the sample irradiated with UV light to polymerize the monomer.[9] Then, the formed

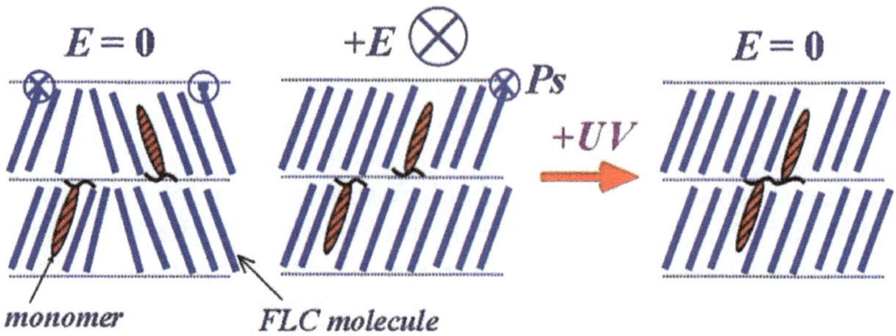

Figure 9.2 Schematic model of a polymer-stabilized FLC photocured under application of DC electric field in SmC* phase.

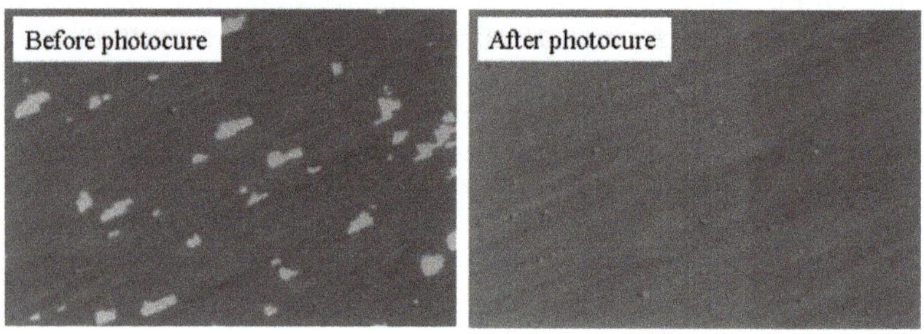

Figure 9.3 Microscopic textures of FLCs before and after polymer stabilization in which the photocure process was performed under application of a DC electric field in the SmC* phase.

polymer templates always adopt the aligned structure during polymerization even at the time of electric field removal.

As an example of the resulting behavior, Figure 9.3 shows the polarizing microscopic textures of a sample, in which 2wt% of the monoacrylate monomer UCL-001 (DIC) was added to the ferroelectric liquid crystal FELIX-M4851/100 (Hoechst), before and after the UV photocuring process.[9] Before polymerization, it is found that there are light-dark bistable domains, whereas a single domain is observed after polymerization. Figure 9.4 shows the electrooptic response of the transmitted light intensity with respect to the pulse voltage. When no electric field is applied, before the polymerization, there are two stable states depending on the polarity of the electric field, whereas only one stable state is obtained after polymerization. Therefore, it is confirmed that the FLC medium is monostable. Figure 9.5 shows the transmittance during the application of an electric field for polymer-stabilized FLCs after polymerization. It is found that the electro-optic characteristics of the polymer-stabilized FLCs are a half-V shaped diode-like response. Then, by monostabilization, continuous transmission light intensity modulation is shown against the electric field intensity, and a grayscale display is possible.

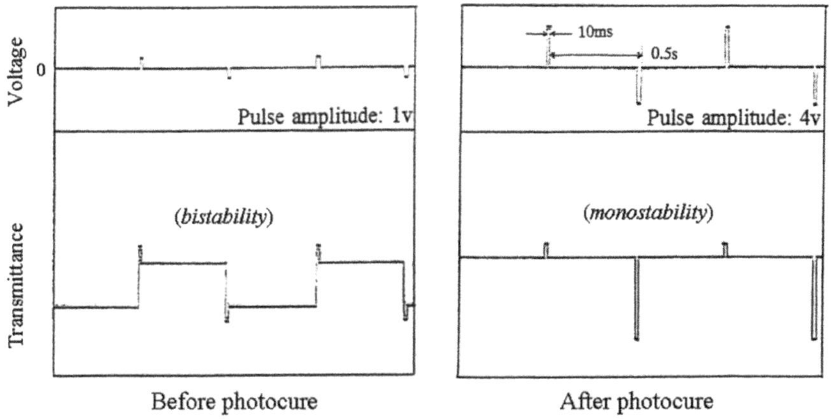

Figure 9.4 Electrooptical characteristics of FLCs before and after polymer stabilization in which photocuring was carried out under application of DC electric field in SmC* phase.

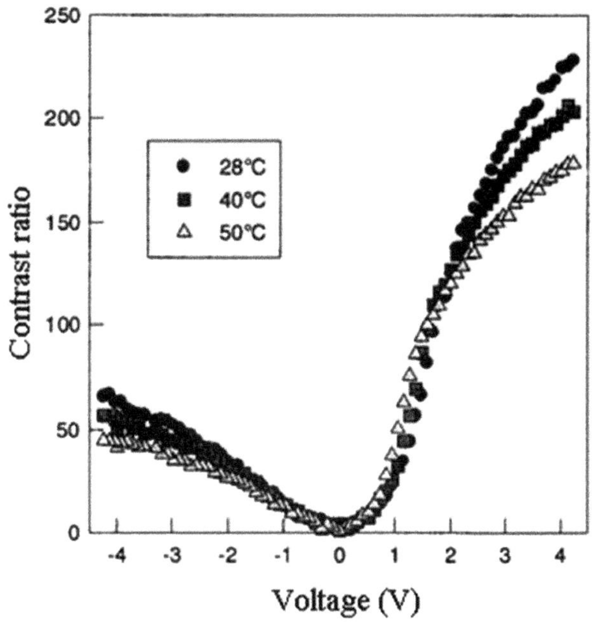

Figure 9.5 Electrooptical characteristics of polymer-stabilized FLC photocured under application of DC electric field in SmC* phase.

9.3 Polymerization in SmC* Phase Under AC Electric Field

In the polymerization under a DC electric field, monostability is easily obtained and half-V-shaped electrooptical characteristics are obtained, but what will happen if the polymerization is carried out under AC electric fields?

Figure 9.6 shows the microscopic textures of a sample in which polymerization was carried out in the same manner as described above under application of a rectangular wave AC electric field (±4 V μm^{-1}, 200 Hz) using FELIX-M4654/100 (Hoechst) as the ferroelectric liquid crystal matrix and UCL-001 (DIC) as a reactive monomer.[17,18] Bistability is exhibited up to a polymer concentration of approximately 4wt%, but the ferroelectric domain size decreases with increasing polymer concentration. If the domain size is uniform and much smaller than the display pixel size, grayscale displays

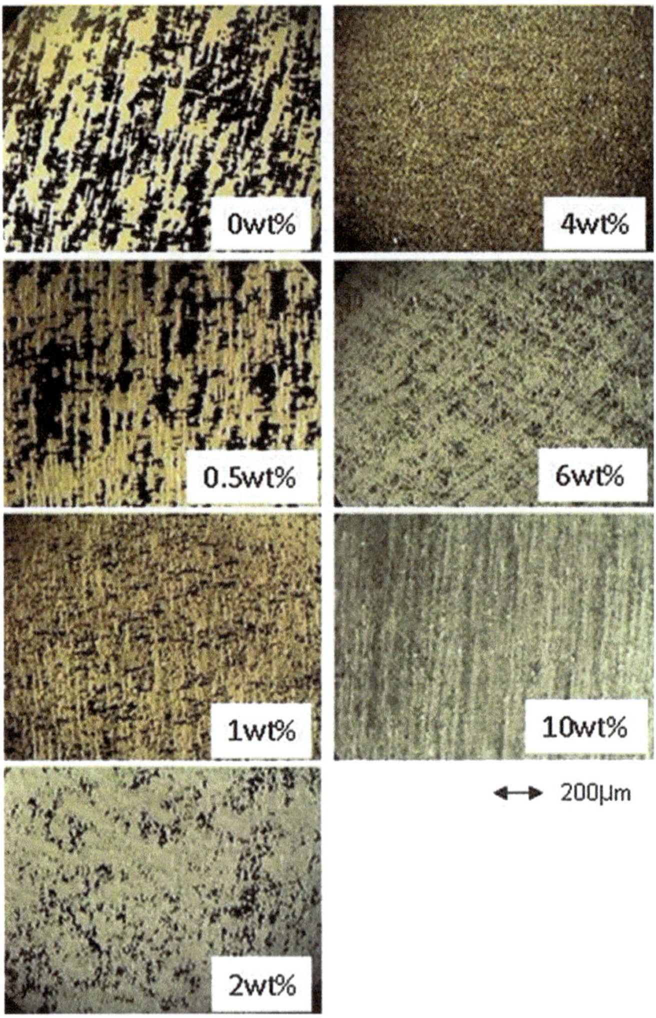

Figure 9.6 Polymer concentration dependence of microscopic texture in a polymer-stabilized FLC photocured under application of an AC electric field in SmC* phase.

become possible by using the volume ratio of the light and dark domains while utilizing the bistable memory property. Figure 9.7 shows the results of measuring the electrooptical characteristics in the memory state after removing the electric field at a polymer concentration of 4wt%.

It can be seen that light and dark can be switched with a minute domain size and the transmitted light intensity modulation can be continuously performed by the applied electric field intensity. On the other hand, when the polymer concentration is increased to 10wt%, the cell becomes monostable as a whole. Figure 9.8 shows the electrooptical characteristics (when an electric field is applied) at 10wt% polymer. The optical axis direction of the medium coincides with the rubbing direction, and by making the rubbing direction coincident with the polarization direction at the time of no applied electric field, a dark state can be obtained. When an electric field is applied, a V-shaped transmitted light intensity modulation, symmetrical with respect to the polarity of the electric field, is obtained and clear domain switching is not observed by polarizing microscopy. The light intensity changes over the whole cell. This V-shaped response is more suitable for display applications than the half-V-shaped response. The ferroelectric liquid crystal exhibiting a V-shaped response by polymer stabilization as described above is called PSV-FLC.

The characteristic changes in the FLC performance due to the above-described polymer concentration can be explained as follows. As shown in Figure 9.9, polymerization is carried out under an alternating electric field, so that the resulting polymer is formed which stabilizes one of the two alignment directions, according to the electric field polarity at the time of polymerization. Consequently, competition in the stabilizing orientation direction occurs for FLC molecules. As a result, at low concentration, it is considered that bistability is maintained in the region far from the polymer although it is monostabilized near the polymer. For the polymerization under DC electric fields discussed in the previous section, monostability was obtained even at a polymer concentration of 2wt%, but the system is not monostabilized at the same concentration for polymerization under an AC electric field. As the polymer concentration is increased, the bistable region shrinks proportionally, so that the domain size becomes minute, and if the bistable region disappears by further increasing the concentration, it is suggested that the whole medium monostabilizes. In this monostabilized state, the monostable domains in two orientation directions coexist, but if the domain size is smaller than the visible light wavelength and the proportions of these domains are equal, the average optical axis of the overall medium coincides with the rubbing direction. Thus, a PSV-FLC obtained by polymerization in SmC* phase under AC electric field is realized by microscopic coexistence of monostable domains in two orientation directions.[17,18]

In the SmC* phase, generally, the tilt angle of the FLC molecule shows a strong dependence on temperature. In addition, the AC electric field has many variable parameters such as amplitude, frequency, and waveform. Furthermore, UV light also has the variable parameters of intensity,

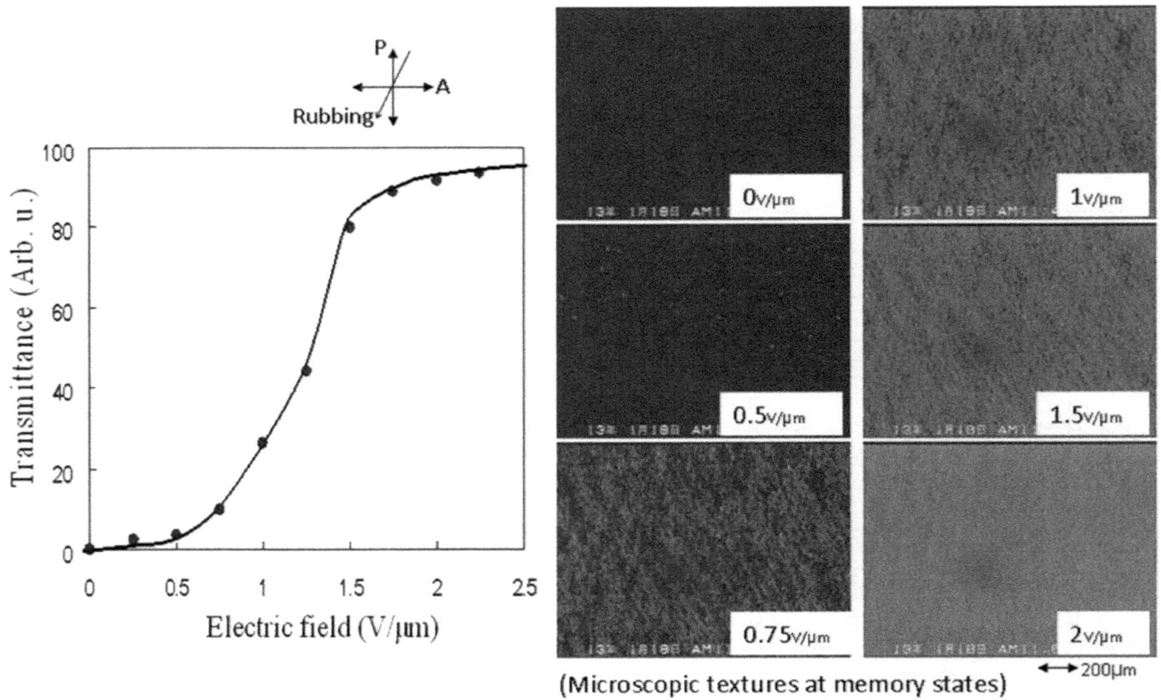

Figure 9.7 Electrooptical characteristics of a polymer-stabilized FLC fabricated using 4wt% monomer; the transmittance data were measured at the memory state ($E = 0$).

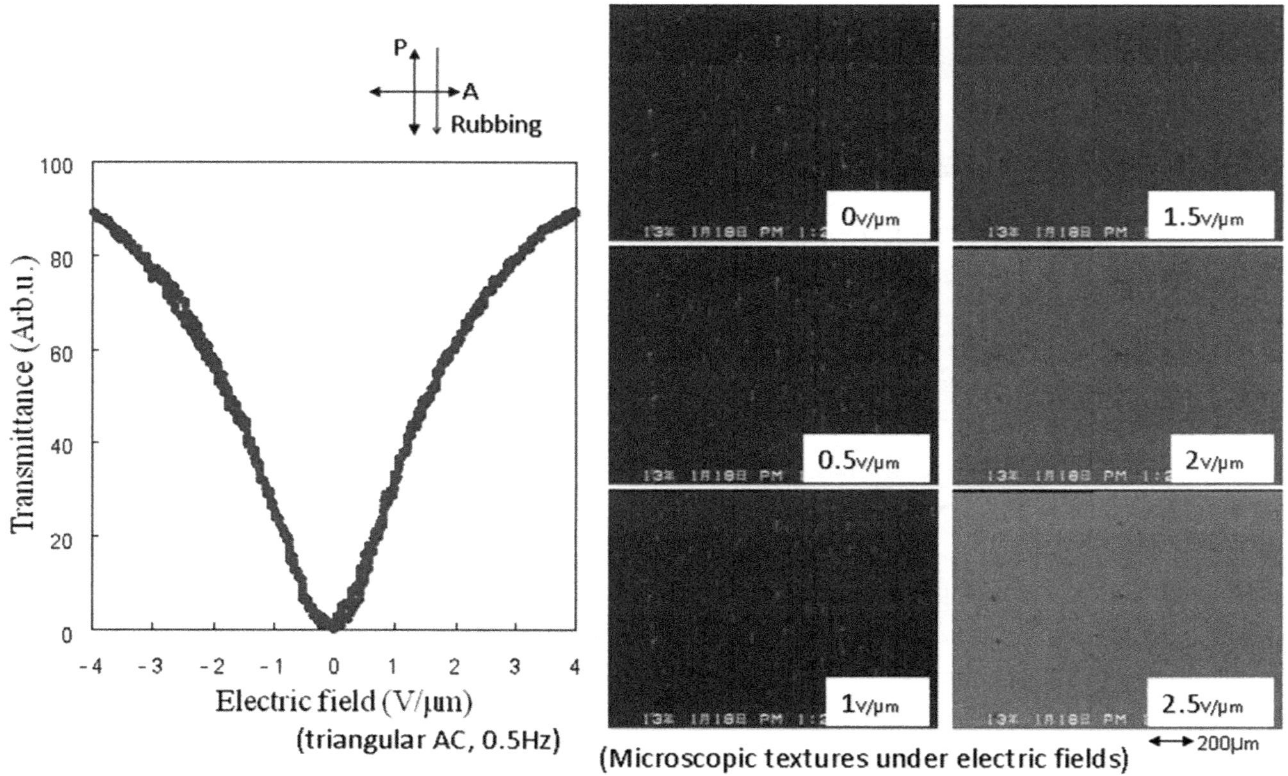

Figure 9.8 Electrooptical characteristics of a polymer-stabilized FLC fabricated using 10wt% monomer; the transmittance data were measured under application of an electric field.

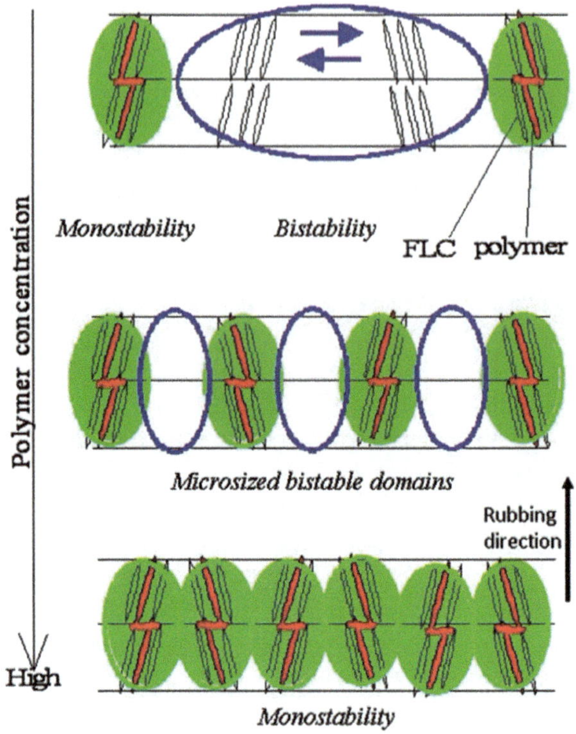

Figure 9.9 Schematic model of the alignment structure in polymer-stabilized FLCs photocured under application of an AC electric field in the SmC* phase.

irradiating time, wavelength, and the like. Of course, the liquid crystal and the monomer material are also of paramount importance. Therefore, the characteristics of the polymer-stabilized FLC fabricated under the application of AC electric field strongly depend on the selection of all of these conditions.

9.4 Polymerization in SmA Phase Under Zero Field Condition

To prepare a PSV-FLC having V-shaped electrooptical characteristics which is monostable by polymer stabilization and symmetrical to the polarity of the electric field, a method of polymerizing in the SmA phase under zero electric field has been proposed.[19] As shown in Figure 9.10(b), if the molecular orientation direction of the SmA phase is made monostable, the orientation direction changes symmetrically depending on the polarity of the applied electric field, so that V-shaped characteristics become possible. In the SmC* case in the preceding paragraph, application of an electric field is indispensable, and control of frequency, amplitude, waveform, *etc.* are also necessary, which may be an obstacle to fabricating a liquid crystal display.

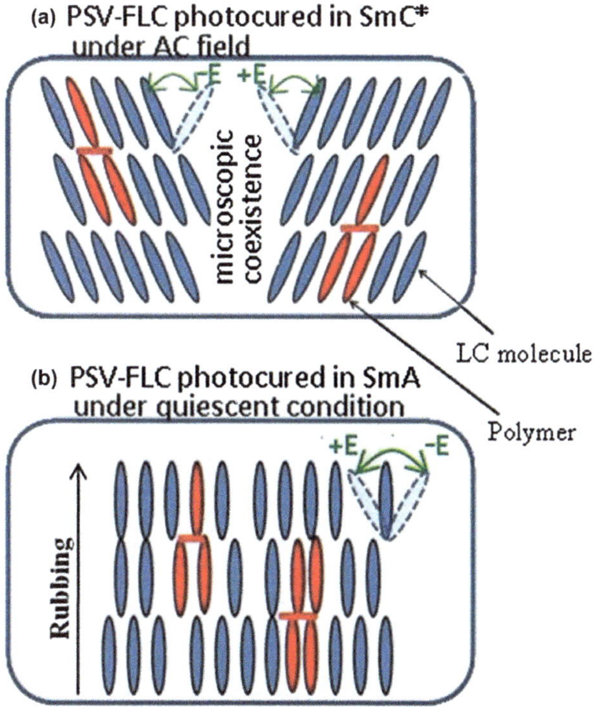

Figure 9.10 Schematic models of alignment structures in polymer-stabilized V-mode FLCs (PSV-FLCs) photocured (a) in SmC* phase under AC electric field and (b) in SmA phase under zero field condition.

With respect to productivity from the viewpoint of practical use, there are advantages in SmA phase polymerization as it does not require the application of an electric field at the time of polymerization and the manufacturing process is relatively simple. The basic principle of FLC monostabilization by SmA phase polymerization is that the molecular orientation direction in the SmA phase during polymerization is maintained in the SmC* phase, but since the phase transition between them is accompanied by smectic layer structure deformation, the mechanism is not simple.

In the case of the horizontally oriented cell, in the SmA phase the smectic layers take a bookshelf structure as shown in Figure 9.11.

As the temperature lowers and the transition to the SmC* phase occurs, the molecules in the layer tilt, thereby narrowing the layer spacing. Then, the layer changes to a chevron structure to suppress the decrease of layer volume.[20] For this structural deformation, the liquid crystal molecules in the bulk region must move in the bending direction of the layer, and the corresponding energy is required. In ordinary FLC cells, the layer does not tilt by as much as the molecule tilts so as to lower this energy, and empirically the bend angle of the layer (chevron angle, δ) is about 80% of the molecular tilt angle (θ) to the layer normal. The difference between them

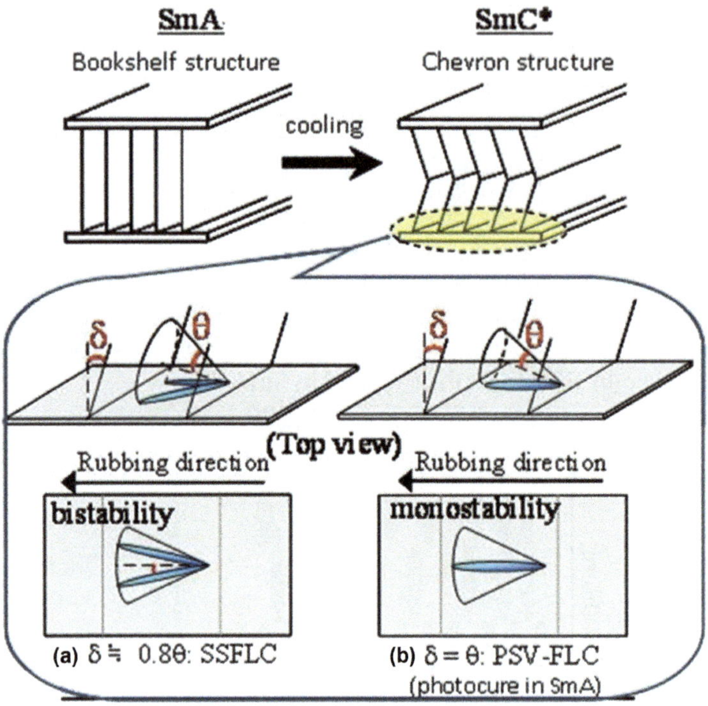

Figure 9.11 Chevron structuresshowing (a) bistability and (b) monostability.

becomes the inclination of the molecule from the rubbing direction in the cell plane (Figure 9.11(a)). As a result, since two stable orientation directions are generated when no electric field is applied, bistability is developed. Therefore, if we can construct the situation of $\delta = \theta$, monostability can be obtained (Figure 9.11(b)). Polymer stabilization by SmA phase polymerization prevents the inclination of molecules in the cell plane by strongly stabilizing the molecular orientation along the rubbing direction in the SmA phase.

References

1. R. B. Meyer, L. Libert and L. Strzelecki, *J. de Phys.*, 1975, **36**, L69.
2. N. A. Clark and S. T. Lagerwall, *Appl. Phys. Lett.*, 1980, **36**, 899.
3. H. Hasebe and S. Kobayashi, *SID Dig. Tech. Papers*, 1985, **16**, 81.
4. T. Tanaka, H. Hasebe and S. Kobayashi, *Proc. Jpn. Disp*, 1986, **86**, 360.
5. T. Makino, Y. Kiyota, T. Yoshihara, H. Shiroto and H. Inoue, *Liquid Crystal Symp. Jpn. Liquid Crystal Soc.*, 1998, 204.
6. T. Uchida, K. Saitoh, T. Miyashita and M. Suzuki, *Proc. IDRC*, 1997, **97**, 37.
7. T. Takahashi, H. Furue, M. Shikada, N. Matsuda, T. Miyama and S. Kobayashi, *Jpn. J. Appl. Phys.*, 1999, **38**, L534.

8. M. Shikada, H. Furue, T. Takahashi and S. Kobayashi, *Mol. Cryst. Liq. Cryst.*, 2001, **368**, 223.
9. H. Furue, T. Miyama, Y. Iimura, H. Hasebe, H. Takatsu and S. Kobayashi, *Jpn. J. Appl. Phys.*, 1997, **36**, L1517.
10. H. Furue, Y. Iimura, H. Hasebe, H. Takatsu and S. Kobayashi, *Mol. Cryst. Liq. Cryst.*, 1998, **317**, 259.
11. H. Furue, T. Takahashi and S. Kobayashi, *Jpn. J. Appl. Phys.*, 1999, **38**, 5660.
12. H. Kikuchi, M. Yokota, Y. Hisakado, H. Yang and T. Kajiyama, *Nat. Mater.*, 2002, **1**, 64.
13. Y. Tanabe, H. Furue and J. Hatano, *Mater. Sci. Eng., B*, 2005, **120**, 41.
14. H. Furue, K. Ikeda and Y. Yamazaki, *Jpn. J. Appl. Phys.*, 2007, **46**, 7132.
15. H. Furue, K. Ikeda and Y. Yamazaki, *J. Photopolym. Sci. Technol.*, 2007, **20**, 19.
16. K. Hibi, S. Kobayashi and H. Furue, *J. Photopolym. Sci. Technol.*, 2012, **25**, 309.
17. H. Furue, H. Yokoyama and S. Kobayashi, *Jpn. J. Appl. Phys.*, 2001, **40**, 5790.
18. H. Furue, T. Takahashi, S. Kobayashi and H. Yokoyama, *Jpn. J. Appl. Phys.*, 2002, **41**, 7230.
19. H. Furue, Y. Koizumi, J. Hatano and H. Yokoyama, *Mol. Cryst. Liq. Cryst.*, 2005, **437**, 195.
20. Y. Ouchi, H. Takano, H. Takezoe and A. Fukuda, *Jpn. J. Appl. Phys.*, 1987, **26**, L21.

CHAPTER 10

Electropolymerisation of (Meth)acrylic Mesogenic Monomers

E. A. SOTO-BUSTAMANTE

University of Chile, Department of Chemistry, Santiago, Chile
Email: esoto@ciq.uchile.cl

10.1 Introduction

Electropolymerisation is the process of obtaining a polymeric material from a monomer by passing an electric current by the application of an electric field. The best known electropolymerisation corresponds to the electrochemical polymerisation of heterocyclic monomers *via* oxidative coupling, to produce intrinsically conductive polymers (ICP). Although less known, electrochemical polymerisation has also been carried out with vinylic monomers. Most of the publications to-date use only small molecule monomers. All electrochemical processes use an electrolyte dissolved either in the monomer or in a monomer/solvent mixture. Depending on the monomer/electrolyte/solvent system, the polymer forms at the anode, cathode or both.

In this chapter we discuss a new electropolymerisation process applied to (meth)acrylic mesogenic monomers.[1] Unlike regular electropolymerisation techniques, our electropolymerisation process is carried out in bulk, both in the isotropic and liquid crystalline states of (meth)acrylic monomers without added electrolytes or initiators. The polymerisation is attained potentiostatically by applying electric fields in the range of 0.5–15 MV μm^{-1} and with

Soft Matter Series No. 8
Polymer-modified Liquid Crystals
Edited by Ingo Dierking

Published by the Royal Society of Chemistry, www.rsc.org

observed currents in the order of tens to hundreds of nA. The monomer can be confined in glass sandwich cells, both with and without superficial alignment layers.

We have not found any record of electrochemical polymerisation of liquid crystals (LC) in solution or bulk. There are cases where the solvent is an LC, which gives a defined morphology to the polymerised monomer.[2–4] In these cases, the monomers used are conjugated heterocycles to obtain ICPs. While there are many systems where an LC monomer is polymerised under the influence of an electric field, the field applied has only been described as necessary to align the LC monomers but has not been credited as involved in the polymerisation induction. The polymerisation is carried out either thermally or photochemically.[5–8]

For a better understanding of our electropolymerisation process and the previously studied ones, it is pertinent to present and discuss the current state of the art.

10.2 Polymerisation Mechanisms

10.2.1 Plasma Polymerisation

While electrochemical polymerisation is directly associated with electrochemical systems, we find it worthwhile to introduce an older type of electrically induced polymerisation, the plasma polymerisation, which has been nicely reviewed by J. Friedrich.[9,10] This polymerisation process dates back to 1796 with N. Bondt *et al.* who used a plasma to produce a polymer from mineral oil. Later, in the 1860s, P. de Wilde,[11] M. Berthelot[12] and P. A. Thenard[13] obtained polymers as sub-products when transforming methane to acetylene in a discharge arc. Many years later, in the 1950s, Schüler and Reinebeck[14–16] revisited this synthesis. In the 1960s, research focused on the kinetics of the process. The preferred regime was glowing discharge.[17] Several mechanisms were proposed, such as an atomic mechanism (monomer destruction and recombination of atoms), and a cationic and radical mechanism, in function of the frequency regime of the plasma.[18] The polymers are obtained as polymeric powders, rigid films, oily films and oils.[19] The polymerisation can be carried out both in gaseous monomers and in condensed phase monomers, depending on the reaction's operational parameters such as voltage, frequency, pressure, carrying gases, monomer/gas flow rate, *etc.* Both powder and films show extensive crosslinking, which makes them insoluble and behave as thermosets. On the other hand, the oily fractions are easily soluble, being primarily oligomers, which also show a high degree of branching.[20] Since the films are highly crosslinked and show no holes, they are quite interesting for technological applications such as textiles,[21] microelectronics,[22] tire reinforcement[23] and osmosis membranes,[24] *etc.* However, the fact that these processes are carried out in batches is a drawback that has limited their industrial application, but the functionalizing of surfaces makes them attractive in the biotechnological field.[25–27] This polymerisation process uses harsh conditions which directly ionize the monomers and even destroy them at atomic level to further rearrange into polymeric chains.[10] While our polymerisation process uses high voltage intensities, it is carried out below the dielectric breakdown of the monomer, hence we expect the mechanism to differ greatly from plasma polymerisations.

10.2.2 Electrochemical Polymerisation

The more well-known and properly-named electropolymerisation system is electrochemical polymerisation. This is carried out in solution. Like most electrochemical systems, it needs a solvent and an electrolyte. The monomer may be added to the solution, or may even be used as the solvent. The electrolyte has the role of conducting the current, and it can also undergo electrochemical reactions, acting itself as an initiator species.

10.2.2.1 Electrochemical Polymerisation of Vinylic Monomers

The electroinitiated polymerisation of vinylic monomers has been extensively studied, and has been reviewed by many authors.[28–33] This field was started in 1948 by Rembold *et al.* who studied the polymerisation of methyl methacrylate (MMA) in an aqueous medium with sodium sulfate as electrolyte.[34] Soon other groups studied this process using many other monomer/electrolyte/solvent systems. Wilson *et al.*,[35] Parravano *et al.*[36] and Kern and Quast[37] polymerised other vinylic monomers in aqueous media using strong mineral acids as electrolyte and mercury and lead as electrodes. The Kolbe reduction was also used to produce carboxylate anion-radicals, then used to polymerise vinyl acetate (VA) and methyl methacrylate (MMA).[38] Tetraalkylammonium salts were also used in aprotic media, polymerising MMA, acrylonitrile (ACN) and styrene (ST), when both anionic and cationic polymerisation in the cathodic and anodic compartment, respectively, were observed.[39] While initially it was thought that the monomer was inert electrochemically and that the initiator species was generated from electrolyte electrolysis, Funt demonstrated that the monomer can also be directly electrolysed, as exemplified in the anionic polymerisation of ACN.[40] In the next years many systems were studied, which were compiled by Breitenbach[41] and Olaj.[31]

In most of these systems, the polymer is obtained homogeneously in solution and heterogeneously precipitated or deposited on the surface of the electrodes. In the latter case it acts as an insulator, passivating the electrodes and hampering both polymerisation control and the study of the kinetics and mechanism of the process.[42]

In radical cases, the yields were low, which was explained by formation of the initiator species on the surface of the electrodes and the consequent annihilation of the radicals near to each other.[43] In ionic polymerisation, the propagating species are not annihilated since they are electrostatically repelled. This allows the migration of these species to the bulk where they propagate. The termination is carried out by chain transfer to the monomer or counter ion of the propagating ionic pair, which explains the low molecular weight observed.[44] This can be controlled by choosing an adequate counter ion for the electrolyte used.[45]

The difficult control of the process and the low electrochemical yield obtained made this process unattractive in comparison with traditional polymerisation techniques such as bulk, solution or emulsion polymerisation as an alternative molecular weight distribution tuning tool. However, since in many cases the polymer gets deposited on the electrodes, this technique can be used as a coating technique to protect metallic substrates against corrosion or to strengthen the bond of the substrate with coatings.[46] However, the films obtained are porous, did not cover the electrodes completely and had poor adhesion.

Relating this kind of electropolymerisation to our system, we will focus on methyl methacrylate electrochemical polymerisation since this is more structurally related to our monomers. Electropolymerisation of methyl methacrylate can occur both in aqueous and aprotic media and form from cathodic and anodic reactions.

10.2.2.2 Electropolymerisation of Methyl Methacrylate at the Cathode

Methyl methacrylate has been widely studied and polymerised at the cathode. In the first studies, Wilson *et al.*,[35] Parravano,[36] Kern and Quast,[37] Das and Palit[47] and Tsvetkov and Glotova[48,49] studied it in an acidic aqueous system, reaching the conclusion that the process occurs by a radical mechanism, initiated by hydrogen radical formation from oxidation of protons. The formation of a cathodic polymer in an aqueous medium has also been reported in the H_2SO_4/K_2SO_8 system (Cram *et al.*[50–52]). The described mechanism was radical and attributed to the reduction of persulfate anions into sulfate radicals.

The cathodic electropolymerisation of methyl methacrylate has also been carried out in aprotic media. Methyl methacrylate has been polymerised in the presence of quaternary ammonium salts, observing both a radical and an anionic mechanism.[40,53] The radical mechanism is initiated by the formation of alkyl radicals coming from the quaternary ammonium cation reduction, while the anionic mechanism is caused by direct monomer reduction and dimerisation.

10.2.2.3 Electropolymerisation of Methyl Methacrylate at the Anode

Anodic electropolymerisation of methyl methacrylate has been observed *via* the reduction of the carboxylate ion (Kolbe reaction) in aqueous[54] and aprotic media.[43] There are also reports of MMA electropolymerisation in a sulfuric acid/methanol system.[55] The polymerisation was also radical, and the initiator species was hydrogen sulfate anions (HSO_4^-) which oxidize to hydrogen sulfate radicals ($HSO_4^\bullet$). Polymerisation *via* direct monomer oxidation at the anode should not occur due to its structure.

10.2.2.4 Electrochemical Polymerisation of Intrinsic Conductive Polymers

The most known and currently studied electrochemical polymerisation is the polymerisation of conjugated heterocyclic monomers to obtain conductive polymers. It started with polypyrrole synthesis by Kanazawa in 1980,[56] and it was further expanded to more monomers such as thiophene, aniline and many other structurally related monomers.[57–59] While is it possible to obtain polymers through anodic and cathodic reactions, anodic oxidative coupling is the most used reaction. The mechanism proposed is a step polymerisation of radicals produced electrochemically.[60] This polymerisation has been extensively reviewed[61–66] and while this area has been widely studied due to its technological applications, its mechanisms and conditions show great differences with our system, both in media and monomer structure and will not be further discussed.

10.2.2.5 Electrografting

The interest in electropolymerising vinylic monomers was renewed by the electrografting process reported in 1981 by Lecayon *et al.* They obtained a covalently deposited polymeric film by cathodic electrolysation of acrylonitrile on nickel electrodes in aprotic media.[67] The polymerisation occurs in 2 steps, as pictured in Figure 10.1. In the first peak, chemical bonding of the monomer to the electrode occurs (−1.7 V). At the second peak, the polymer detaches from the electrode and further polymerises in the bulk of the electrolytic cell.[68]

Studies of the mechanism of the process[70] imply that electrografting happens through an anion-radical formation at the electrode surface at the first potential peak. At the second potential peak, the polymers detach as a dianion which propagates in the bulk[69] (Figure 10.2).

This process was extended to several monomers such as ethyl acrylate, methyl methacrylate, trimethylsilyloxy methacrylate and dimethylaminoethyl methacrylate.[71] While a polymer was always observed by electrolysing the system, for a successful electrografting (polymer covalently bound to the substrate), the solvent, electrolyte and monomer chosen were of great importance. A correlation was found between the donor–acceptor character of the monomer and solvent pair.[70] Since the polymer obtained is insulating and the polymer is insoluble in the solvent, the thickness of the films obtained range between 50 and 200 nm.

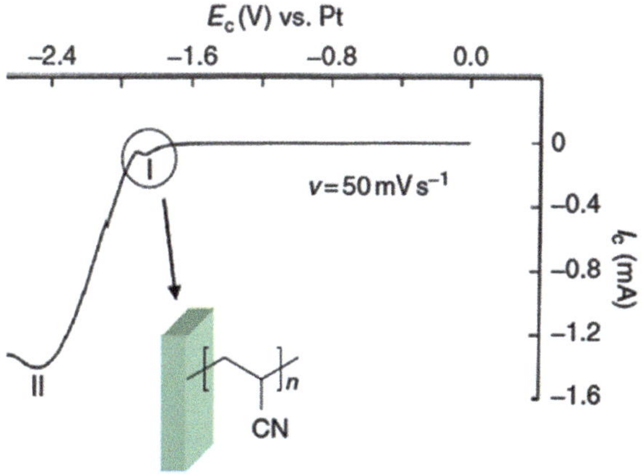

Figure 10.1 Voltammogram of acrylonitrile (AN) reduction with tetramethylammonium perchlorate (TMAP) and nickel cathode. (I) Electrografting polymerisation peak, (II) peak of diffusion polymerisation.[69]
Reproduced from ref. 69 with permission from Elsevier, Copyright 2010.

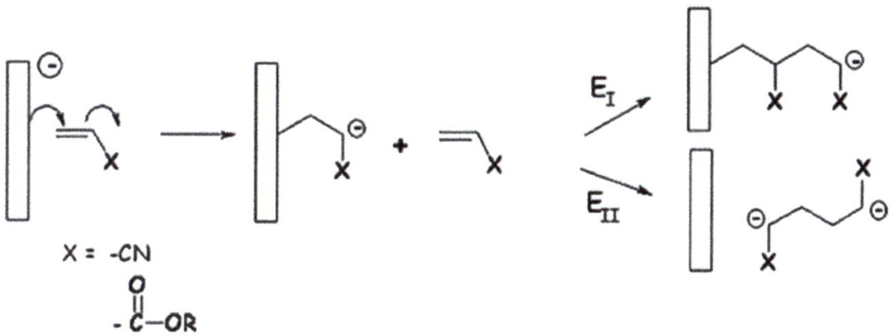

Figure 10.2 Cathodic electropolymerisation mechanism of ACN and MMA according to ref. 69.
Reproduced from ref. 69 with permission from Elsevier, Copyright 2010.

The electrografting process occurs in aprotic media and in the absence of oxygen and water, similar to our bulk system. The initiation is caused by monomer reduction and not by electrolyte reduction, which could make it a similar system, however, in the electrografting system the electrolyte cation has an important role in stabilizing the anion.

While electrografting, being anionic, requires an aprotic medium, strictly dry and without oxygen,[67] electrode coating can also be obtained both in protic and aprotic conditions, as described by Subramanian *et al.*,[72] Tidswell and Mortimer,[73,74] Mengoli *et al.*[75] and Cram *et al.*[50–52] among others, yielding insoluble polymers in known solvents, even though methyl methacrylate should form a linear and soluble polymer and does not suffer from crosslinking. This suggests a possible grafting to the electrode surface. Adsorption of radicals is expected on metallic substrates due to a deficiency of an electron pair.[29]

While methyl methacrylate and related monomers have been electropolymerised in many systems, both protic and aprotic, we expect that in our case the polymerisation mechanism is similar to aprotic systems, having only the organic molten monomer. The mechanism could be either radical, anionic or both. While the polymerisation system has no added initiators, it cannot be ruled out that adventitious species such as oxygen or water among others could act as initiators even in trace amounts.

10.2.2.6 Initiatorless Electropolymerisation in Bulk

Our polymerisation,[1] as mentioned at the start of the chapter, is carried out in the molten monomer, either liquid crystalline (smectic A, C or nematic) or in the isotropic state. The monomer is confined between 2 conductive substrates 3–20 μm apart which could have or not have an alignment inducing layer *e.g.* polyimide. We achieve polymerisation of this system by applying an electric field potentiostatically. For all the monomers studied, we observe a systematic induction time which has been attributed to adventitious dissolved oxygen. The polymer is obtained as both a soluble and an insoluble fraction adhered to the substrate surface of the electrode. The insoluble polymer is obtained at the cathode in cells with alignment inducing layers and on the anode in cells without alignment layers. The process is highly favored by the application of a DC field, occurring only scarcely with an AC field.[76]

While we have studied many (meth)acrylic thermotropic calamitic monomers (shown on Figure 10.3), the most studied system is the acrylic phenyl benzoate family, Figure 10.3c.

Most monomers tried have been polymerised in the isotropic phase of the monomer, in glass sandwich cells with ITO electrodes and planar alignment layers. We observe the formation of a birefringent material in the electrode area, corresponding to a monomer/polymer mixture which phase separates when it reaches a threshold concentration. The material is partially soluble and has been determined to be polymeric in nature by ^{1}H-NMR, FTIR and GPC. XRD studies on these polymers show a behavior consistent with previously studied polymers and monomer/polymer mixtures, obtained by radical means.[77] One of the first cases studied was M6R8 ((E)-6-(3-hydroxy-4-(((4-(octyloxy)phenyl)imino)methyl)phenoxy)hexyl methacrylate), which was shown to be an SmC_2 phase with a molecular length about 1.4 times the monomeric length.

While most of the Schiff bases studied have the imine group internally chelated with the hydroxyl group, they still hydrolysed with time which brought complications to the studies. To avoid these issues, we synthetized two acrylic azobenzene families (Figure 10.3b),[77] which did not electropolymerise as readily as the previously-used Schiff bases. The cause of this

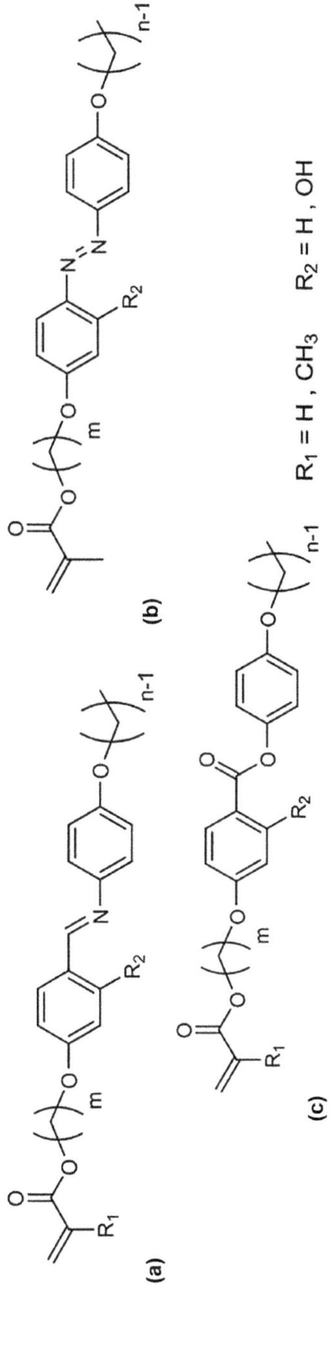

Figure 10.3 General structures of the monomers studied.

will be considered in future studies and we consequently synthesized a family of acrylic phenyl benzoates which we polymerised satisfactorily. With this family we explored systematically the electropolymerisation system.

10.2.2.7 Electropolymerisation of Acrylic Phenyl Benzoates (A6En)

Now we will focus on the electropolymerisation of the acrylic phenyl benzoate family of 4-(alkyloxy)phenyl 4-((6-(acryloyloxy)hexyl)oxy)benzoate, with an *n*-carbon terminal tail with $n = 1$ to 12 (A6En) (Figure 10.3c). Members of this acrylic family have been partially studied before by Portugall, Ringsdorf and Zentel[78] and Horváth, Cser and Hardy.[79,80] Yoshida and Kakuchi also reported the monomers A6E1, A6E2 and A6E6 for use in UV-curable coatings.[81] Only details of members A6E1, A6E2, A6E4 and A6E6 have been published. The rest of the monomers are new. Their polymers have been extensively studied, but are primarily obtained through thermal radical polymerisation.[78–80]

As mentioned before, we know that the polymer deposits on a preferred electrode, as seen in IPS cells and in open sandwich glass cells. This behavior would be in agreement with an anionic polymerisation of the methacrylate moiety as observed in its electrochemical polymerisation in aprotic media, however, we observe an inverse behavior in cells without alignment layers.

Furthermore, it is noteworthy that the polymerisation is carried out at high temperatures and, as will be addressed later, the molecular weight obtained is high (~1–2 MDa) which would agree more with a radical process. As we mentioned before, the induction time observed and the clear inhibition by oxygen on monomer/air interfaces, also agree with a radical process.

To rule out the effect of the cell's material and construction on the polymerisation process, we have constructed cells of glass, calcium fluoride (CaF_2) and zinc selenide (ZnSe) with gold sputtered electrodes. We have also used germanium and vitreous carbon as conductive electrodes. In all cases, both polymerisation and the formation of insoluble polymer on the electrodes was observed.

We have also polymerised ethyl acrylate, dodecyl methacrylate and polystyrene on ITO coated glass cells. In all cases we have obtained insoluble polymer (using tetrahydrofuran and dichloromethane as solvents).

Going back to our most used system, we electropolymerised the whole family in the monomer isotropic state 80 °C/85 °C in INSTEC sandwich glass cells of 8.0 μm gap and antiparallel planar alignment, with a DC voltage of 30 V.

The polymers obtained show the phase diagram seen in Figure 10.4 obtained by DSC and XRD measurements. Smaller monomers (A6E1 up to A6E5) form a nematic phase and a monolayer SmA/SmB phase. Longer monomers (A6E5 up to A6E12) are strictly smectogenic, showing a d-spacing of 1.4–1.5 times the molecular length, which could correspond to a SmA_d phase or a SmC_2 phase.

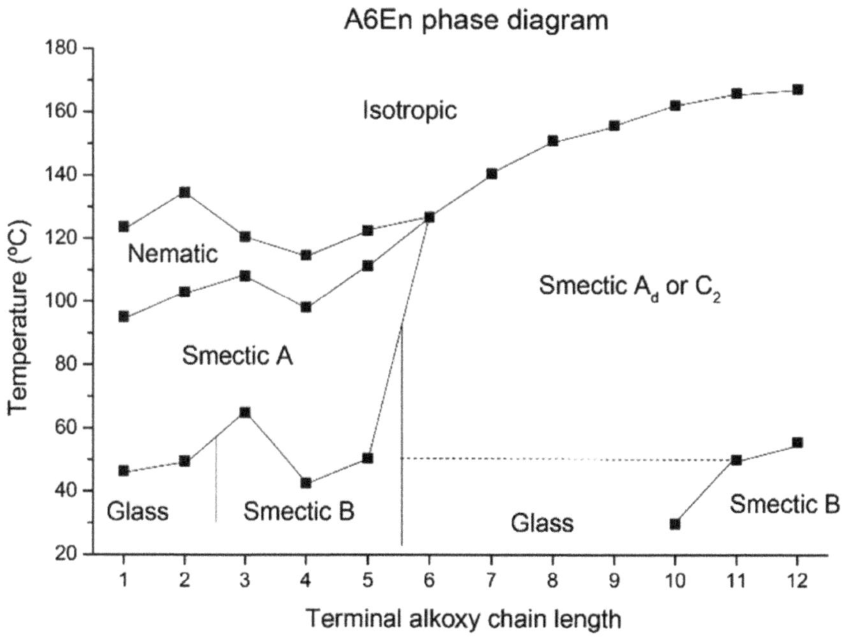

Figure 10.4 Phase diagram of the electropolymerised monomers A6En, obtained by DSC and PXRD.

10.3 Characterization Techniques

10.3.1 Polarized Optical Microscopy (POM)

In all cases, the polymer forms as a birefringent phase which appears from the isotropic monomer. The alignment of this polymer is clearly superior in lower members, however, the best alignment is observed for the monomers A6E3 and A6E4. Figure 10.5 shows the alignment of 3 representative monomers, A6E1, A6E4 and A6E7. We see a worse transmission and extinction of A6E1 and A6E7 compared with A6E4. A6E1 and A6E2 show a disordered nematic phase, for which we are still not clear of the cause. Polymers A6E5 up to A6E12 produce only a smectic phase. The absence of a nematic phase in between the isotropic and smectic phase produces a deficient alignment, showing the typical grainy texture of a smectic phase for polymers.

We have focused the rest of the study on the A6E4 monomer due to its availability and the better alignment of the polymer obtained.

A6E4 has been polymerised in its nematic phase, Figure 10.6, and isotropic phase. In the latter, it has been polymerised at temperatures where the polymer is obtained in its smectic phase at 85 °C (Figure 10.7), nematic phase at 120 °C (Figure 10.8) and isotropic state at 140 °C (not shown).

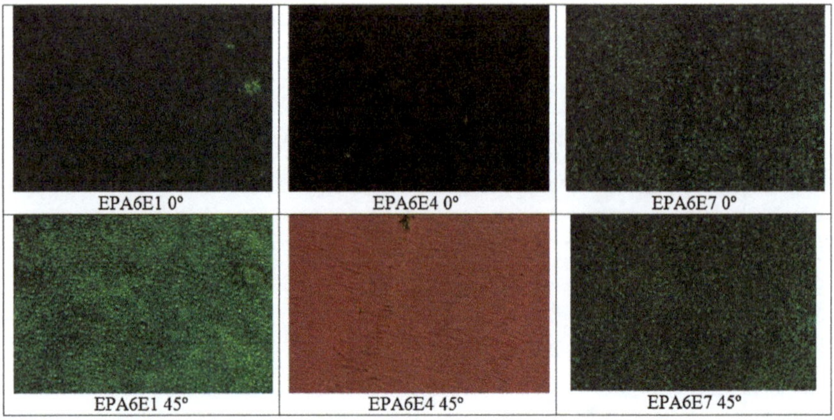

Figure 10.5 POM microphotographs of A6E1, A6E4 and A6E7 electropolymerised in the isotropic state at 85 °C and 30 V. Top row, cells with the rubbing direction parallel to the polarizers. Lower row, cells with the rubbing direction at 45° with the polarizers.

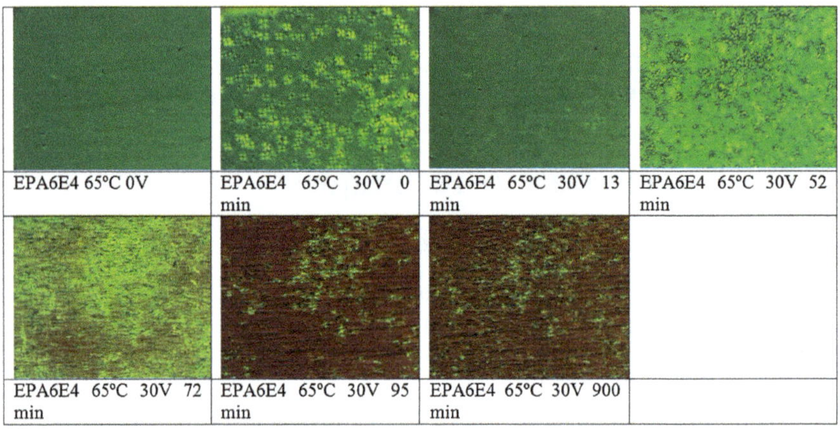

Figure 10.6 POM microphotographs of A6E4 electropolymerised at 65 °C at 30 V for different times of DC electric field applied.

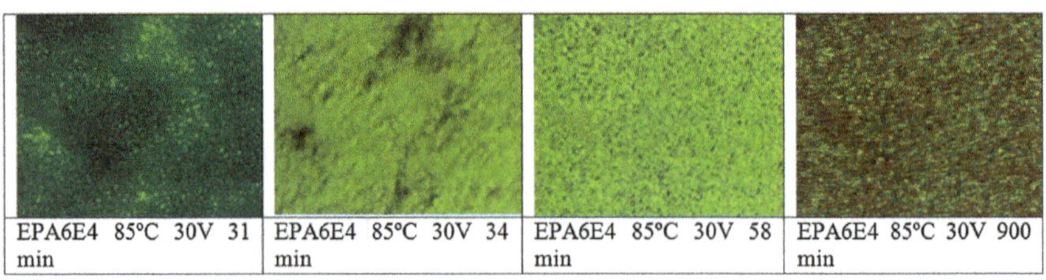

Figure 10.7 POM microphotographs of A6E4 electropolymerised at 85 °C at 30 V for different times of DC electric field applied.

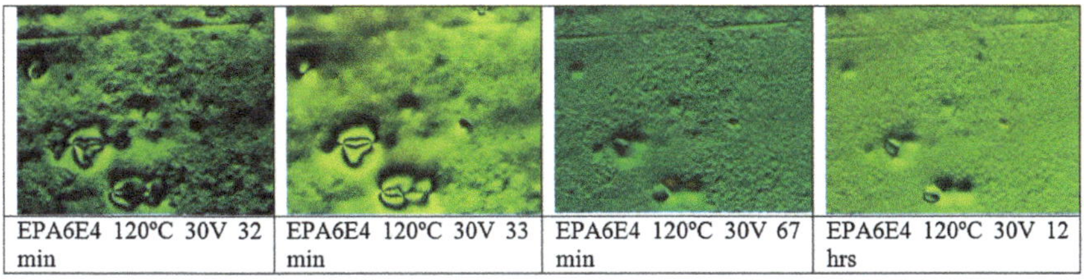

Figure 10.8 POM microphotographs of A6E4 electropolymerised at 120 °C at 30 V for different times of DC electric field applied.

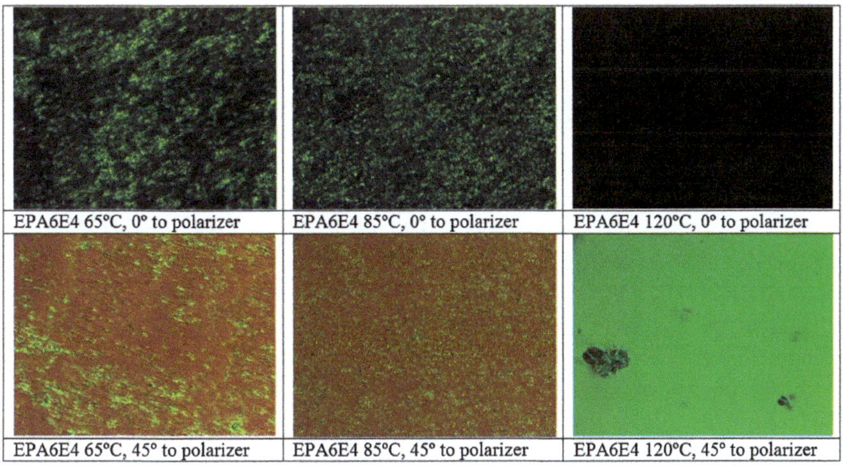

Figure 10.9 POM microphotographs of A6E4 electropolymerised in the nematic (65 °C) and isotropic state (85 °C and 120 °C) at 30 V. Top row: cells with the rubbing direction parallel to the polarizers. Lower row: cells with the rubbing direction at 45° to the polarizers.

By electropolymerising in the nematic phase (Figure 10.6), the monomers distort their alignment due to the effect of the field (electrohydrodynamic effect) but they readily stabilize after some minutes. The nematic phase changes to a smectic A phase. A clear planar-aligned focal conic texture forms with time. With increasing electropolymerisation time, this smectic texture shows an increasing amount of defects.

Polymerising in the polymer smectic phase at 85 °C shows a similar texture (Figure 10.7) however the alignment is more disordered than in the 65 °C case.

Polymerising the polymer nematic phase, we obtained a more aligned phase as shown in the sequence in Figure 10.8. The phase has less defects than in the 65 °C and 85 °C cases, however, it is more disordered than the monomeric nematic phase (Figure 10.6).

Figure 10.9 shows POM images of the polymer obtained at 65 °C, 85 °C and 120 °C. It can clearly be seen that the best alignment is obtained at the temperature where the polymer will be nematic (120 °C), as evidenced by the amount of visible defects and the brightness and darkness of the micrographs when the samples are parallel and at 45° with respect to the polarizers.

At this point it is necessary to mention that a small fraction of the polymer obtained is insoluble and well bound to the substrate as a rather thin film. Polymers obtained by radical polymerisation are totally soluble in tetrahydrofuran (THF) and dichloromethane (DCM). In our case, we always obtain a fraction of the polymer deposited on the cathode which is insoluble and has the ability to maintain its alignment even after being washed and swollen in THF and DCM.

We are not sure if this insolubility is caused by a very high molecular weight, a partial crosslinking or a grafting reaction to the surface of the substrate. We have polymerised this monomer with 365 nm UV light in absence of an initiator, a procedure which has been described by Lee *et al.* for similar monomers.[82] While the polymer, its texture and its molecular weight are similar to that obtained by electropolymerisation, we observe an absence of insoluble material in the UV initiated cells.

We also polymerised A6E4 and A6E12 radically in THF with azobisisobutyronitrile (AIBN), obtaining a polymer fully soluble in THF, DCM and toluene. These polymers show considerably lower molecular weight than the bulk electro- and photopolymerised samples. To ascertain if the insoluble polymer is crosslinked/grafted or just has a very high molecular weight, further experiments are necessary.

We tried to follow the kinetics of this polymerisation method through a range of techniques, but this was unsuccessful. The fact that part of the polymer is insoluble, the difficulty in removing the material from the cells and the fact that the induction time was not very reproducible, made it impossible to plot yield *vs.* time curves. We tried using transmission FTIR with germanium cells, however there were many effects which interfered with the meaningful integration of the 810 cm^{-1} C=C deformation signal of the acrylate, as the signal was modified by the phase transition of the isotropic monomer to the mesogenic polymer. Also, the change in refraction index from the monomer to the polymer continuously modified the baseline interference pattern of the germanium cell.

To extract information about the polymerisation mechanism, we studied the molecular weight distribution by changing parameters such as the intensity and duration of the electric field applied and the temperature.

10.3.2 Molecular Weight Study by Gel Permeation Chromatography (GPC)

Since, as previously mentioned, our studies focus on obtaining a highly aligned side chain liquid crystalline polymer, we studied the A6E4 monomer, since its polymer yielded the best alignment in planar cells. Experiments were performed to see the effect of the electric field, time and temperature on the molecular weight of the obtained polymer.

The molecular weight does not depend greatly on time (Figure 10.10a), which would be a consequence of a radical process. We see a broadening of the molecular weight distribution with a considerably longer polymerisation time (Figure 10.10b), which could be explained by side reactions of the remaining monomer with the already-formed polymer (high weight increase) and polymerisation of remaining monomers with lower molecular weight due to monomer starvation.

The molecular weight does not change greatly with the voltage applied except for at the lower value (10 V) (Figure 10.11a). In the literature, the behavior is mixed. Funt *et al.*[83] and[84] found a decrease of molecular weight with current for MMA, while Márquez *et al.* found an increase in molecular weight with current.[85] The molecular weights obtained are similar and are high, which increases the error in the measurements. Also, the polymerisation is occurring in the phase separation regime (which will be discussed below), which increases the molecular weight obtained by protecting the propagating radical from termination analogous to the Trommsdorff effect.

The most important effect is observed with temperature as shown on Figure 10.11b. We see a sharp molecular weight decrement with increasing temperature. While the monomer is isotropic at 70 °C and above, the mesophase of the obtained monomer/polymer mixture changes. The first big change is observed at 120 °C. At this temperature the monomer/polymer mixture is isotropic for a long time until there's enough polymer to make it turn nematic. At 140 °C the monomer/polymer mixture is isotropic at all compositions.

From Figure 10.11b we see a mixed behavior of the molecular weight distribution (MWD) for samples polymerised at high temperature, which could be caused by the purely thermal behavior of viscosity or due to the order of the mesophases present. To study the effect of temperature and the mesophase on the MWD of the polymer obtained by electropolymerisation, we chose the monomer A6E12 which only shows one mesophase (smectic).

It was found that both temperature and the phase in which the polymer is obtained govern the molecular weight distribution (see Figure 10.12).

At low temperatures (70 °C–100 °C), the polymer segregates directly from the isotropic monomer, producing exclusively a high molecular weight.

At higher temperatures (120 °C and above), the polymer is partially soluble in its isotropic monomer, producing an isotropic polymer. At low polymerisation times, only isotropic polymer is obtained, with a broad and low MWD. By increasing the polymerisation time, the excess polymer formed segregates in a smectic phase and the high molecular weight fraction appears to show a bimodal MWD.

The observed behavior has been previously seen and described for radical polymerisation carried out in the mesogenic state, both thermally[86–88] and photoinitiated.[89,90] The molecular weights for polymers obtained for both

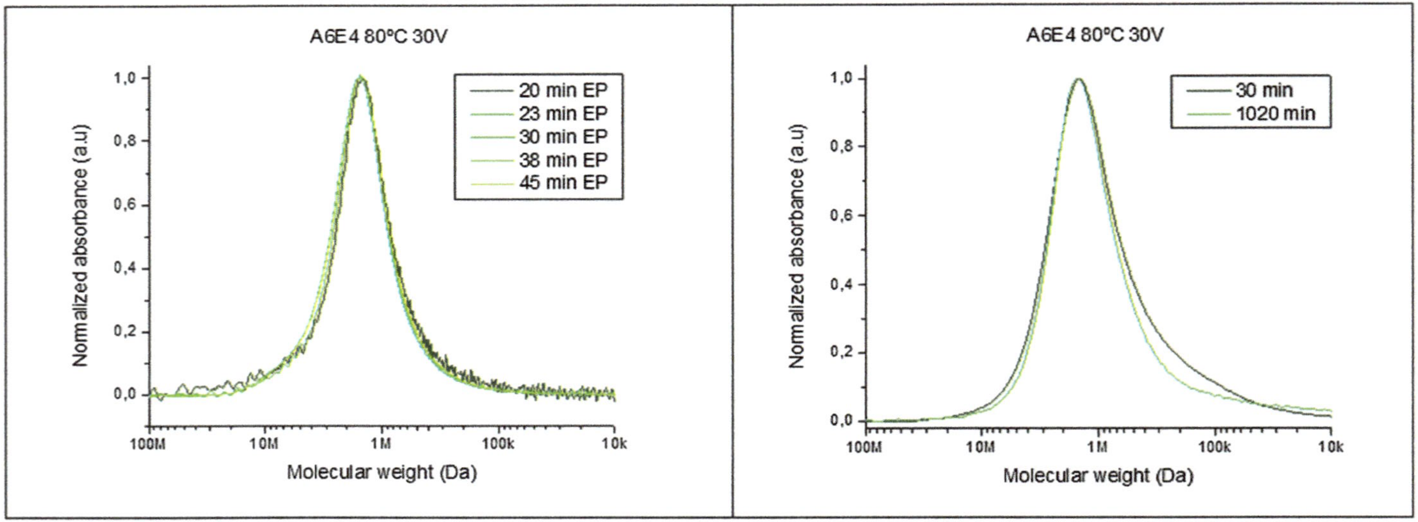

Figure 10.10 Gel permeation chromatograms for A6E4 electropolymerised at 80 °C and 30 V at different times; (a) short time dependence and (b) long time dependence of the molecular weight obtained.

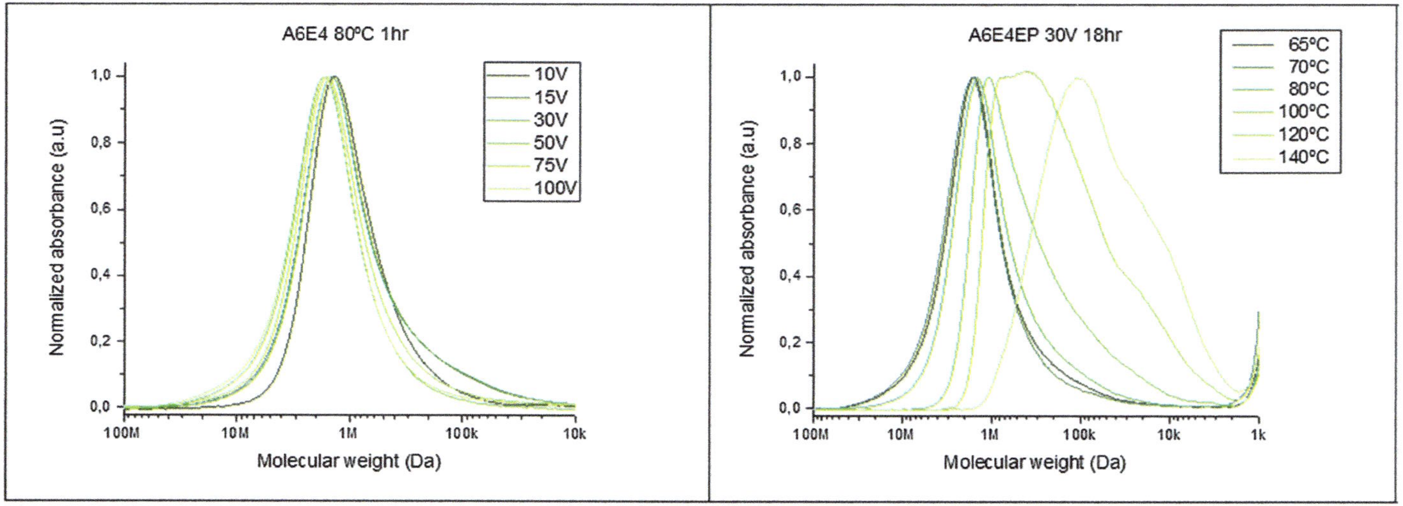

Figure 10.11 Gel permeation chromatograms for A6E4 electropolymerised at (a) 80 °C and different voltages for one hour and (b) different temperatures at 30 V for 18 hours.

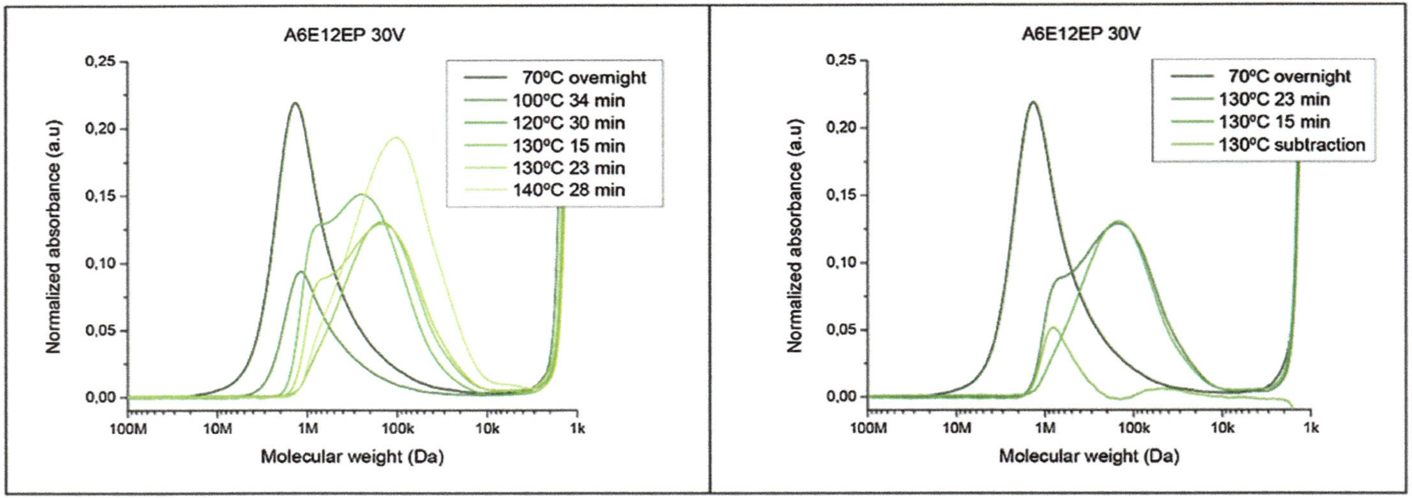

Figure 10.12 (Left) GPC chromatograms for electropolymerised A6E12 at 120 °C and 130 °C at different times. (Right) chromatograms for the 130 °C samples before and after phase separation and the subtraction of both chromatograms.

the nematic and the smectic phase are considerably higher than for the isotropic state.

This behavior has been explained by the segregation of the polymerisable moieties in the smectic layers, which would facilitate chain propagation, and also avoid termination reactions *via* recombination and chain transfer to the mesogenic moiety.

Phase separation in the polymerisation of common acrylic monomers is well-known for trapping radicals, protecting them from termination reactions and thus accelerating polymerisation rates. The propagating radicals being protected from termination increases the molecular weight.[91]

The high molecular weight, which is independent of the polymerisation time, together with a high inhibition by oxygen, implies a radical mechanism. Oxygen inhibition has been observed by polymerisation in open cells with interdigitated electrodes, and at the air–monomer interface on partially filled cells.

Unexpectedly, the molecular weight is not greatly affected by the strength of the field applied, which could be explained by a surface limited initiation process. The process shows an induction time which is attributed to adventitious dissolved oxygen in the sample. This is consistent with another observation: after the process has started, it can be paused by removing the field. By reapplying the field the process can be instantaneously resumed, not showing another induction time.

Finally, we address the mesophases observed in the polymers obtained, and their alignment with the substrates.

10.3.3 X-ray Diffraction

10.3.3.1 Powder X-ray Diffraction (PXRD)

X-ray powder diffraction from samples removed from the cells shows that the polymer obtained from short tail monomers exhibits a nematic, smectic A and smectic B phase, as pictured in Figure 10.13a, while longer polymers from longer tail monomers yield only smectic phases (Figure 10.13b).

A6E5 is a limiting case, and shows both phases are obtained (Figure 10.14a). Once it is isotropized, it only shows a monolayer smectic A/B structure by cooling (Figure 10.14b).

10.3.3.2 Grazing Incidence Wide-angle X-ray Scattering (GIWAXS)

To gain insight both on the alignment of the samples and to properly identify the phases of the polymer obtained, we carried out Synchrotron GIWAXS on two samples. The measurements were carried out on open cells on the cathode substrate, for beam incidence perpendicular and parallel to the rubbing direction of cells with polyimide (PI) layers which induce planar alignment.

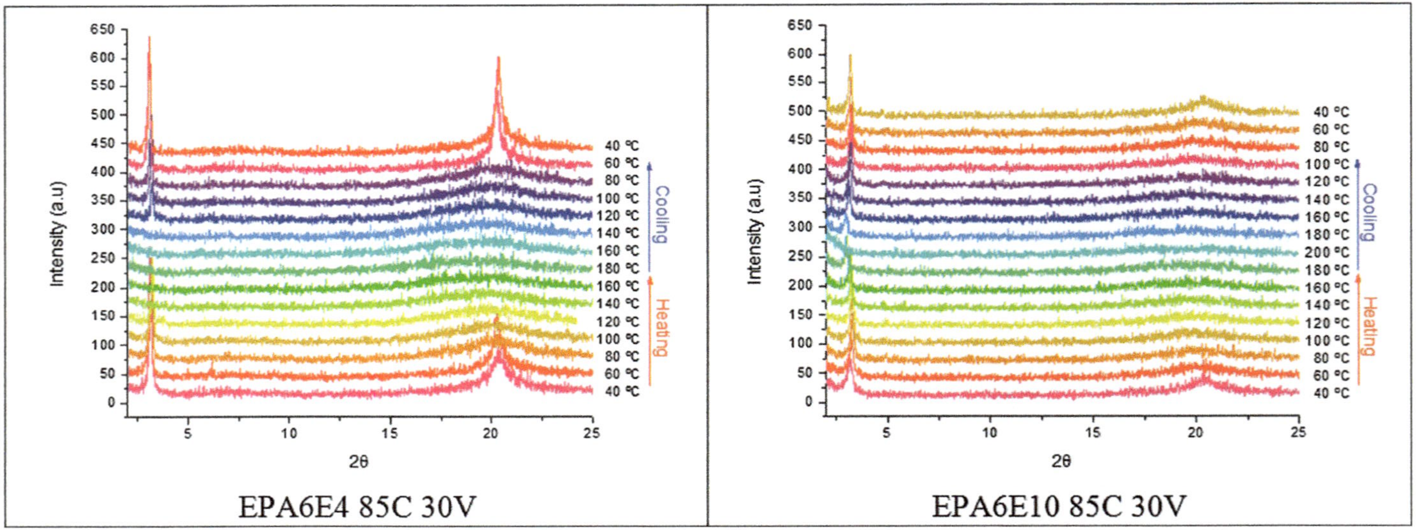

Figure 10.13 PXRD patterns with temperature by heating and cooling for (a) EPA6E4 and (b) EPA6E10.

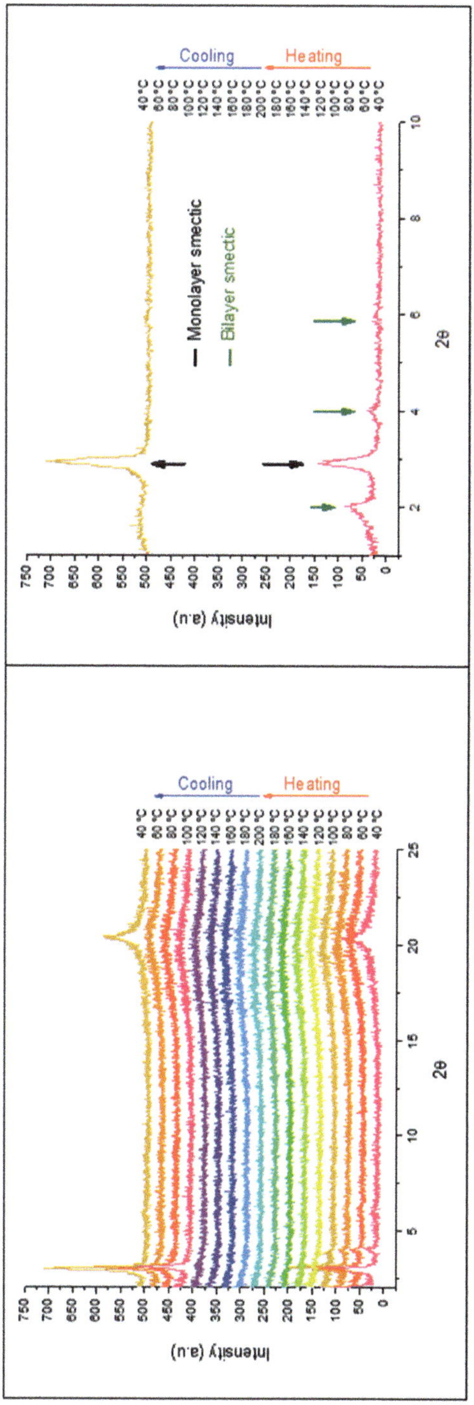

Figure 10.14 (a) PXRD patterns with temperature by heating and cooling for A6E5; (b) emphasis on the low angle area with a mixed phase at the start and an exclusive monolayer phase by cooling.

Figure 10.15a shows the GIWAXS image obtained for the electropolymerised A6E4 sample with the beam perpendicular to the rubbing direction. It shows a clear smectic B phase at room temperature with many low angle reflexes (up to the 006 order visible). The high angle reflex is perpendicular to the lamellar reflexes, as is expected for orthogonal phases.

Figure 10.15b shows the GIWAXS scattering with beam incidence parallel to the rubbing. We observe almost no lamellar reflexes, which shows the good alignment of the smectic phase induced by the rubbing direction of the PI layers. The high angle distance shows an isotropic arc, suggesting that there are no correlations between the layers, and the d-spacing agrees with that observed in Figure 10.15a. The 2D pattern agrees well with the study of Davidson[92] on a methacrylic homolog (PM6AOC4H9).

For the EPA6E12 sample, Figure 10.16a with beam incidence perpendicular to the rubbing direction shows fewer reflexes at low angles; the signals of 002, 003 and 004 are visible and are tilted with respect to both the substrate and the high angle distance reflex (010). The 001 reflex is buried under the beam stopper.

The tilt angle observed from GIWAXS of the EPA6E12 is 28° which is too small to account for the whole interlayer d-spacing of 1.5 times the molecular length of the monomer, which would need a tilt angle of 41°. Hence the layer thickness observed is caused both by a tilted phase and interdigitation of the terminal alkoxy tails.

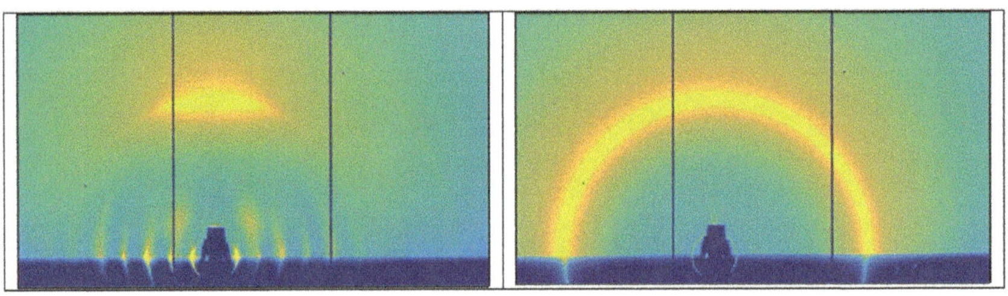

Figure 10.15 GIWAXS of EPA6E4 with incidence perpendicular and parallel to the rubbing direction.

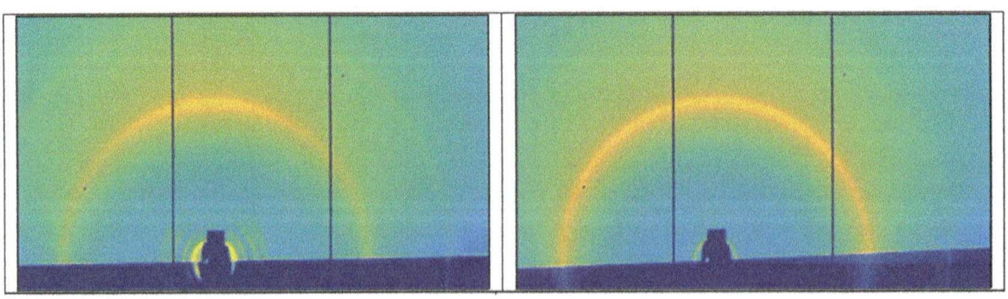

Figure 10.16 GIWAXS of electropolymerised A6E12 with incidence perpendicular and parallel to the rubbing direction.

Figure 10.16b, with the beam incidence parallel to the rubbing direction, shows slightly stronger low angle reflexes, due to a poorer alignment compared with the A6E4 sample. The high order reflex is isotropic which shows that there are not interlayer correlations or that the sample is disordered enough for them not to be visible.

With this technique it is clear that the polymers of members A6E1 up to A6E5 of this monomeric family form a monolayer orthogonal smectic phase with good alignment. A6E6 up to A6E12 members show a tilted smectic phase, which has the molecules parallel to the rubbing direction and smectic layers diagonal across the substrates, as schematized in Figure 10.17.

On further examination of the high angle region of the GIWAXS diffractograms (Figure 10.18) for EPA6E4 and EPA6E12, it is possible to see a coexistence of a high order smectic phase (sharper peak) together with a flexible one (diffuse halo). This behavior resembles semicrystalline polymers, with the coexistence of a crystal phase with amorphous material.

10.4 Electropolymerisation of Mesogenic Acrylic Monomers in Liquid Crystalline Hosts

In this part, we continue the work with a family of azo compounds, which are more stable than the chelated Schiff-bases M6R8. Two families of monomers were synthesized as shown in Figure 10.19.[77]

These families did not polymerize as well as the Schiff bases (Figure 10.3a). They need very high voltages which were near the dielectric breakdown, and longer polymerisation times. The polymer was mostly obtained bound to the cell surface instead of the bulk. This hampered the study of the molecular weight of the polymer obtained and even the identification by ^{1}H-NMR. We used the best polymerising monomer of this series, A6OA12 (Figure 10.19b with a dodecyloxy terminal chain). The hydroxy group in the meta-position to the azo group prevents *E–Z* isomerization, facilitating the optical studies of the polymerisation.

This study focuses on surveying this electropolymerisation process as a technique for *in-situ* polymer formation, as in polymer stabilization vertical alignment (PS-VA) or polymer stabilized ferroelectric liquid crystals (PS-FLC), which are primarily prepared photochemically. We studied the effect of electrolysing the A6OA12 monomer as a host in both positive and negative dielectric anisotropy nematic mixtures. In the ferroelectric case, we used the FELIX M4581-050 mixture. The effect of the polymer formed on the alignment and switching of these LC hosts was explored. We also observed the polymer formed at the electrodes with scanning electron microscopy and studied its morphology.

10.4.1 Electropolymerisation in a Nematic Host

Several mixtures were prepared to evaluate the monomer used in polymer stabilisation in LCs. For the study of the polymerisation in the nematic

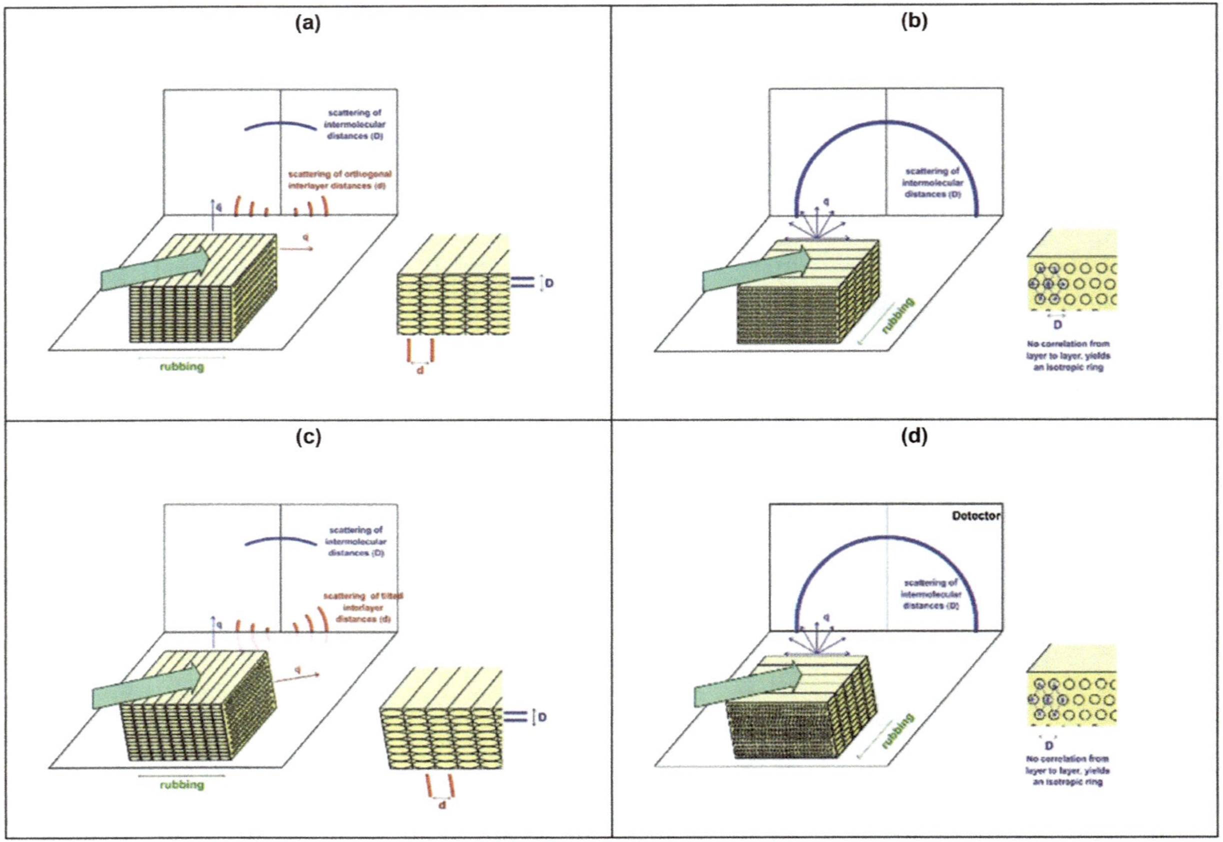

Figure 10.17 Diagram of the scattering produced by the interlayer and intralayer distances of the molecules for EPA6E4 with a beam incidence (a) perpendicular and (b) parallel to the rubbing direction and EPA6E12, with a beam incidence (c) perpendicular and (d) parallel to the rubbing direction.

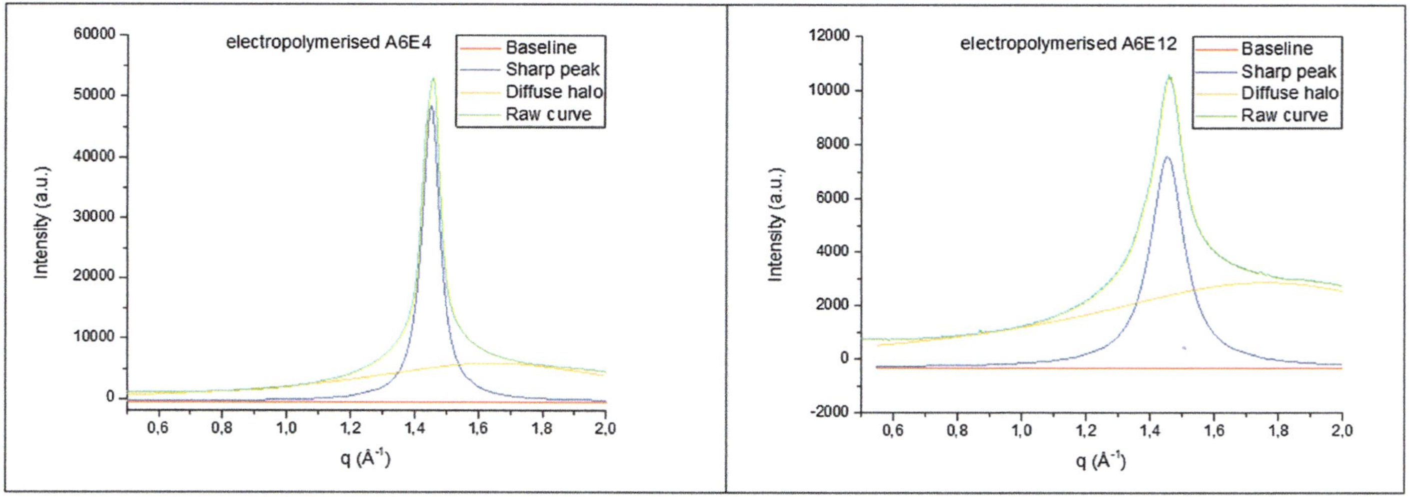

Figure 10.18 Line cuts of the 2D GIWAXS diffractograms of (a) electropolymerised A6E4 and (b) electropolymerised A6E12. Both the raw line cut and Lorentz fits are shown.

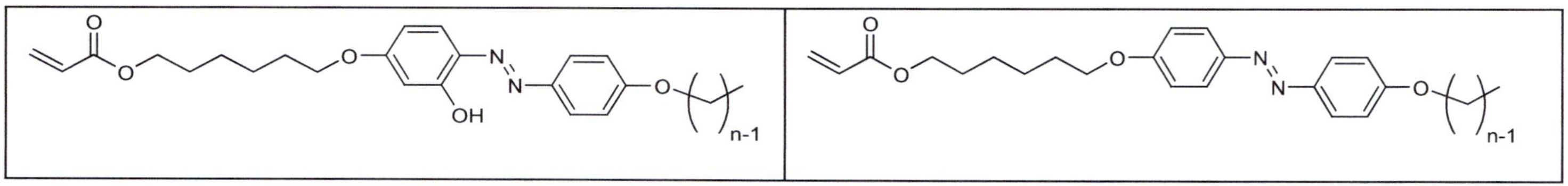

Figure 10.19 Structure of the two acrylic azobenzene families synthesized. (a) (*E*)-6-(4-((4-alkyloxyphenyl)diazenyl)-3-hydroxyphenoxy)hexyl acrylate (A6OAn) and (b) (*E*)-6-(4-((4-alkyloxyphenyl)diazenyl)phenoxy)hexyl acrylate (A6An).

phase, 2 mixtures were used, ZLI1132 with positive dielectric anisotropy and ZLI2806 with negative dielectric anisotropy. The monomer was mixed at 2.5 and 5% in weight and was electropolymerised in the nematic and isotropic phases.

For the ZLI1132 in a 5 μm planar cell, polymerisation was carried out at 30 V in both phases. In the nematic at 65 °C, the polymer formed on the surface of the electrode and acted as a new alignment layer. As the DC field was applied to the mixture forming the polymer, this changed the alignment of the cell from planar to homeotropic, the nematic host being homeotropic when applying the field, acts as a template for the polymer. The monomer was polymerised also in the isotropic state, however when cooling to 65 °C neither a pure homeotropic nor planar texture was observed, but instead an intermediate state (Figure 10.20).

The director realignment was faster in cells with higher concentration (12 hours for 2.5% wt and 3–6 hours for 5% wt) and it also occurred noticeably faster in a thicker self-made cell, with a PVA alignment layer, which was attributable to lower anchoring forces.

The polymer was observed in all A6OA12/ZLI1132 cells polymerised, appearing on both electrodes as a corrugated or rope-like structure, with a periodicity between 500–750 nm and usually found near the spacer used for the cell construction. On a control cell with pure ZLI1132 subjected to the same polymerisation treatment, there were not any structures on the substrate. Since the width of the strands is of the same magnitude as observed in other polymerisation mechanisms in LCs,[93] this suggests that it is due to the polymer growth kinetics (Figure 10.21).

The fact that the change from fully planar to fully homeotropic is gradual, with time of polymerisation, could be used to make tunable tilt retardation layers.

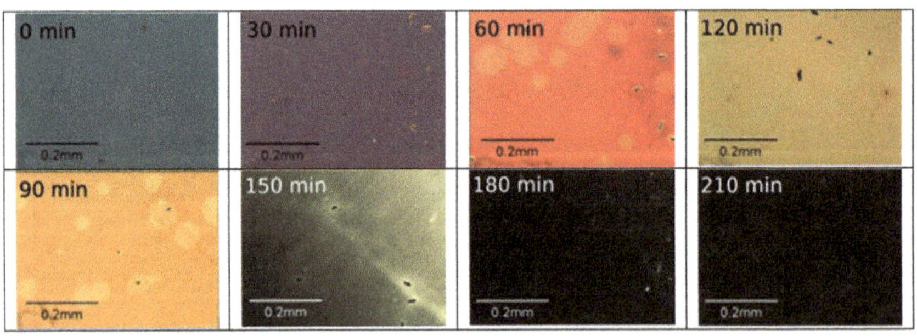

Figure 10.20 Polarizing optical microscopy photographs of a 5% by weight mixture of A6OA12 in a 5 mm AWAT cell with planar alignment, with no applied voltage as the picture was taken but after the cell had been exposed to a DC voltage of 30 V for a time given by the label in the corner of the photos. The alignment direction is 45° to the polarizers and the horizontal.
Reproduced from ref. 96 with permission from the Royal Society of Chemistry.

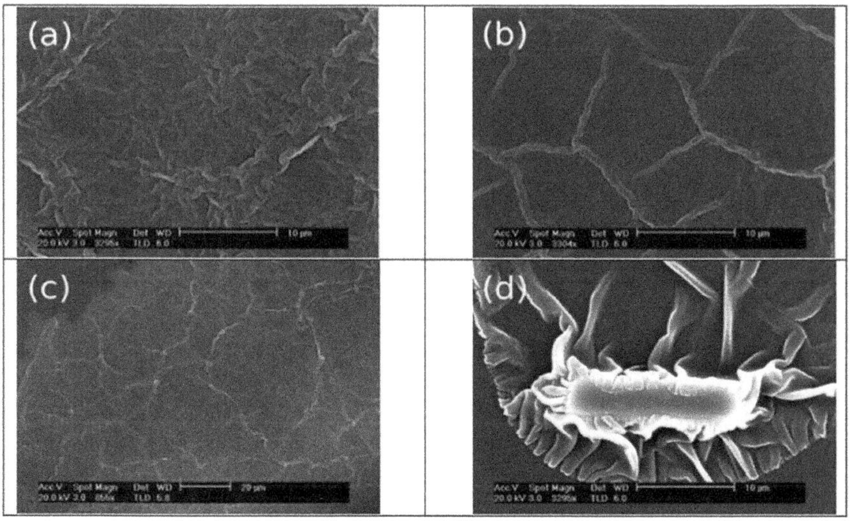

Figure 10.21 SEM images of the polymer network polymerised in the nematic phase of a mixture of 5% by weight A6OA12 in ZLI1132, left under voltage for 150 minutes so that the nematic director was partially realigned. Structures on samples polymerised for longer did not appear to be clearly different to these. Images (a), (b) and (d) are from the anode, and (c) is from the cathode. Image (d) includes a spacer bead.
Reproduced from ref. 96 with permission from the Royal Society of Chemistry.

Regarding the polymerisation of the ZLI2806 mixture in a homeotropic cell, in which a degenerate planar state was forced by the field on, this did not show any sign of polymer formation, either by director realignment or by SEM on the electrode surfaces, in agreement with the observed behavior of the polymerisation of Schiff bases.[76]

10.4.2 Electropolymerisation in Smectic A* and C* (Ferroelectric) Hosts

The monomer was also mixed at 5% wt in the ferroelectric mixture FELIX M4581-050 and was polymerised in both the SmA* (68 °C) and the SmC* (45 °C) phase in planar cells. At both temperatures an effect was observable. Polymerising at 68 °C in the SmA* phase (50 V, 16 hours) produced some streaks which were present in the nematic and in the SmC* phase after heating and cooling, explained by the stabilizing effect of the polymer formed (Figure 10.22).

Polymerising in the SmC* phase at 45 °C (100 V for 16 hours) produced zigzag defects on the electrode area, which remained visible when heating at the SmA*, Ch* and isotropic phases (Figure 10.23).

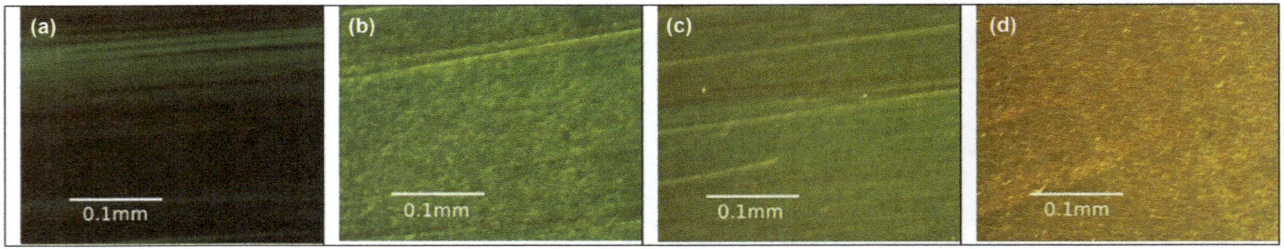

Figure 10.22 Polarizing optical microscopy photographs of a sample of 5% by weight A6OA12 mixed with FELIX M4581-050 ferroelectric liquid crystal. (a) Before polymerisation at 68 °C in the smectic-A phase. (b) At 72 °C, in the nematic phase, after polymerisation for 16 hours at 50 V and 68 °C in the smectic-A phase. (c) The same sample after it was then cooled to 71 °C, still in the nematic phase. (d) The same sample after then being cooled to 30 °C, in the smectic-C* phase, with 10 V DC applied across the cell.
Reproduced from ref. 96 with permission from the Royal Society of Chemistry.

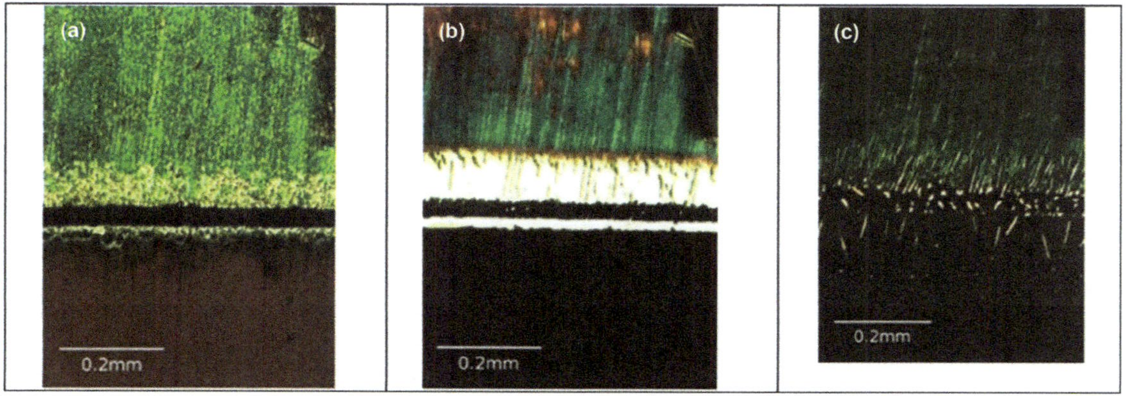

Figure 10.23 Polarizing optical microscopy photographs of the texture of the 5% mixture of A6OA12 monomer in FELIX M4581-050 after polymerisation at 100 V at 45 °C. At the top of the pictures is the area between the ITO electrodes, and at the bottom is the unpolymerised area outside the electrodes. Pictures were taken at (a) 45 °C in the smectic-C* phase, (b) at 71 °C in the chiral nematic phase and (c) at 80 °C in the isotropic phase.
Reproduced from ref. 96 with permission from the Royal Society of Chemistry.

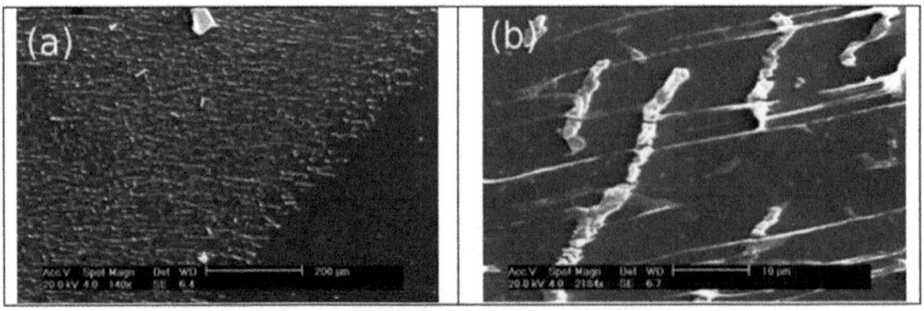

Figure 10.24 SEM images of the polymer network resulting from a 5% by weight mixture of A6OA12 monomer in FELIXM4581-050. (a) Image from the cathode and showing the boundary of the electrode (*i.e.* of the layer of conductive indium tin oxide). (b) Image from the anode, at higher magnification.
Reproduced from ref. 96 with permission from the Royal Society of Chemistry.

The polymer texture seen by SEM differs greatly from that observed in ZLI1132 system. The polymerisation in the SmA* and SmC* phases formed strands aligned with the director of the liquid crystal host with a length of 100–200 μm and 1–2 μm thickness. These thick strands align with the smectic layers in the host LC (around 30° at RT), while some faint streaks are visible in the direction of rubbing (Figure 10.24).

10.5 Conclusions and Outlook

The discussed process seems to be radical in nature, according to the behavior of the polymer formed, as assessed by GPC on variation of time, temperature and voltage. The locus of this electropolymerisation is still not clear. There is the possibility that it forms onto the electrode surface and gets immobilized there as has been described for the electrografting process. It could be also formed in the bulk and get further electrodeposited and immobilized onto the electrodes.

As Lee *et al.*[82] pointed out, for the photoinitiated process, no prior examination has been documented for the initiatorless polymerisation of liquid crystalline materials, up to 2016. However initiatorless polymerisation of these materials may have been observed but not described.[94,95] The same occurs with our described process. Since it has been demonstrated that the application of a DC field is capable of initiating the polymerisation of a reactive monomer, it could be possible that the common photopolymerisation used for instance in polymer stabilized vertical alignment,[8] is not purely photochemically induced. It may be concurrent with a certain degree of electropolymerisation.

The electropolymerisation of mixtures shows that the polymer morphology is affected by the phase and alignment of the host matrix, as shown with nematic mixtures of positive and negative dielectric anisotropy and a ferroelectric smectic C* host. The polymer is formed onto the substrates, and does not increase the switching voltage of the host, which is consistent with a surface polymerisation and not a bulk network through the host material.

The fully planar to fully homeotropic alignment rearrangement with polymerisation time is gradual, which could be used to make tunable tilt retardation layers. The fact that an important part of the polymer is obtained ordered and insoluble after long times of polymerisation, could be used to obtain optical retarder films without added crosslinking agents.

References

1. E. A. Soto-Bustamante and V. H. Trujillo-Rojo, *In Situ Polymerisation Process for Obtaining an Electro-optical Apparatus, Said Polymer and Electro-optical Apparatus, and Uses Thereof*, Universidad De Chile, European, 2013.
2. L. Huang, *et al.*, Polyaniline nanowires by electropolymerization from liquid crystalline phases, *J. Mater. Chem.*, 2002, **12**(2), 388–391.
3. H. Goto, Liquid crystal stepwise electropolymerization – an approach to create insect photonic structure, *RSC Adv.*, 2013, **3**(18), 6347–6355.
4. H. Goto and S. Nimori, Liquid crystal electropolymerisation under magnetic field and resultant linear polarised electrochromism, *J. Mater. Chem.*, 2010, **20**(10), 1891–1898.
5. L. He, *et al.*, In-situ photopolymerization of an oriented chiral liquid crystal acrylate in an electric field, *Polym. Bull.*, 1995, **34**(1), 7–12.
6. N. Yohei, *et al.*, Relation between Monomer Structure and Image Sticking Phenomenon of Polymer-Sustained-Alignment Liquid Crystal Displays, *Jpn. J. Appl. Phys.*, 2011, **50**(5R), 051702.
7. S. Kurihara, K. Iwamoto and T. Nonaka, Preparation and structure of polymer networks by polymerization of liquid crystalline monomers in DC electric field, *Polymer*, 1998, **39**(15), 3565–3569.
8. C. Y. Huang, W. Y. Jhuang and C. T. Hsieh, Switching of polymer-stabilized vertical alignment liquid crystal cell, *Opt. Express*, 2008, **16**(6), 3859–3864.
9. J. Friedrich, *The Plasma Chemistry of Polymer Surfaces: Advanced Techniques for Surface Design*, Wiley-VCH Verlag & Co., Weinheim, Germany, 2012.
10. J. Friedrich, Mechanisms of Plasma Polymerization – Reviewed from a Chemical Point of View, *Plasma Processes Polym.*, 2011, **8**(9), 783–802.
11. M. P. Wilde, Vermischte Mittheilungen, *Ber. Dtsch. Chem. Ges.*, 1874, **7**(1), 352–357.
12. M. Berthelot, Sur l'absoption de l'azole libre par les matières organiques à la température ordinaire, *C. R. Hebd. Séances Acad. Sci.*, 1876, **82**, 1283.
13. P. Thenard and A. Thenard, Acétylène liquéfié et solidifié sous l'influence de l' effluve électrique, *C. R. Hebd. Séances Acad. Sci.*, 1874, **78**, 219.
14. H. Schüler and L. Reinebeck, Über neue Spektren in der Glimmentladung mit Naphthalindampf, *Z. Naturforsch.*, 1951, **6a**, 271.

15. H. Schüler and L. Reinebeck, Über Diacetylen in der Glimmentladung, *Z. Naturforsch.*, 1954, **9a**, 350.
16. H. Schüler and L. Reinebeck, Chemische Reaktionen in der positiven Säule einer Glimmentladung, *Z. Naturforsch.*, 1960, **15a**, 308.
17. H. Kobayashi, A. T. Bell and M. Shen, Effects of Monomer Flow Rate, Flow Configuration, and Reactor Geometry on the Rate of Plasma Polymerization, *J. Macromol. Sci., Chem.*, 1976, **10**(3), 491–500.
18. S. Morita, G. Sawa and M. Ieda, Thin Polymer Films by the Glow-Discharge Indirect Method, *J. Macromol. Sci., Chem.*, 1976, **10**(3), 501–517.
19. H. Kobayashi, M. Shen and A. T. Bell, Effects of Reaction Conditions on the Plasma Polymerization of Ethylene, *J. Macromol. Sci., Chem.*, 1974, **8**(2), 373–391.
20. M. Shen and A. T. Bell, A Review of Recent Advances in Plasma Polymerization, in Plasma Polymerization, *ACS Symp. Ser.*, 1979, 1–33.
21. A. Zille, F. R. Oliveira and A. P. Souto, Plasma Treatment in Textile Industry, *Plasma Processes Polym.*, 2015, **12**(2), 98–131.
22. R. K. Sadhir and H. E. Saunders, Plasma polymerization: A new process for electronics, in 1985 EIC 17th Electrical/Electronics Insulation Conference, 1985.
23. E. L. Lawton, Adhesion improvement of tire cord induced by gas plasma, *J. Appl. Polym. Sci.*, 1974, **18**(5), 1557–1574.
24. A. F. Stancell and A. T. Spencer, Composite permselective membrane by desposition of an ultrathin coating from a plasma, *J. Appl. Polym. Sci.*, 1972, **16**(6), 1505–1514.
25. J. Yang, *et al.*, A High-Throughput Assay of Cell-Surface Interactions using Topographical and Chemical Gradients, *Adv. Mater.*, 2009, **21**(3), 300–304.
26. F. Brétagnol, *et al.*, Direct Nanopatterning of 3D Chemically Active Structures for Biological Applications, *Adv. Mater.*, 2007, **19**(15), 1947–1950.
27. D. Shi, *et al.*, Quantum-Dot-Activated Luminescent Carbon Nanotubes via a Nano Scale Surface Functionalization for in vivo Imaging, *Adv. Mater.*, 2007, **19**(22), 4033–4037.
28. B. L. Funt, Electrolytically controlled polymerizations, *J. Polym. Sci., Part D: Macromol. Rev.*, 1967, **1**(1), 35–56.
29. G. S. Shapoval and A. V. Gorodyskii, Electrochemical Initiation of Polymerisation, *Russ. Chem. Rev.*, 1973, **42**(5), 370.
30. R. V. Subramanian, *Electroinitiated Polymerization on Electrodes*, Springer, Berlin, Heidelberg, 1979.
31. O. F. Olaj, Electrolytically initiated vinyl polymerization, *Makromol. Chem., Macromol. Symp.*, 1987, **8**(1), 235–254.
32. J. W. Breitenbach and C. Srna, Electrochemical initiation of polymerization, *Pure Appl. Chem.*, 1962, **4**(2–4), 245.
33. G. D. Guerra, 24 – Electroinitiated Polymerization A2 – Allen, Geoffrey, in *Comprehensive Polymer Science and Supplements*, ed. J. C. Bevington, Pergamon, Amsterdam, 1989, pp. 453–458.

34. E. A. Rembold, *A Study of Electrochemical Polymerization*, Ohio State University, 1947.
35. E. Dineen, T. C. Schwan and C. L. Wilson, Electrolytic Reduction of Organic Compounds: V. Cathodic Initiation of Vinyl Polymerization in Acrylic Acid, Ethyl Acrylate, and Methacrylate. Carbonyl Reduction in Crotonic Acid, *J. Electrochem. Soc.*, 1949, **96**(4), 226–233.
36. G. Parravano, Polymerization Induced by Hydrogen in Metals, *J. Am. Chem. Soc.*, 1951, **73**(2), 628–630.
37. V. W. Kern and H. Quast, Über die auslösung der polymerisation ungesättigter verbindungen durch aktivierten wasserstoff, *Die Makromol. Chem.*, 1953, **10**(1), 202–220.
38. H. Z. Friedlander, S. Swann and C. S. Marvel, Attempts at the Electrolytic Initiation of Polymerization, *J. Electrochem. Soc.*, 1953, **100**(9), 408–410.
39. V. J. W. Breitenbach, C. Srna and O. F. Olaj, Polymerisationsanregung durch elektrolyse. 2. anionische, kationische und radikalische mechanismen, *Die Makromol. Chem.*, 1960, **42**(1), 171–176.
40. B. L. Funt and F. D. Williams, Electroinitiated anionic polymerization of acrylonitrile, *J. Polym. Sci., Part A: Gen. Pap.*, 1964, **2**(2), 865–880.
41. J. W. Breitenbach, O. F. Olaj and F. Sommer, *Polymerisationsanregung durch Elektrolyse*, Springer, Berlin, Heidelberg, 1972.
42. M. M. Baizer, Recent developments in organic synthesis by electrolysis, *Tetrahedron*, 1984, **40**(6), 935–969.
43. O. F. Olaj, J. W. Breitenbach and F. Sommer, Die Kinetik der durch Acetationen-Entladung angeregten radikalischen Polymerisation von Methylmethacrylat im nichtwäßrigen Medium, *Die Makromol. Chem.*, 1971, **148**(1), 135–154.
44. J. W. Breitenbach and C. Srna, Electrochemical initiation of polymerization, *Pure Appl. Chem.*, 1962, **4**(2–4), 245–254.
45. B. L. Funt and T. J. Blain, Effects of pulses of current on the cationic electropolymerization of isobutylvinylether, *J. Polym. Sci., Part A-1: Polym. Chem.*, 1971, **9**(1), 115–127.
46. B. M. Tidswell, Electrochemical polymerization, in *Macromolecular Chemistry*, ed. A. D. Jenkins and J. F. Kennedy, 1980, The Royal Society of Chemistry, vol. 1, pp. 74–80.
47. M. N. Das and S. R. Palit, A note on polymerization initiated by free radicals obtained by electrolysis, *Sci. Cult.*, 1950, **16**, 34.
48. N. S. Tsvetkov, The polymerization of methyl methacrylate by cathodic hydrogen, *Polym. Sci. U.S.S.R.*, 1962, **3**(4), 570–576.
49. N. S. Tsvetkov and Z. F. Glotova, Effect of the electrolyte composition on electrochemical polymerization, *Polym. Sci. U.S.S.R.*, 1964, **5**(1), 49–54.
50. S. L. Cram, *Electrochemically Polymerised Acrylics as Adhesion Promoting Layers at a Metal/Polymer Interface*, Department of Materials Engineering, University of Wollongong, 2000.
51. S. L. Cram, *et al.*, Mechanism of electropolymerisation of methyl methacrylate and glycidyl acrylate on stainless steel, *Electrochim. Acta*, 2002, **47**(12), 1935–1948.

52. S. L. Cram, *et al.*, Electrochemical polymerization of acrylics on stainless steel cathodes, *J. Appl. Polym. Sci.*, 2003, **87**(5), 765–773.
53. N. Yamazaki, I. Tanaka and S. Nakahama, Electroinitiated Anionic Polymerization of Vinyl Monomers, *J. Macromol. Sci., Chem.*, 1968, **2**(6), 1121–1137.
54. W. B. Smith and H.-G. Gilde, The Kolbe Electrolysis as a Source of Free Radicals in Solution II. The Polymerization of Some Vinyl Monomers, *J. Am. Chem. Soc.*, 1960, **82**(3), 659–661.
55. A. Aurizi, V. Filippeschi and G. Pistoia, Anodic polymerization of methyl methacrylate in $CH_3OH–H_2SO_4$ solutions, *J. Appl. Electrochem.*, 1977, 7(2), 139–145.
56. K. Keiji Kanazawa, *et al.*, Polypyrrole: An electrochemically synthesized conducting organic polymer, *Synth. Met.*, 1980, **1**(3), 329–336.
57. W.-S. Huang, B. D. Humphrey and A. G. MacDiarmid, Polyaniline, a novel conducting polymer. Morphology and chemistry of its oxidation and reduction in aqueous electrolytes, *J. Chem. Soc., Faraday Trans. 1*, 1986, **82**(8), 2385–2400.
58. J. Simonet and J. Rault-Berthelot, Electrochemistry: A technique to form, to modify and to characterize organic conducting polymers, *Prog. Solid State Chem.*, 1991, **21**(1), 1–48.
59. G. Tourillon and F. Garnier, New electrochemically generated organic conducting polymers, *J. Electroanal. Chem. Interfacial Electrochem.*, 1982, **135**(1), 173–178.
60. S. Sadki, *et al.*, The mechanisms of pyrrole electropolymerization, *Chem. Soc. Rev.*, 2000, **29**(5), 283–293.
61. R. J. Waltman and J. Bargon, Electrically conducting polymers: a review of the electropolymerization reaction, of the effects of chemical structure on polymer film properties, and of applications towards technology, *Can. J. Chem.*, 1986, **64**(1), 76–95.
62. M. D. Imisides, *et al.*, The use of electropolymerization to produce new sensing surfaces: A review emphasizing electrode position of heteroaromatic compounds, *Electroanalysis*, 1991, **3**(9), 879–889.
63. W. Plieth, 11 – Intrinsically Conducting Polymers, in *Electrochemistry for Materials Science*, Elsevier, Amsterdam, 2008, pp. 323–363.
64. J. Heinze, B. A. Frontana-Uribe and S. Ludwigs, Electrochemistry of Conducting Polymers—Persistent Models and New Concepts, *Chem. Rev.*, 2010, **110**(8), 4724–4771.
65. E. Schab-Balcerzak, *Electropolymerization*, InTech, Rijeka, Croatia, 2011.
66. G. Inzelt, Chemical and Electrochemical Syntheses of Conducting Polymers, in *Conducting Polymers: A New Era in* Electrochemistry, Springer, Berlin, Heidelberg, 2012, pp. 149–171.
67. G. Lecayon, *et al.*, Grafting and growing mechanisms of polymerised organic films onto metallic surfaces, *Chem. Phys. Lett.*, 1982, **91**(6), 506–510.
68. C. Jérôme, 4.36 – Electroinitiated Polymerization A2 – Matyjaszewski, Krzysztof, in *Polymer Science: A Comprehensive Reference*, ed. M. Möller, Elsevier, Amsterdam, 2012, pp. 903–918.

69. S. Gabriel, R. Jérôme and C. Jérôme, Cathodic electrografting of acrylics: From fundamentals to functional coatings, *Prog. Polym. Sci.*, 2010, **35**(1–2), 113–140.
70. N. Baute, *et al.*, "Bonding" electropolymerization: A new electron-transfer strategy for the tailoring of performant polymer-based composites, *Macromol. Symp.*, 1998, **134**(1), 157–166.
71. N. Baute, *et al.*, Electrografting of Acrylic and Methacrylic Monomers onto Metals: Influence of the Relative Polarity and Donor–Acceptor Properties of the Monomer and the Solvent, *Eur. J. Inorg. Chem.*, 1998, **1998**(11), 1711–1720.
72. B. K. Garg, R. A. V. Raff and R. V. Subramanian, Electropolymerization of monomers on metal electrodes, *J. Appl. Polym. Sci.*, 1978, **22**(1), 65–87.
73. B. M. Tidswell and D. A. Mortimer, Electropolymerization: direct film formation on metal substrates-I, *Eur. Polym. J.*, 1981, **17**(7), 735–744.
74. B. M. Tidswell and D. A. Mortimer, Electropolymerization: direct film formation on metal substrates II locus of polymerization, *Eur. Polym. J.*, 1981, **17**(7), 745–754.
75. G. Mengoli, *et al.*, The inhibition of the corrosion of mild steel in aqueous acids by in situ polymerization of unsaturated compounds, *Corros. Sci.*, 1991, **32**(7), 743–753.
76. P. Romero-Hasler, *Electropolymerization Process and Its Applications on Light Switching Devices*, University of Chile, Chile, 2015.
77. E. A. Soto Bustamante, *et al.*, Antiferroelectricity in achiral mesogenic mixtures of organic materials, *Chem. Phys. Lett.*, 2000, **322**(1), 45–50.
78. M. Portugall, H. Ringsdorf and R. Zentel, Synthesis and phase behaviour of liquid crystalline polyacrylates, *Die Makromol. Chem.*, 1982, **183**(10), 2311–2321.
79. F. C. Judit Horváth and G. Hardy, Polimerizáció folyadékkristályban, XXII. Hexiloxi távtartó csoportot tartalmazó polimerek mezomorf fázisának vizsgálata, *Magyar Kémiai Folyóirat*, 1986, **92**, 481–488.
80. K. N. Judit Horváth, F. Cser and G. Hrady, Polimerizáció folyadékkristályban, XVII. Á mezogén oldallánc lágy szubsztituensek hatása a mezogén polimerek tulajdonságaira, *Magyar Kémiai Folyóirat*, 1984, **90**, 270–280.
81. K. Yoshida and T. Kakuchi, UV curable coatings with mesogenic side groups, *Prog. Org. Coat.*, 2005, **52**(3), 165–172.
82. K. M. Lee, *et al.*, Initiatorless Photopolymerization of Liquid Crystal Monomers, *ACS Appl. Mater. Interfaces*, 2016, **8**(41), 28040–28046.
83. B. L. Funt and K. C. Yu, Electroinitiated polymerization of methyl methacrylate in a homogeneous medium, *J. Polym. Sci.*, 1962, **62**(174), 359–367.
84. G. Pistoia and O. Bagnarelli, Electrochemical bulk polymerization of methyl methacrylate in the presence of nitric acid, *J. Polym. Sci., Polym. Chem. Ed.*, 1979, **17**(4), 1001–1013.
85. J. Márquez, *et al.*, Electrochemically controlled molecular weight of polystyrene production, *Polym. Bull.*, 1990, **24**(4), 451–458.

86. Y. B. Amerik, *et al.*, Polymerization of p-methacryloxybenzoic acid in the liquid crystalline state, *Polym. Sci. U.S.S.R.*, 1967, **9**(12), 2931–2938.
87. B. A. Krentsel and Y. B. Amerik, Radical polymerization in anisotropic media under conditions of weak intermolecular interaction, *Polym. Sci. U.S.S.R.*, 1971, **13**(6), 1526–1546.
88. E. M. Barrall and J. F. Johnson, A Review of the Status of Polymerization in Thermotropic Liquid Crystal Media and Liquid Crystalline Monomers, *J. Macromol. Sci., Part C*, 1979, **17**(1), 137–170.
89. C. E. Hoyle, *et al.*, Liquid crystallinity: Medium effects on photopolymerization rates, *Polym. Eng. Sci.*, 1992, **32**(20), 1490–1493.
90. C. E. Hoyle, *et al.*, Efficient polymerization of a semi-fluorinated liquid crystalline methacrylate, *Polymer*, 1993, **34**(14), 3070–3075.
91. P. Hayden and H. Melville, The kinetics of the polymerization of methyl methacrylate. II. The crosslinked and heterogeneous reaction, *J. Polym. Sci.*, 1960, **43**(141), 215–227.
92. P. Davidson, Liquid Crystalline Polymers Part 1X-ray diffraction by liquid crystalline side-chain polymers, *Prog. Polym. Sci.*, 1996, **21**(5), 893–950.
93. C. V. Rajaram, S. D. Hudson and L. C. Chien, Morphology of Polymer-Stabilized Liquid Crystals, *Chem. Mater.*, 1995, 7(12), 2300–2308.
94. W. Zheng and C. C. Chan, Two-photon absorption in diacrylate mesogens at 632.8 nm wavelength, *Europhys. Lett.*, 2011, **93**(1), 13001.
95. S.-T. Wu, Absorption measurements of liquid crystals in the ultraviolet, visible, and infrared, *J. Appl. Phys.*, 1998, **84**(8), 4462–4465.
96. N. Kasch, I. Dierking, M. Turner, P. Romero-Hasler and E. A. Soto-Bustamante, *J. Mater. Chem. C*, 2015, **3**, 8018–8023.

CHAPTER 11

Polymer-stabilized Antiferroelectric Liquid Crystals and Their Applications

PER RUDQUIST

Microtechnology and Nanoscience, Electronics Materials and Systems Laboratory, Chalmers University of Technology, 412 96 Gothenburg, Sweden
Email: per.rudquist@chalmers.se

11.1 Introduction

Antiferroelectric liquid crystals (AFLCs) generally have smectic structures, with antipolar arrangement of molecular layers, each exhibiting a spontaneous polarization density **P** normal to the smectic layer normal **z**. The application of a sufficiently high electric field **E** normal to **z** switches the material from the antiferroelectric state into a ferroelectric state where the polarization is oriented in the same direction in all layers. When the field is switched off the material relaxes back to the antiferroelectric state.

Until the mid-1990s, only *anticlinic* smectic phases of *chiral* molecules had been shown to exhibit antiferroelectricity. The most important, and most studied, example of anticlinic AFLCs is the so-called smectic C_a* (SmC_a*) phase,[1,2] which soon after its discovery also attracted great interest for use in display screens.[3–5] Later, liquid crystal antiferroelectricity was reported also in phases of (achiral) bent-core mesogens,[6,7] and today synclinic, anticlinic, as well as non-tilted antiferroelectric bent-core smectic materials have been found.[8] The origin of polar order is different for rod-like and bent-core

Soft Matter Series No. 8
Polymer-modified Liquid Crystals
Edited by Ingo Dierking

Published by the Royal Society of Chemistry, www.rsc.org

mesogens; in the former, polar order is a result of tilt of chiral molecules, whereas in the latter the mesogens pack more effectively when ordered in a polar way within a smectic layer.

In this chapter, we focus on SmC_a*-type polymer-stabilized AFLCs, also referred to as *AFLC gels*.[9–11] Recent work on polymer-stabilized bent-core AFLCs[12] indicates that many of the polymer-stabilization effects observed for rod-like molecules in the SmC_a* phase can be expected also for bent-core mesogens.

In sample cells and display-type devices polymer-stabilization has been proven to influence both static and dynamic performance of AFLC materials. In particular it can

(i) stabilize helical as well as helix-suppressed *structures*
(ii) stabilize or destabilize anticlinic and synclinic *states*
(iii) enhance electrooptic grey-scale capabilities
(iv) dramatically increase the electrooptic working speed.

Whereas (i) to (iv) are related to stabilization of *director structures*, AFLC gels have also been discussed in terms of stabilization of the *bookshelf smectic structure* in AFLC devices and hereby make it less sensitive to the impact of mechanical shock and temperature variations.[9,11]

Empirically, the concept of polymer-stabilized liquid crystals is most effective when the reactive monomer and the liquid crystal have very similar molecular structures and the polymerization is carried out *in situ*.[13] It should be pointed out, however, that little is still known regarding the microscopic nature of the polymer network, especially when formed in the anticlinic state. The studied AFLC–monomer combinations are still very few and due to the limited statistical basis, one should be careful when trying to establish general rules for polymer-stabilized AFLCs. We will limit ourselves to some of the observed effects in these few studied AFLC–monomer systems, rather than trying to provide in-depth knowledge on the detailed features of the networks themselves.

The structure of the present chapter is as follows. In Section 11.2 we give an overview of the AFLC display device, together with a brief summary of some seemingly established effects of polymer stabilization of the AFLC structure in device cells. In Section 11.3 we briefly discuss various strategies of polymer-stabilization with the main focus on *in situ* photopolymerization. In Section 11.4 we go into more detail regarding polymer-stabilized AFLCs for display applications. In addition to the examples from the literature, some complementary experiments have been carried out for further illustration and discussion. In Section 11.5 we focus on the influence of polymer-stabilization on some physical properties of AFLC systems *e.g.* the dielectric and effective elastic properties of AFLCs[14] as well as on the influence on Bragg scattering from the polymer-stabilized helical AFLC structures.[15–17] The chapter ends with a brief discussion in Section 11.6.

11.2 The Antiferroelectric Liquid Crystal Display

11.2.1 The Smectic C_a* Phase

In the SmC_a* phase, schematically depicted in Figure 11.1a, the director **n** is tilted the angle θ with respect to the layer normal **z**, with opposite tilt direction in adjacent layers (anticlinic arrangement). Molecular chirality in combination with director tilt makes each smectic layer polar with an electric polarization density **P** normal to the tilt plane spanned by **z** and **n**. The steric coupling between θ and **P** makes the anticlinic structure *antipolar* and *antiferroelectric*. In addition, the molecular chirality leads to the formation of a helical superstructure along **z**. The helix pitch p is often about one micrometer or less in optically pure SmC_a* materials, leading to Bragg reflection of circularly polarized light,[2] analogous to selective reflection in cholesteric liquid crystals.[18]

A typical SmC_a* material, like the prototype material MHPOBC and similar molecules, *cf.* Figure 11.2, may exhibit several smectic phases, but generally no nematic phase, in the phase sequence. On heating, the SmC_a* phase is often directly followed by the SmC* and/or the SmA* phase before the isotropic phase is reached. Some AFLC materials, *e.g.* MHPOBC, also exhibit one or more sub-phases in between these phases, with ferri-, or antiferroelectric behaviour.[2] AFLCs developed for display applications are generally multicomponent mixtures having the SmC_a* phase over several tens of degrees, including room temperature.

11.2.2 The AFLC Display Geometry

The AFLC display device is schematically illustrated in Figure 11.1b. The smectic layers are uniformly arranged perpendicular to the bounding glass plates in the so-called bookshelf geometry.[†] The layer normal **z** – which is the slow axis of the macroscopic optical indicatrix of the anticlinic structure – is oriented along one of the crossed polarizers which gives a dark state for $\mathbf{E}=0$. An electric field $\pm\mathbf{E}$, applied normal to the cell plane, can switch the system from the anticlinic antiferroelectric (*AF*) dark state into the two synclinic, oppositely polarized ferroelectric ($\pm F$) bright states, having **P** along **E**, *cf.* Figure 11.1c. In the $\pm F$ states **n**, and hence the new slow axis, is tilted $\pm\theta$ away from the polarizer axis, depending on the sign of E, and P. The two field-induced $\pm F$ states provide equal brightness, see Figure 11.1c. Maximum brightness in the $\pm F$ states is achieved when $\theta = 45°$ and when these states constitute a half-wave plate, *i.e.* when $d\Delta n = \lambda/2$, where λ is the wavelength of light in vacuum, and $\Delta n = n_{\parallel} - n_{\perp}$ is the birefringence. Here $n_{\parallel}$ and $n_{\perp}$ are the indices of refraction measured parallel and perpendicular to **n** in the F-state, respectively.

[†]The term "bookshelf geometry" is motivated by the fact that as the smectic layers are oriented with respect to the glass plates as books on shelves.

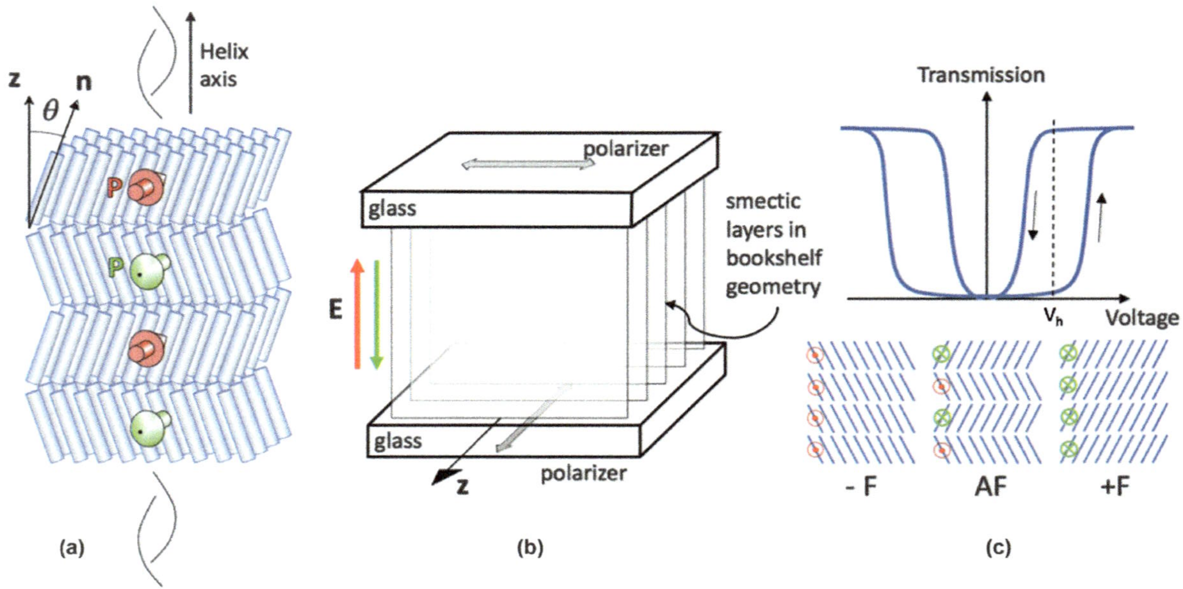

Figure 11.1 (a) Schematic illustration of the antiferroelectric SmC_a* phase with its anticlinic arrangement of rod-like molecules, here represented by cylinders. (b) The basic AFLC device structure. (c) The electrooptic transmission–voltage is characterized by a double hysteresis loop and both the *AF* and $\pm F$ states are stable at the holding voltage $\pm V_h$.

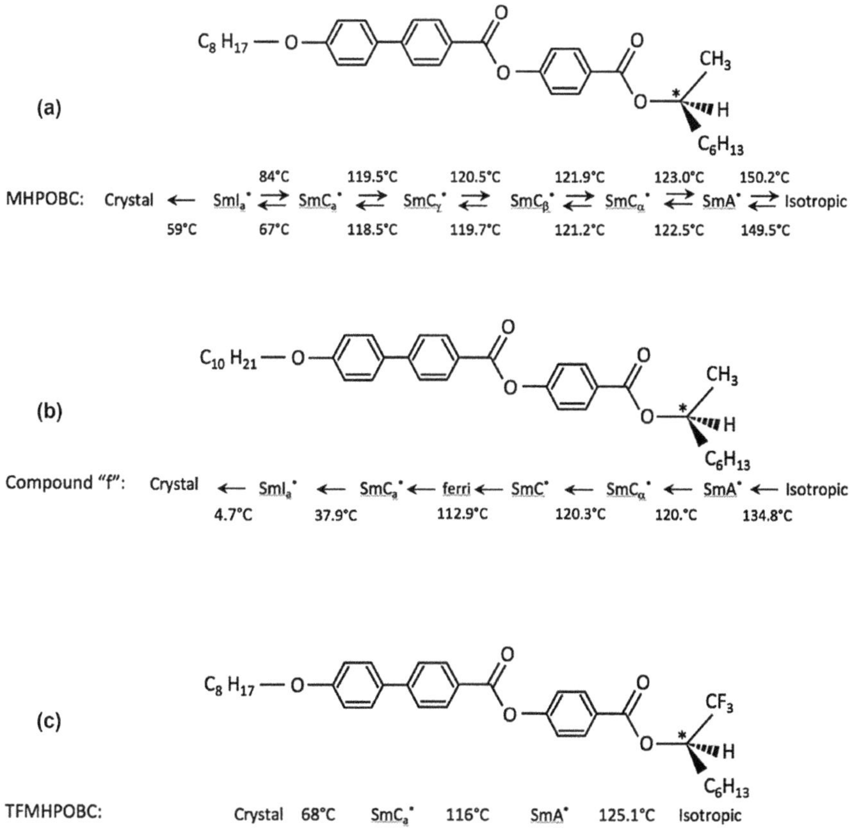

Figure 11.2 (a) Molecular structure and phase sequence of 4-(1-methylheptyl-oxy-carbonyl)phenyl 4′-octyloxybiphenyl-4-carboxylate (MHPOBC), the most studied and most important reference AFLC. Data from ref. 19. A majority of SmC_a* materials have structures more or less similar to the one of MHPOBC, in particular the methyl heptyloxycarbonyl tail which seems to stabilize anticlinic order.[5] Examples are "Compound f" (b) used in ref. 11, TFMHPOBC (c) used in ref. 58, and the orthoconic antiferroelectric mixtures shown in Figure 11.3.

The symmetric double hysteresis transmission–voltage curve (Figure 11.1c) allows for simple DC-compensated drive, with subsequent frames written with opposite sign of voltage. The fact that the electrooptic switching takes place in the form of finger-like bright *F*-domains appearing in the dark *AF* domains allows for grey level generation. After a certain grey level has been written with the data pulse, the grey level set by the fraction of bright and dark domains can be secured over the frame time by the application of a holding voltage V_h, for which both the *AF* and *F* states are stable.

Many AFLC prototype displays have been developed[5] but no AFLC displays have yet been commercialized. The main reason is that the AFLC display technology still suffers from some well-known problems which are primarily related to contrast and switching speed. Interestingly, it seems that many of these problems could possibly be ruled out or solved by means of polymer-stabilization of the AFLC structure.

11.2.3 Contrast

The absence of a nematic phase in the phase sequence makes AFLCs difficult to align by means of conventional techniques such as rubbed polymer surface layers. Therefore, the bookshelf structure generally has inhomogeneities and defects which result in *static light leakage* in the dark state. The bookshelf structure is also sensitive to mechanical shock, just as in ferroelectric LCDs. Moreover, there is a tendency for formation of horizontal chevrons, *i.e.* kinks of the smectic layers in the cell plane, when the AFLC cell is switched by means of electric fields.[20] Hence, the alignment quality also tends to decrease under addressing of the device, causing a further increase in light leakage. One proposed way to overcome the problem of alignment instabilities, just as in FLC devices, is *polymer-stabilization of the bookshelf structure*.[21–24]

In addition, the so-called pretransitional effect,[25–27] giving a small field-induced rotation of the slow axis at moderate applied fields below the *AF–F* transition, gives *dynamic light leakage* under application of the holding voltage. Yu *et al.* demonstrated a suppression of the pretransitional effect and improved contrast in passive matrix addressing conditions in polymer-stabilized AFLC systems.[28]

Both the static and dynamic light leakage can be significantly reduced by using so-called orthoconic antiferroelectric liquid crystals (OAFLCs).[29–32] Examples of OAFLC structures developed by the Dabrowski group are shown in Figure 11.3. In these materials $\theta \approx 45^\circ$ and when surface-stabilized, *i.e.* when the helix is suppressed, and the tilt plane is everywhere parallel to cell surfaces, they become uniaxial with the optic axis normal to the cell substrates. This *orthoconic state* ideally gives full extinction between crossed polarizers, irrespective of smectic layer misalignment. But as we shall see, in OAFLCs, polymer-stabilization is often also required for high dynamic contrast performance.

11.2.4 Switching Speed

In AFLCs the switching from dark to bright ($AF \rightarrow F$) is driven by the electric field E and can be made very fast, of the order of 10 or 100 μs. The relaxation from bright to dark ($F \rightarrow AF$) is generally a much slower process. In some cases, the field-induced F-states can even become metastable which severely reduces both working speed and dynamic contrast. It is well-known that the $F \rightarrow AF$ switching can be made faster by means of so-called reverse pulse voltages,[33] which, however, make matrix addressing significantly more complex. An efficient strategy to overcome the problems of F metastability and slow $F \rightarrow AF$ relaxation is to *polymer-stabilize the (surface-stabilized)*

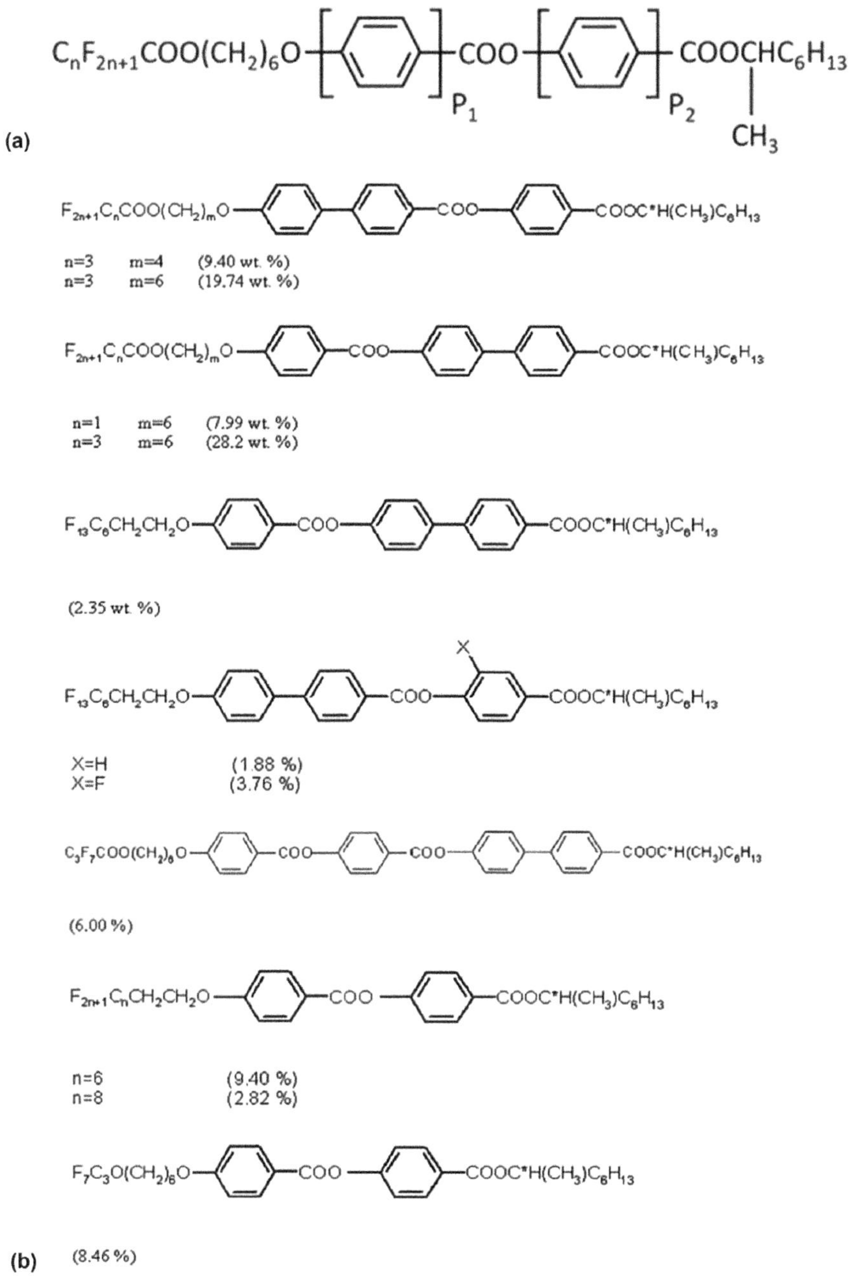

Figure 11.3 (a) Types of molecules used for the orthoconic AFLC mixtures W182.[32] P1 and P2 can take the values 1 and 2, and 2 and 1, respectively with (P1 + P2 = 3). (b) Composition of the 11 component orthoconic AFLC mixture W193B. All individual components have the MHPOBC-tail stabilizing anticlinic order. In addition, the opposing tails are all partly fluorinated.
Part B reproduced from ref. 34 with permission from IOP Publishing.

anticlinic state.[35] Such stabilization has proven to reduce the effective relaxation times from hundreds of milliseconds (or even longer in the case of metastable $\pm F$ states) to a few milliseconds or even sub-milliseconds in orthoconic AFLCs.[36,37]

11.2.5 Memory Type Devices

By performing the polymer-stabilization in the field-induced ferroelectric state, the balance of ferroelectricity and antiferroelectricity is shifted and photo-optical memory type devices based on this effect have been proposed. These architectures exhibit different stability of the $+F$ and $-F$ states[11,38] and sometimes even a stable F state in the absence of applied electric fields (see Section 11.4).

11.3 Polymer-stabilization

11.3.1 *In-situ* Photopolymerization

The common strategy used to include a low concentration polymer network in the AFLC comprises *in-situ* photopolymerization of reactive molecules dissolved into the liquid crystal. A general description of *in-situ* photopolymerization in liquid crystals[39,40] is given in Chapter 3 of this book. Typically, less than about 5% of a photoreactive monomer, together with a small amount of suitable photo-initiator, is added to the low molar mass AFLC material before filling of the device cell. When the desired structure and alignment has been obtained through the action of *e.g.* boundary conditions (surface alignment layers) and external electric or mechanical fields, the sample is irradiated with ultraviolet light of a suitable wavelength for excitation of the photo-initiator and activation of the photopolymerization. In the process, the monomers are cross-linked and form a network inside the liquid crystal.

A common picture of such liquid crystal gels, also confirmed by electron microscopy in some cases,[24] is that the polymer forms an anisotropic network of strands separated from the LC monomer and oriented along the director orientation present during illumination. After cross-linking, the liquid crystal director field in turn tends to follow the fixed oriented strands, which constitutes the basis for the stabilization of the liquid crystal structure. In other words, the anisotropic network works as a distributed "alignment surface" inside the bulk. This picture is certainly plausible for nematic, smectic A and smectic C liquid crystals, where the director field is slowly varying in space. However, one could argue that networks formed in the anticlinic SmC_a* phase, where the director abruptly changes on a nanometer scale between neighbouring layers, could be different, also on a molecular level. But there is no evidence that the networks created in the SmC_a* phase itself are either (even partly) smectic and/or anticlinic.

Artal *et al.* studied systems of "compound **f**" (Figure 11.2b) and three different cross-linkers and also monitored the polymerization process itself with time.[11] Two of the monomers used in their work are mesogenic themselves and one of these, "compound **b**" (Figure 11.4b), has the same core as the host "compound **f**". The third cross-linker has no aromatic rings and is not mesogenic. The cross-linking was carried out in the SmC_a*, SmC*, and SmA* phases and the polymerization rates *vs.* conversion were measured using differential scanning calorimetry (DSC). In all three cases, the polymerization rate increased as the percentage of cross-linker increased and this was attributed to a hindering of the mobility of macroradicals at a higher degree of cross-linking. The non-aromatic cross-linker exhibited about 4 times higher polymerization rate compared to the mesogenic ones. This was attributed to its shorter molecular length (providing higher mobility), and to the fact that the non-aromatic cross-linker, in principle, could also display a different type of segregation within or between smectic layers. These measurements were carried out in open DSC pans in a nitrogen atmosphere, and not in thin sample cells. In Section 11.4 we will instead focus on polymer-stabilized structures in thin AFLC display-type cells and devices.

Most studies of AFLC gels have comprised the use of rod-like mono- and di-acrylate liquid crystalline monomers, exhibiting nematic and/or smectic liquid crystal phases by themselves. There are also mono- and di-acrylate monomers that themselves exhibit the anticlinic smectic phase (*cf.* compound **b** in ref. 11), and antiferroelectric SmC_a*.[41] The mesogenic SmC_a* compounds in ref. 41 have *e.g.* been utilized for the formation of thermodynamically stable pyroelectric polymers, in which all molecules are

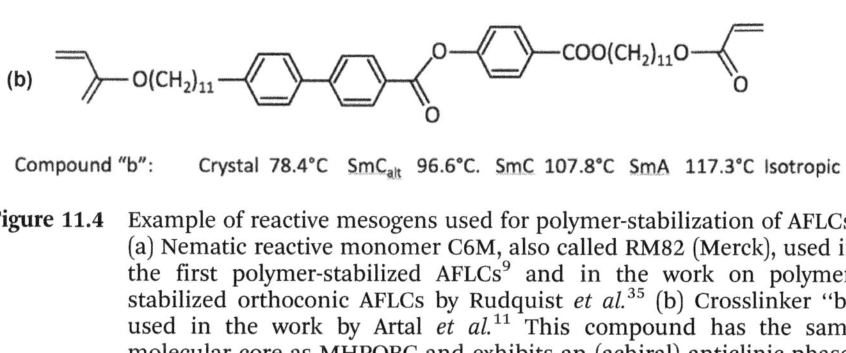

Figure 11.4 Example of reactive mesogens used for polymer-stabilization of AFLCs. (a) Nematic reactive monomer C6M, also called RM82 (Merck), used in the first polymer-stabilized AFLCs[9] and in the work on polymer-stabilized orthoconic AFLCs by Rudquist *et al.*[35] (b) Crosslinker "b" used in the work by Artal *et al.*[11] This compound has the same molecular core as MHPOBC and exhibits an (achiral) anticlinic phase.

cross-linked in the field-induced ferroelectric state. The macroscopic polarization of these pyroelectric materials cannot be switched after polymerization, as the mesogenic parts are all locked-in by the polymer network.

11.3.2 Addition of Polymers

Another strategy for "polymer-stabilization" of low-molecular mass antiferroelectric liquid crystals is to mix them with their related polymers. Nishiyama and Goodby found that in certain chiral acrylate SmC* materials, polymerization has the effect of stabilizing anticlinic/antiferroelectric ordering at the expense of synclinic/ferroelectric ordering.[42] One possibility for this stabilization of anticlinic order is that the side-chain mesogens are believed to be arranged in a zigzag fashion, *cf.* Figure 11.5, which could stabilize anticlinic order at the smectic interphases. The studied polymers were found to be miscible with low molar mass materials and, as also pointed out by the authors, this opens up for the possibility of inducing and stabilizing the antiferroelectric SmC_a* phase in mixtures with low molar mass smectic and AFLC materials.[42] Stabilization of anticlinic order by means of doping with such *anticlinicity-promoting polymers* might also constitute a means to speed up the $F \rightarrow AF$ relaxation in some cases, *cf.* Section 11.4.5, and if so, has important implications for device applications.

Related to this type of stabilization of anticlinic order is the case of induced antiferroelectricity in liquid crystals through the addition of polymers reported by Soto-Bustamente and coworkers.[43,44] They observed antiferroelectric polarization hysteresis behaviour in mixtures of achiral side-chain polymers with their monomers, even though neither of the two counterparts exhibited antiferroelectric behaviour themselves. In this case, the spontaneous polarization was *parallel* to the tilt plane. This is often referred to as

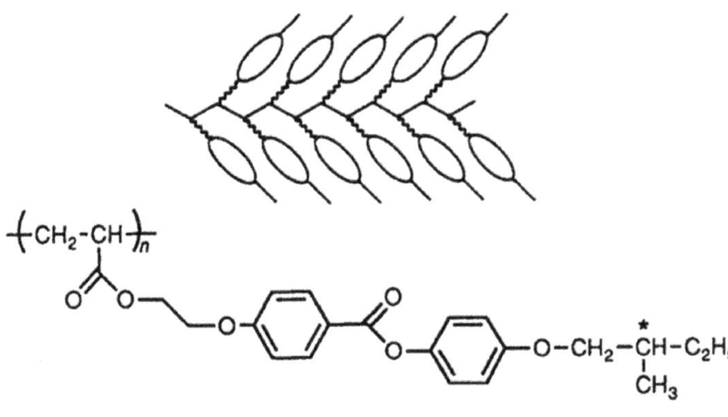

Figure 11.5 Zigzag conformational ordering of polymer liquid crystals may stabilize anticlinic/antiferroelectric structures when mixed with low molar mass (chiral) smectic materials.
Reproduced from ref. 42 with permission from The Royal Society of Chemistry.

"longitudinal polarization" to distinguish it from the "transversal polarization" in chiral SmC^*_a materials. The mixtures of Soto-Bustamente *et al.* exhibit a smectic structure with alternating synclinic and anticlinic arrangement of the mesogens. Freely suspended film experiments on this system later revealed that the tilt order originates from the anticlinic arrangement of the side-chain mesogens at the polymer backbone.[45] The field-induced switching of this longitudinal polarization in sample cells does not, however, give any electrooptical response, as the dielectric tensors have the same shape and orientation in both field-induced ferroelectric states.

11.4 Examples of Polymer-stabilized AFLC Devices

11.4.1 Bookshelf Structure

It is well known that the bookshelf structure, Figure 11.1b, is sensitive to external disturbances and from mechanical shock. In nematic liquid crystal displays disturbances in the director field due to mechanical shock generally heal out by themselves. But this is usually not the case in smectic LCDs. Even a light accidental touch of the display panel can cause the smectic layers to be reoriented over relatively large areas and the uniform bookshelf alignment is hereby destroyed. The first Canon FLC display even had a transparent protective barrier sheet placed a few millimeters in front of the display panel to prevent damage from mechanical shock to the FLC layer. The bookshelf alignment may also deteriorate due to temperature variations, which cause the smectic layer thickness to change leading to the formation of layer chevrons and other inhomogeneities in the bookshelf structure. Changes in temperature might also induce phase transitions to other phases, after which the bookshelf geometry is not fully restored after going back to the SmC_a^* phase. A third effect disrupting the bookshelf structure could be field-induced layer rotation which may occur under asymmetric driving conditions also in AFLCs.[46]

Polymer-stabilization of the AFLC bookshelf structure was first reported by Strauss and Kitzerow in 1996.[9] They used a commercial AFLC mixture CS4000 (Chisso Corporation, Japan) with the phase sequence Cryst −10 °C SmC_a^* 82 °C SmC* 84 °C SmA* 100 °C Iso combined with the reactive liquid crystalline di-acrylate monomer CM6, *cf.* Figure 11.4a. They found that a highly cross-linked polymer network indeed increased the stability of the bookshelf alignment of the liquid crystal. Observations of such polymer enhanced stability of the bookshelf alignment of AFLCs was also reported *e.g.* by Artal *et al.*,[11] Caillaud *et al.*,[36,37] Rudquist *et al.*,[35] and by Atorf *et al.* for bent-core AFLCs.[12]

11.4.2 Surface-stabilization and Helix Suppression

In SmC* materials the helix can be unwound by the surface action which produces a surface-stabilized ferroelectric liquid crystal (SSFLC).[47] As a rule of thumb, in FLC materials surface-stabilization can be realized when the cell gap is similar to, or smaller than, the helix period. Similarly, an SmC_a^* material can become surface-stabilized in small cell gaps. However, in the anticlinic AFLC, the effective period of the helical structure is not the pitch p, but half the pitch $p/2$. Thus, with a similar rule of thumb, the cell gap must be smaller than about $p/2$ to suppress the formation of a helix through surface-stabilization. A theoretical analysis of the conditions for surface-stabilization of AFLCs is found in ref. 48. Moreover, the anticlinic antipolar arrangement of molecules of the SmC_a^* phase is not compatible with any known surface condition. In fact, surfaces generally promote synclinic, synpolar order. Therefore, surface anchoring promoting the SmC_a^* tilt plane being parallel to the surface is inherently weak. This means that the surface-stabilized state is easily lost; a small decrease in the value of the intrinsic SmC_a^* pitch due to a temperature variation might ruin the surface-stabilization. The surface-stabilization can also be lost after switching of the structure; after relaxing back from the *F* state we might get unwanted *AF* states in which the tilt-plane is vertical, or where there is even a partial helical structure.[49] By performing the polymer stabilization in the surface-stabilized state, the formation of unwanted helical and vertical tilt plane states can be prevented, as discussed in ref. 32. The idea is then to carry out the cross-linking at a temperature where the intrinsic pitch is large enough to allow for surface-stabilization. After *in situ* photopolymerization, the polymer network prevents the helix from forming and the helix-free state is secured far away from the cross-linking temperature, also where the pitch should have been too small for surface-stabilization. Obviously, this strategy of polymer-stabilization of a surface-stabilized AFLC is especially attractive for OAFLC devices where the helix-free orthoconic state must be secured in a broad temperature interval.

11.4.3 Polymer-stabilized States

Figure 11.6 shows a microphotograph of a pixelated OAFLC cell between crossed polarizers. The OAFLC material (W193B) is mixed with 4% of a nematic reactive monomer (RM82). A circular area (lower left) has been cross-linked in the surface-stabilized orthoconic state. After heating to the isotropic phase and cooling back to the SmC_a^* phase, the black surface-stabilized *AF* state is immediately restored. Outside the illuminated area the sample is in metastable $\pm F$-states. This reveals that the polymer network stabilizes the anticlinic structure present during cross-linking. Between the pixels in the illuminated area the material was in metastable $\pm F$ states during UV-illumination and here the synclinic states became memorized by the network. Hence, the network stabilizes the tilt- and polar order, *AF*, $+F$, *or* $-F$, that was present during cross-linking. These asymmetric switching and storage effects was first observed by Glossmann *et al.*, who studied the AFLC mixture CS4000 with 8.5% of the reactive monomer C6M, *cf.* Figure 11.7.[38] When the polymerization was carried out under the

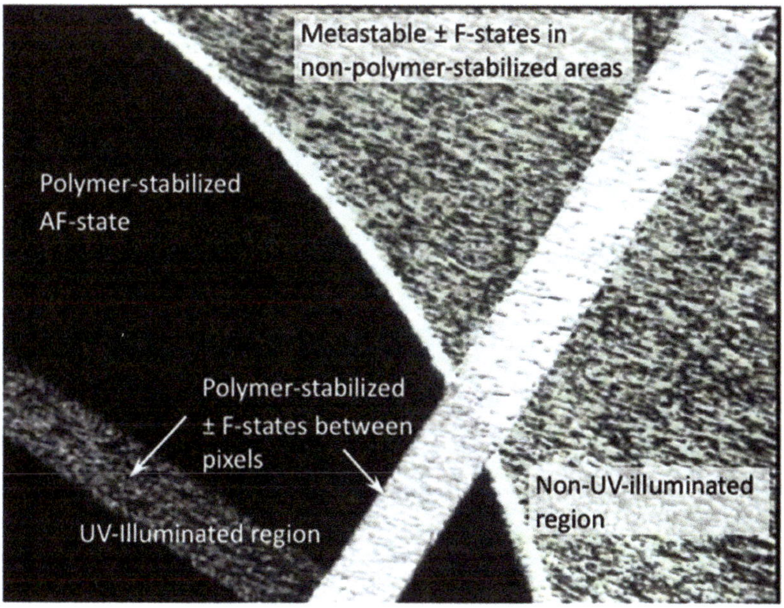

Figure 11.6 Polymer-stabilized orthoconic AFLC device cell.[35] The microphotograph shows a device cell filled with the orthoconic mixture W193B and 4% of the reactive monomer RM82. After filling, the cell was first electronically addressed until the pixel areas obtained the helix-free surface-stabilized orthoconic state, with the tilt-plane parallel to the cell plane. Between the pixels (non-addressed areas), metastable $\pm F$ domains were still present. The electric field was then switched off and a circular region of the cell was illuminated with UV-light. The microphotograph is taken after heating the sample to the isotropic phase and cooling down to the antiferroelectric phase. In the polymer-stabilized region (left) both the surface-stabilized *AF* state (in pixels) and the $\pm F$-states (between pixels) are stabilized. In the non-polymerized area (right) the material is in metastable *F* domains. The experiment shows that both *AF* and *F*-states can be polymer-stabilized in AFLCs. The distance between adjacent pixels is 100 μm.

application of an electric field keeping the material in the $+F$ state, the lobes of the double hysteresis transmission–voltage curves were subsequently shifted towards negative voltages. Similar observations were made by Artal *et al.* using a smaller concentration of the polymer network, *cf.* Figure 11.8.[11] In both these studies, the *AF* state was stabilized when the cross-linking was carried out in the zero field *AF* state.

The polymer-stabilization of states with different polar and tilt-order becomes even more striking in very thin surface-stabilized cells. Figure 11.9 shows transmission–voltage curves of a 0.8 μm thick bookshelf cell with the orthoconic W193B, mixed with 6% of RM82 under the application of a slowly varying (0.1 Hz) triangular voltage. Before cross-linking (b) the field-induced *F* states are metastable as the surfaces promote synclinic order. For

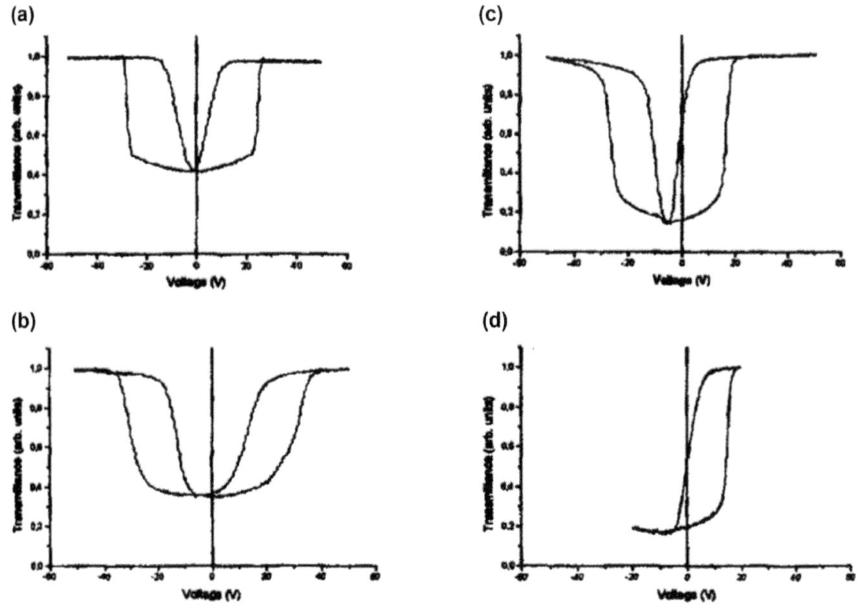

Figure 11.7 Transmission–voltage curve for (a) pure CS4000, (b) CS4000 with 8.5wt% C6M polymerized at $E=0$, (c) and (d) CS4000 with 8.5wt% C6M, polymerized at 23.8 V/2 μm.
Reproduced from ref. 38 with permission from Taylor & Francis.

amplitudes below the *AF–F* threshold (~5 V/0.8 μm) there is essentially no switching and the sample stays in the dark *AF* state. When the amplitude is increased above the *AF–F* threshold, the sample switches between the metastable synclinic ferroelectric states, with only fractions of the sample relaxing to the *AF* state when the field passes through zero. The fraction of domains reaching the *AF* state decreases with faster rate of change of the triangular voltage. The surface-mediated metastability of field-induced *F*-states in AFLC devices and their influence on the electrooptic response are discussed in detail in ref. 50 and 33.

After cross-linking at zero field *i.e.* in the anticlinic *AF* state, Figure 11.9a, the metastability of the *F*-states is gone. Now the whole sample relaxes back to *AF* from the ±*F* states in every half cycle of the voltage and a well-developed double hysteresis loop is obtained. After polymerization the threshold for *AF–F* switching has also increased. Hence, the formed polymer network strongly stabilizes the surface-stabilized dark *AF* state.

When we instead perform the cross-linking with the AFLC material in a field-induced *F*-state (c), the network seems to stabilize the +*F*-state as also observed in ref. 11 and 38. The sample can still be switched between the three states +*F*, *AF*, and −*F*, but the transmission–voltage curve is strongly asymmetric with regards to the applied voltage. In other words, it is easier to induce the +*F* state, in which the cross-linking was carried out, than to induce the opposite −*F* state. Moreover, the +*F* and −*F* states no longer have

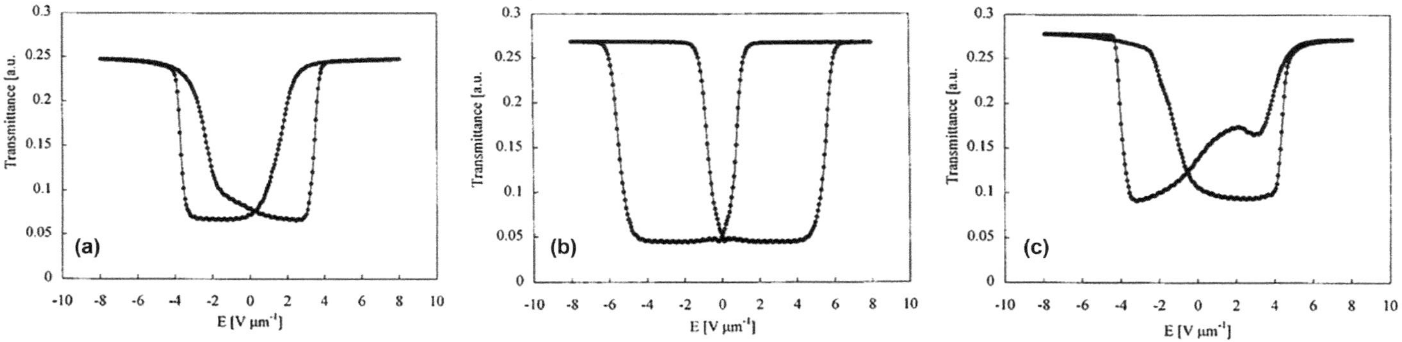

Figure 11.8 Transmission voltages of the MHPOBC homologue "Compound f" and gels of Compound f and 2% anticlinic smectic "Crosslinker b", *cf.* Figure 11.2. (b) Pure Compound f with well-developed AFLC double hysteresis loop. (a) Gel cross-linked in the *AF* state and (c) gel cross-linked under application of electric field. In (a) the *AF* state is stabilized *vs.* the $\pm F$ states. In (c) the TV-curve becomes asymmetric w.r.t. $E = 0$.
Reproduced from ref. 11 with permission from American Chemical Society, Copyright 2001.

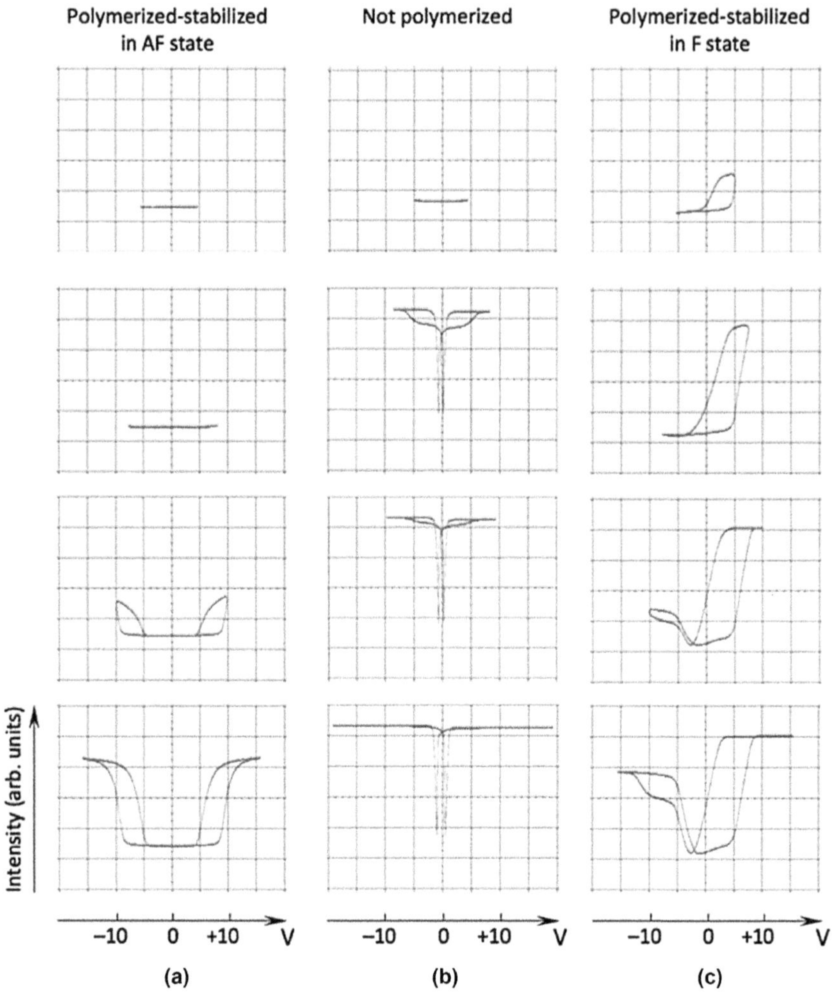

Figure 11.9 Examples of polymer network modified AFLC cells. (a)–(c) show transmission–voltage curves for W193B mixed with 6% RM82 polymer-stabilized under different conditions; (a) polymer-stabilized in the *AF* state. Tri-state AFLC switching with characteristic double hysteresis loop, and full relaxation to the dark *AF* state around zero voltage. (b) Before polymer-stabilization. Strong surface-mediated metastability of the $\pm F$ states. The major part of the sample switches directly back and forth between the two *F*-states and the *AF* state is not achieved during switching. (c) Polymer-stabilized in the (field-induced) $+F$ state. The $AF \rightarrow F$ and $AF \rightarrow F$ switching thresholds are now different, and the TV curve is strongly asymmetric w.r.t. zero volts. Moreover, the effective tilt angle is higher in the "network favoured" $+F$ state than in the "network-disfavoured $-F$ state", which gives different transmission levels in the $+F$ and $-F$ states. Triangular AC voltage, 0.1 Hz. The amplitude of the applied voltage increases in steps between the top and the bottom rows.

the same brightness. This is consistent with the idea that molecules close to the strands are fixed and do not take part in the switching.

In this particular cell with W193B and 6% RM82, we find that the low-frequency threshold voltages (roughly determined from the graphs in Figure 11.9) depend on the polymerization conditions as listed in Table 11.1. Figure 11.10 depicts a similar cell in which the left and right parts were cross-linked in the *AF* and $-F$ states, respectively. Notably, depending on the history of switching, the part cross-linked in the $-F$ state under applied electric fields may adopt the $-F$ or the *AF* state depending on the history of

Table 11.1 Low-frequency threshold voltages measured for the 0.8 μm thick OAFLC cell in Figure 11.9.

Switching	Polymer-stabilized in *AF*	Not polymer stabilized	Polymer-stabilized in $+F$
$AF \rightarrow +F$	+10 V	+5 V	+5 V
$+F \rightarrow AF$	+7 V	~0 V	+2 V
$AF \rightarrow -F$	−10 V	−5 V	−12 V
$-F \rightarrow AF$	−7 V	~0 V	−5 V

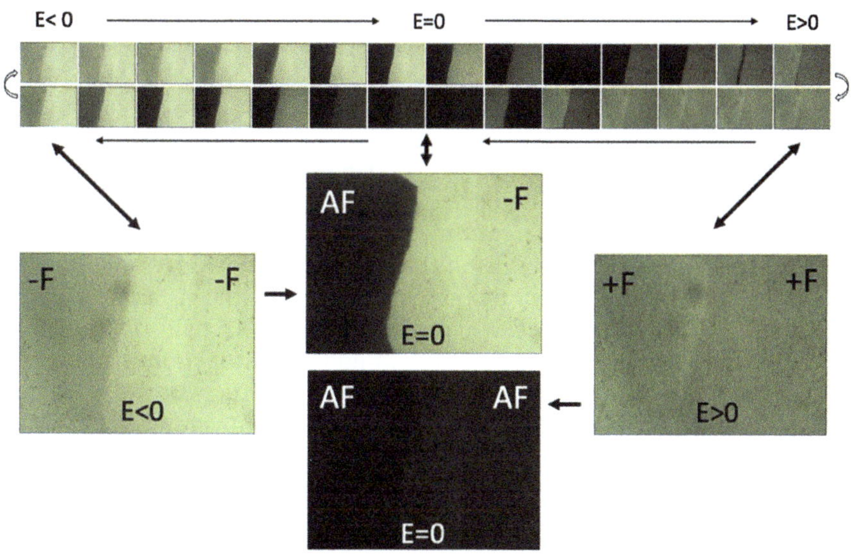

Figure 11.10 Microphotographs of 0.8 μm thick cell of W193B and 6% RM82. The left part of the cell is polymer-stabilized in the *AF* state (at $E=0$) and the right part is polymer stabilized in the $-F$ state. Top: sequence of photographs taken while switching with a 0.1 Hz triangular AC voltage. Bottom: In the right part of the cell, both the $-F$ and the *AF* states are stable at $E=0$ and the adopted state ($-F$ or *AF*) depends on the history of the sample. The left part exhibits a symmetric double hysteresis loop, where the *AF* state is always stable at $E=0$; *cf.* transmission–voltage curves in Figure 11.9 and voltage levels in Table 11.1.

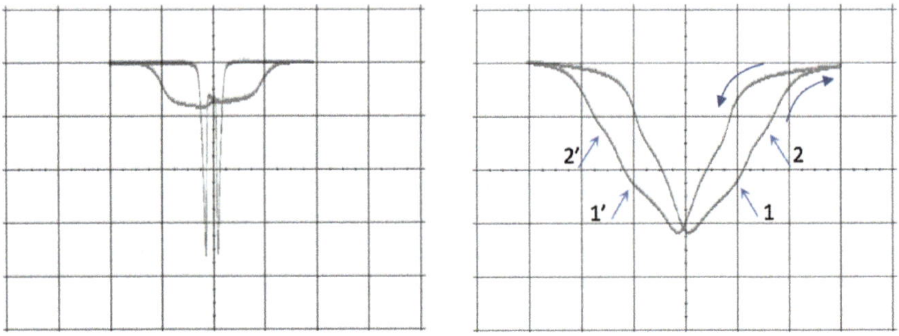

Figure 11.11 Example of cross-linking under AC field: Transmission–voltage curves ($f = 0.1$ Hz) for W193B with 6% RM82 for a triangular voltage before (a) and after (b) cross-linking while applying 100 Hz triangular voltage during UV-illumination. The 1 (1′) and 2 (2′) arrows indicate $AF-F$ switching thresholds for domains likely polymerized in the $+F$ ($-F$) states and in the AF state, respectively; *cf.* Table 11.1. Horizontal divisions represent 5 V, arb. units for transmission (vertical scale). Cell thickness 0.8 μm.

the sample. Hence, both the dark *AF* state and the bright $-F$ state are stable in the absence of applied fields.

The transmission–voltage curves can be modified further by carrying out the cross-linking under various forms of AC voltages. For example, Figure 11.11 shows the electrooptic response of a similar cell cross-linked during the application of a 100 Hz triangular voltage. The peculiar transmission–voltage curve indicates that there is an ensemble of microdomains cross-linked at different director tilt configurations and with different threshold voltages. The kinks in the curve marked with arrows fit well to the threshold voltages for domains cross-linked in the *AF* and $\pm F$ states (Table 11.1). The possibilities of using various types of field treatment and shadow masks during subsequent cross-linking steps open up new types of display-type and memory type polymer-stabilized AFLC devices with a plethora of functionalities, and switching behaviour, also in individual pixels.

11.4.4 Greyscale

The AFLC structure is monostable, but it provides tri-state electrooptic switching, see Figure 11.1c. Grey levels can in principle be obtained by controlling the fraction of bright *F* domains *vs.* dark *AF* domains through pulse-height and/or pulse-width modulation, in combination with a holding voltage, under which both the *AF* and *F* states (and hence the grey levels written) are stable. However, the slope of the transmission–voltage loop is rather steep and a small variation in applied data pulse amplitude or pulse width may give a large difference in the transmission level. The addressing of AFLCDs is also sensitive to temperature variation.

The work by Strauss and Kitzerow[9,10] revealed and highlighted yet another effect from polymer-stabilization. In their system of CS4000 and C6M they observed that the presence of a polymer network reduced the hysteresis and made the optical transmittance due to the $AF - F$ transition more continuous, with implications for grey level control, *cf.* Figure 11.12, and Figure 11.13. Furthermore, they found that in the polymer-stabilized cells the grey levels were due to the appearance of bright domains with a diameter of a few μm, rather than the well-known finger-like domains running along the smectic layers in non-polymer-stabilized AFLCs. A similar effect of polymer-stabilization on grey scale behaviour and domain features was demonstrated by Rudquist *et al.* in OAFLCs,[35] *cf.* Figure 11.14. When increasing the polymer content to 19% in W193B, Caillaud *et al.*, observed that

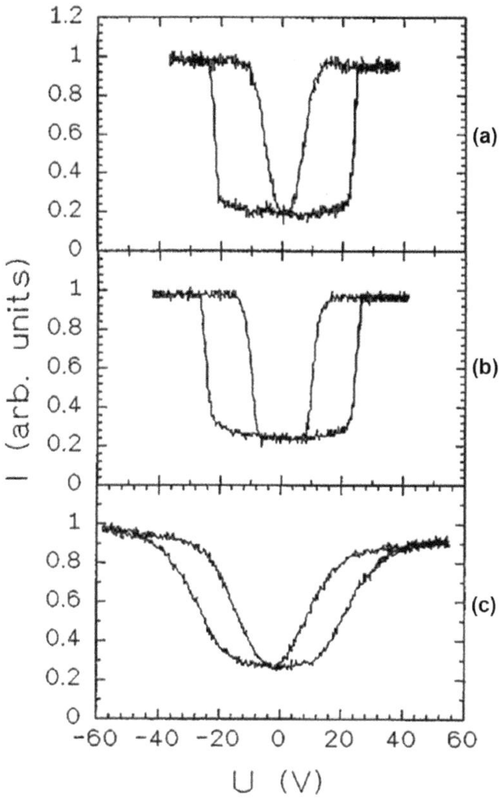

Figure 11.12 Transmission–voltage (T–V) curves for three different AFLC cells with varying amount of polymer. (a) CS4000, no polymer (b) CS4000 with 1.7wt% C6M, (c) CS4000 with 8.3wt% C6M. In (b) the network stabilizes the AF state and both the $AF - F$ and $F - AF$ switching occurs at somewhat higher voltages than for the pure CS4000 sample in (a). In (c) the higher concentration (8.3wt%) of polymer decreases the slopes of the T–V curves, extending the voltage range for greyscale generation. Reproduced from ref. 9 with permission from AIP Publishing.

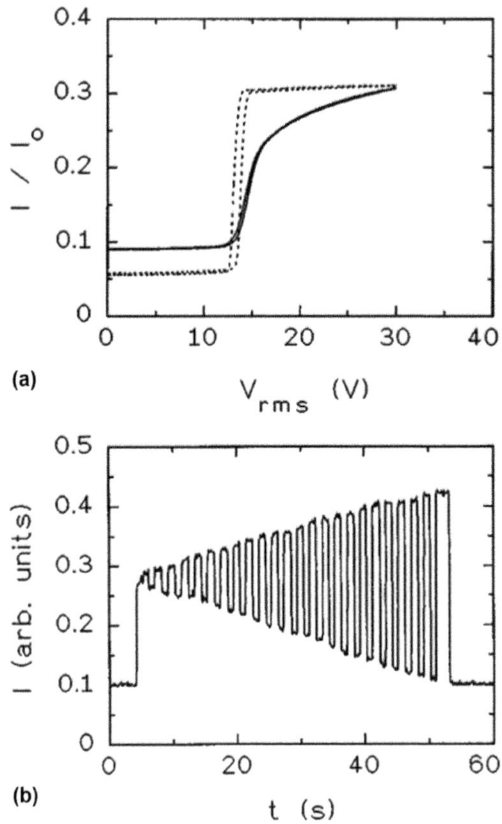

Figure 11.13 Effect of polymer stabilization on the electrooptics of the AFLC mixture CS4000.[9] (a) Average value of the transmitted intensity *versus* rms value of an applied AC voltage sine, frequency 100 Hz. Dashed line: CS 4000 without polymer; solid line: CS 4000 with 1.7wt% C6M. (b) Time dependence of the transmitted intensity of the 1.7wt% C6M sample. Based on the curve presented in (a), the amplitude of the applied voltage was varied with time in order to achieve different equidistant grey levels.
Reproduced from ref. 9 with permission from AIP Publishing.

the characteristic threshold for switching was gone, and instead, a seemingly analogue response was obtained.[36] In the latter case, one could speculate that the very high concentration of the nematic monomer destabilized the SmC_a* and therefore the cross-linking was at least partly carried out in an (V-shaped switching) SmC* structure.

11.4.5 Switching Dynamics

The working speed of AFLC display type devices is mainly limited by the relatively slow relaxation from the field-induced ferroelectric state to the

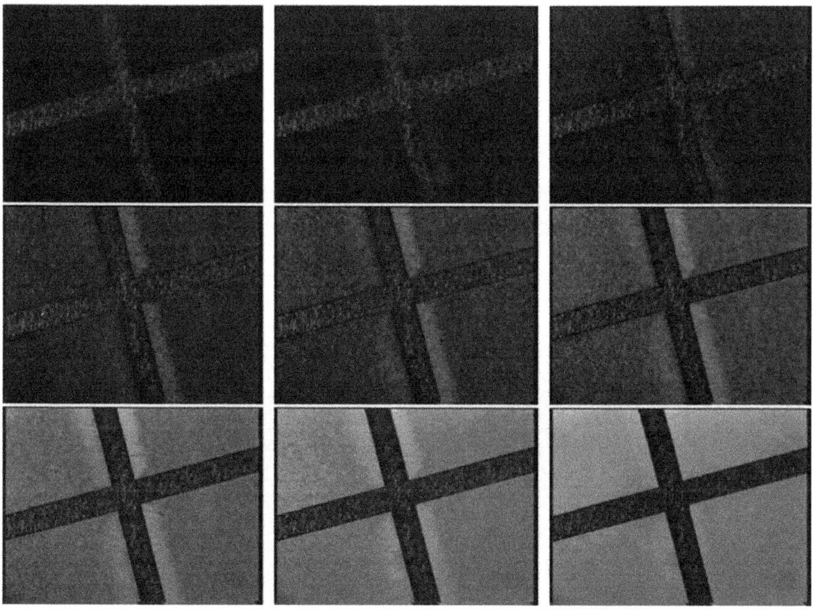

Figure 11.14 Greyscale response in a polymer-stabilized orthoconic AFLC cell subjected to 100 Hz bipolar square wave voltage. Voltage amplitude increases from upper left (0 V) to lower right (12 V). The switching from *AF* to *F* does not occur in finger-like domains but in smaller grain-like domains with different effective threshold voltages for $AF-F$ switching.
Reproduced from ref. 35 with permission from Taylor & Francis.

antiferroelectric state.[33] In order to secure a stable greyscale performance of the device, it is desirable that each pixel switches back to its *AF* dark state before next grey-level is written. At video frame rates the $F \rightarrow AF$ relaxation is generally too slow when using simple voltage pulse driving. However, by using tailored addressing waveforms, incorporating a short reverse voltage pulse just before switching off the voltage, the response times could be shortened drastically. For example, Okada *et al.* demonstrated a reduction in the $F \rightarrow AF$ response time of almost three orders of magnitude for MHPOBC, from 20 ms to 30 μs.[33] But the use of reverse pulse voltages and other tailored driving waveforms in AFLCDs[51] makes electronic addressing significantly more complex.

An alternative approach to decreasing the $F \rightarrow AF$ response times is to polymer-stabilize the *AF* state, which has proven to be particularly effective for surface-stabilized orthoconic AFLCs. Most orthoconic AFLC materials have pitch values of about 1 μm or less, and in cells thin enough to assure surface-stabilization, the field-induced *F*-states are often metastable.[32,35] The $F-AF$ relaxation might then take seconds or minutes, or even much longer. Figure 11.15 shows the electrooptic response of the surface-stabilized orthoconic mixture W193B with 4% of RM82 in a 0.8 μm thick cell subjected to a

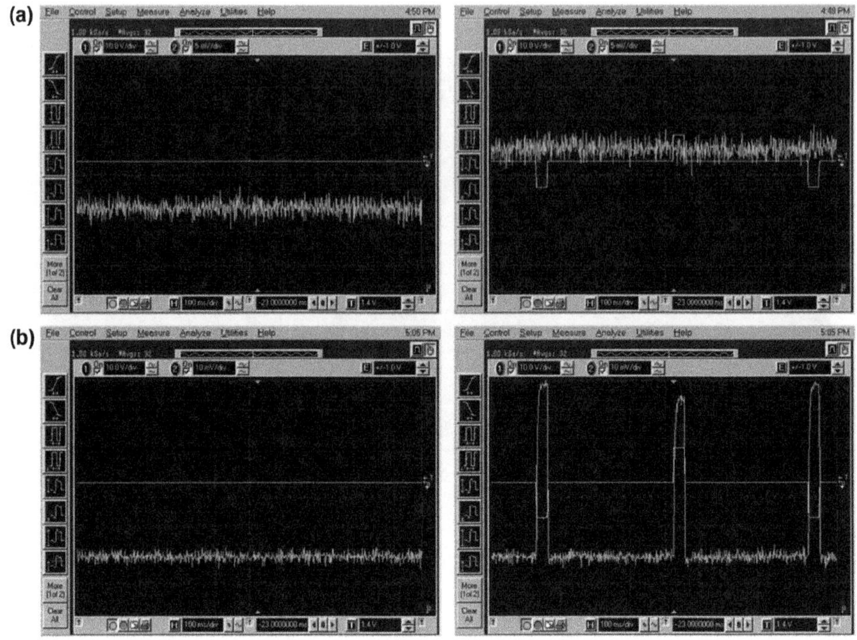

Figure 11.15 Effect of polymer-stabilization in 0.8 μm thick orthoconic AFLC device of W193B with 4% of RM82. Electrooptic response (green) when the cell is subjected to a bipolar pulse train (yellow) before (a) and after (b) cross-linking. Left: $E = 0$, the cell is in the dark state. Right: When 12 V pulses of 30 ms width and 500 ms separation are applied, the OAFLC switches to the bright *F*-states. (a) Without polymer-stabilization the relaxation, however, is extremely slow due to the surface-mediated metastability of the field-induced *F* states and the dark state is never reached between pulses. (b) When the *AF* state is polymer-stabilized, the *F*-states are no longer metastable and the relaxation occurs within a few ms, without any need of special addressing waveforms. Reproduced from ref. 35 with permission from Taylor & Francis.

bipolar pulse train. Before stabilization (a), the cell never relaxes back from the bright $\pm F$ states to the dark *AF* state between pulses. On the other hand, after stabilization (b) the relaxation takes place in a few ms. In Figure 11.16 is shown the electrooptic response for different amplitudes of the driving pulses. For pulse voltages in the range 8.5–12 V the greyscale is perfectly continuous.

Caillaud *et al.* observed that in non-surface-stabilized OAFLC cells, and where the *F*-states are not metastable, the relaxation time was in the sub-millisecond range also without polymer stabilization.[36] Interestingly, in these thicker cells the polymer stabilization, used to stabilize the bookshelf structure in stroboscopic welding visors, slightly increased the relaxation time, *cf.* Figure 11.17. There could be several reasons for this behaviour. One possibility is that when the cell is thick compared to the helix pitch, the surface (synclinic) influence on the bulk is relatively small, and the $\pm F$ states are therefore not metastable. Thus, the relaxation would be fast also in the

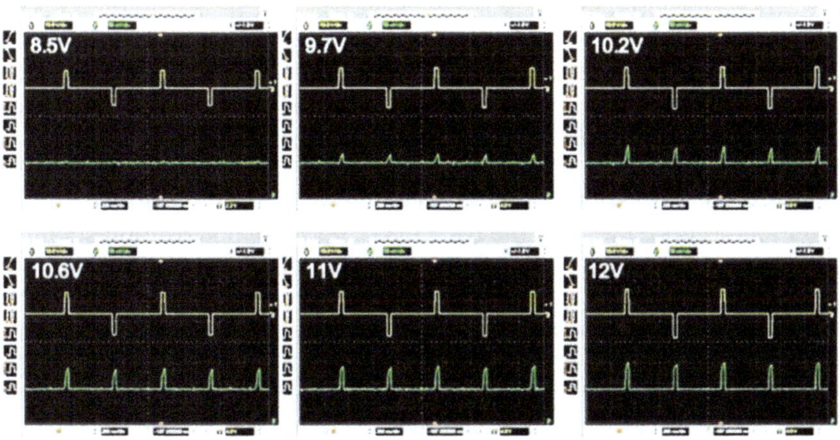

Figure 11.16 Electrooptic response of polymer-stabilized orthoconic AFLC device, for increasing value of the pulse voltage.
Reproduced from ref. 35 with permission from Taylor & Francis.

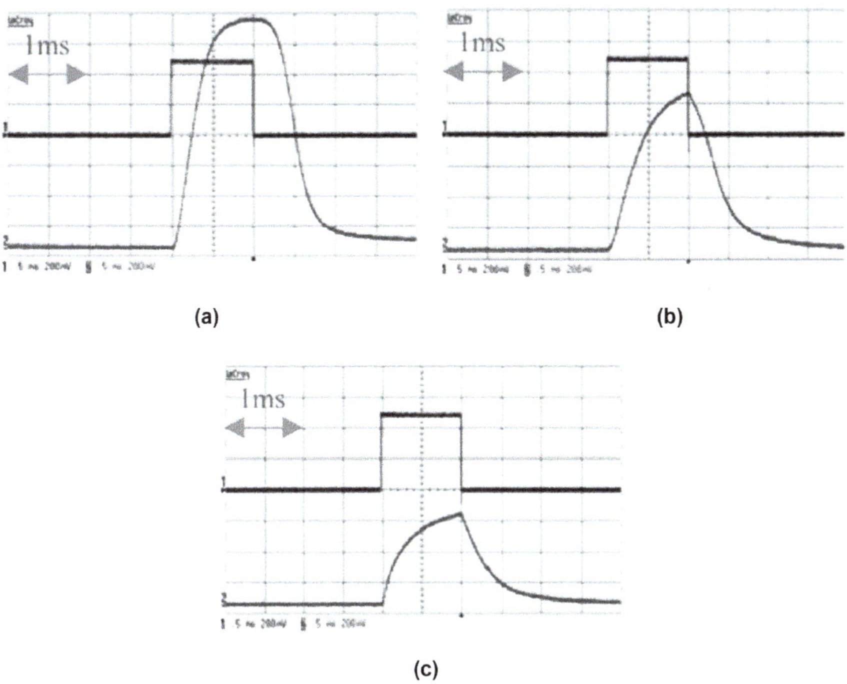

Figure 11.17 Electrooptic response of W193B to a 1 ms pulse of 45 V (a) Pure W193B, (b) with 5% of the reactive monomer RM257, and (c) with 10% of RM257. Cell gap 2 μm, *i.e.* here the short pitch W193B is not surface-stabilized.
Reproduced from ref. 36 with permission from Taylor & Francis.

pure AFLC; the presence of a scarce network only increases the effective viscosity for switching. Another possibility is that the *F*→*AF* relaxation is still as slow, but the electrooptic effect is fast. One could imagine a two-stage scenario in which the bright synclinic *F* state first relaxes into a dark polarization-stabilized twisted SmC* state[52] with **P** parallel to the glass plates, or alternatively, into a dark short-pitch SmC* helix state with the optic axis (collinear with the helix) along the layer normal, before the *F*→*AF* relaxation occurs slowly "in the dark". But, both the polarization-stabilized twisted SmC* state and the helical C* states should be avoided in AFLCDs. The reason is that both these unwanted states give thresholdless "V-shaped switching". Hence, any region of an AFLC pixel exhibiting any of these states will give an immediate increase of transmission when the applied field is increased from zero and, hence, a seemingly large pretransitional effect, and significant light leakage at the holding voltage.

The polymer mediated faster relaxation from *F* to *AF* was also observed by Yu *et al.*, who studied AFLC devices with several AFLC materials under real passive matrix display driving conditions.[28] These authors found that the stabilization of the *AF* state suppresses the pretransitional effect and hereby increases the dynamic contrast of an AFLC display. This conforms well to the scenario of polymer-suppression of analogue V-shaped switching states.

If the polymer- and surface-stabilized OAFLC is combined with short reverse pulses the relaxation time can be of the order of 100 μs. Figure 11.18 shows the electrooptic response of a polymer-stabilized OAFLC cell displaying field-sequential colour at a frequency of 720 Hz.[53] Here the backlight is rapidly switched between red, green, and blue in sequence and the cell synchronously switched between "white" and "black" to produce various colours and combination of colours in the time domain.

The concept of polymer-stabilized OAFLCs have also been demonstrated for ternary phase-only modulation of light[54] and recently for real time phase modulation measurements in liquid crystals.[55]

11.5 Effects on Physical Properties

11.5.1 Phase Sequence

From the limited amount of published research, it seems that a polymer concentration of less than 2–5% does not influence the phase behaviour of the host SmC_a* material significantly. Artal *et al.* noticed a slight shift in phase transitions a few degrees towards lower temperatures when adding 2% of the monomer **b** to the host "compound **f**".[11] After the cross-linking, regardless of the photopolymerization conditions, the SmC_α* subphase was absent and the temperature region in which the SmC* to SmC_a* occurs was extended over several degrees. Singh and Gleeson observed that when the polymerization was carried out in the SmC* phase, this phase was stabilized at the expense of the sub-phases.[16] Singh and Bradshaw observed an approximately linear decrease of the phase transition temperatures with

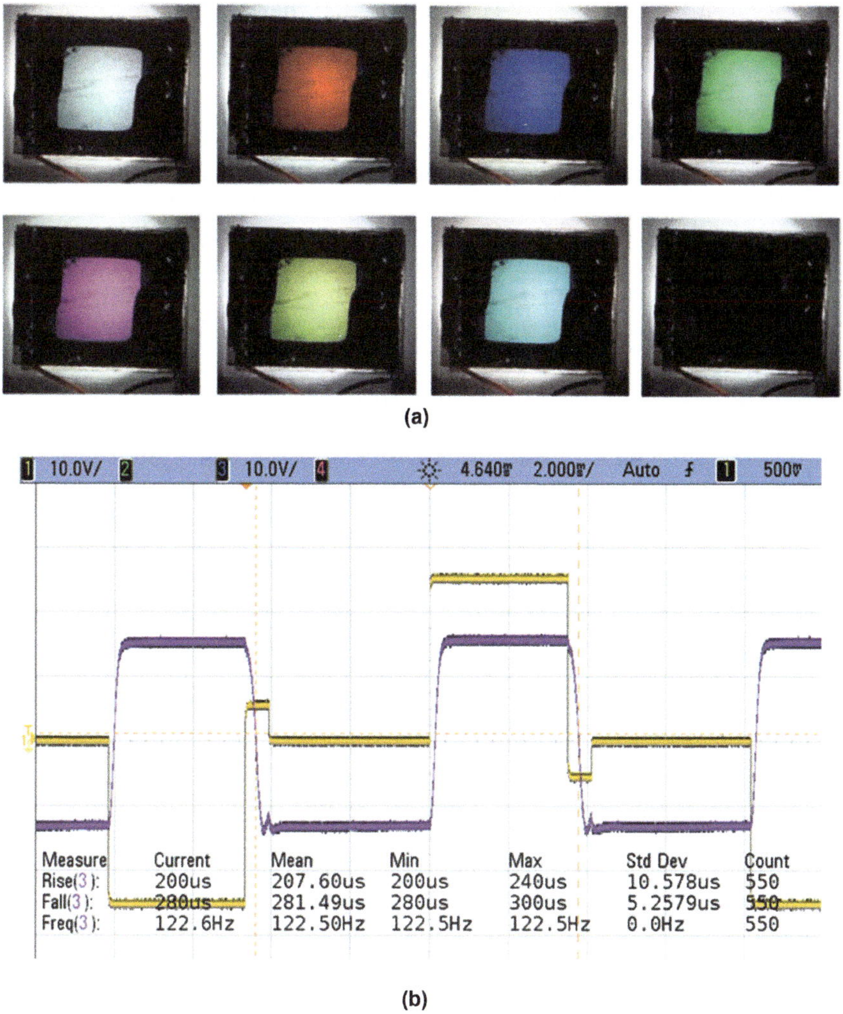

Figure 11.18 (a) A 10 mm×10 mm polymer-stabilized surface-stabilized OAFLC cell displaying field sequential colour switching at 720 Hz. The orthoconic state prevents light leakage in the black state while the polymer network and the short reverse pulse (b) assures a relaxation times of less than 300 μs.
Reproduced from ref. 53 with permission from John Wiley and Sons, Copyright 2012 Society for Information Display.

increasing polymer concentration for polymer concentrations up to 5% in the AFLC AS661 when stabilized in the SmC* phase.[15]

11.5.2 Molecular Tilt Angle

Electrooptic measurements reveal that the apparent molecular tilt of the field-induced ferroelectric states decreases with increasing polymer concentration. Strauss and Kitzerow found that the apparent tilt angle is well described with $\theta \sim (T_c - T)^{1/2}$ in the monomeric as well as polymer-stabilized system of CS-4000 and C6M, and that the tilt decreases approximately linearly with C6M concentration in the concentration range 0–5%.[10] In some cases, the addition of the monomer leads to a slight decrease in tilt even *before* cross-linking. For instance, in the case of the orthoconic AFLC mixtures W193B and W182 the tilt angle measured from the switching angle or tilt angle of the *F* states, decreased from 45° to 43° after adding 4% of the monomer RM82.[35] (A slightly smaller decrease in tilt could be expected if the monomer itself has a similar structure to the host, *cf.* compounds **f** and **b** in ref. 11.) After cross-linking, the switching angle was further reduced, to about ±40°.[36,37] But if the molecules close to the network do not take part in the switching, one would of course also expect the optical switching angle to decrease with increasing polymer content. Hence, the tilt angle in the anticlinic *AF* state could indeed be larger than the tilt measured from electrooptic switching.

11.5.3 Spontaneous Polarization

As the AFLCs have no macroscopic polarization in the absence of applied fields, one usually refers to the value of the measured polarization in the induced ferroelectric state as being a measure of the polarization of the AFLC material itself. This does not necessarily have to be the case; the smectic layer polarization density in the anticlinic state could in principle differ from the value in the synclinic state. When adding a reactive monomer which in itself is not liquid crystalline, or is liquid crystalline but does not have a ferroelectric nor antiferroelectric phase, there is obviously a dilution and we expect the spontaneous polarization to linearly decrease with the increasing monomer concentration. However, the decrease can be strongly non-linear[10] and cannot then be explained solely by dilution. Rather, it indicates that molecules close to the network strands are tightly fixed to the network and do not take part in the switching. Analogously to the case of the tilt angle, the layer polarization in the anticlinic state could possibly be higher than the measured polarization in the electric field-induced switched state. A reduction in (switching) **P** with increasing polymer concentration is observed also in polymer-stabilized ferroelectric liquid crystals.[24]

Strauss and Kitzerow also found that the spontaneous polarization in the polymer-stabilized cells of CS-4000 and C6M followed the $P \sim (T_c - T)^{1/2}$ behaviour where T_c is the SmC*–SmA* phase transition temperature.[10]

11.5.4 Dielectric Spectroscopy

In the $SmC_a{}^*$ phase there are several dielectric relaxation processes. Antiphase fluctuations in the azimuthal orientations in odd and even layers constitute a polar high frequency mode (~1 MHz), whereas rotations about the short axis are believed to give a low frequency (10–100 kHz) polar mode.[56] (The low frequency mode has also been discussed in terms of a collective mode being due to in-phase azimuthal fluctuations.[57]) Lakshmi Mashura *et al.* carried out dielectric spectroscopy on TFMHPOBC (Figure 11.2c) in its pristine form and when stabilized by a polymer network of 10% of RM82.[58] The inclusion of the network led to a significant reduction of the antiphase fluctuation frequency but almost no change in the relaxation frequency of the low frequency mode. The results indicate that the 10% network leads to an effective decrease in the antiferroelectric coupling constant, *i.e.* that the anticlinic state is slightly destabilized by the network. Furthermore, the effects of confinement, *i.e.* restricted geometry effects at large volume fractions of polymer strands, were discussed.

11.5.5 Selective Reflection and Pitch Stabilization

In the SmC_a* phase the helical pitch usually strongly varies with temperature. For instance, MHPOBC even changes its helix handedness a few degrees into the SmC_a* phase and in the process the pitch becomes infinite at a certain temperature. The possibility of *polymer-stabilization of the helix pitch* at a certain value was demonstrated by Furue *et al.*[59] Figure 11.19 shows the center wavelength of the Bragg peak as function of temperature in the AFLC mixture CS-4001. At $T \approx 18$ °C the pitch goes to infinity and the handedness changes from left-handed to right-handed on heating.[59] When doped with a very small amount of the photocurable monomer UCL-001, the overall pitch behaviour is preserved, *cf.* Figure 11.19b. After cross-linking at a temperature 10 °C below the SmC_a*–SmA* transition in the AFLC–monomer mixtures, the temperature for which the reversal of pitch handedness occurs is slightly shifted to lower temperatures (b) and, importantly, the value of the right-handed pitch seems to be preserved closer to the value present during illumination. This effect becomes evident at a polymer concentration of 5%, (c). After photocuring the helix inversion is suppressed and the pitch of stabilized right-handed helix is essentially independent of temperature. Hence, in this system both the handedness and the helical pitch becomes fixed to the values present at cross-linking temperature. Such polymer-stabilization of the SmC_a* helix structure can have important implications for wide-temperature range Bragg-scattering components of AFLCs, and for extension of the surface-stabilized regime in AFLC electrooptic devices, as discussed in Section 11.4.

Bragg-scattering from polymer-stabilized AFLCs has also been studied by Singh and Gleeson.[16,17] They carried out the photocuring at elevated temperatures, in the SmC* and SmA* phases, respectively. When polymerized in the SmC*/SmA* phase, the Bragg peak in the SmC_a* phase was slightly shifted towards longer/shorter wavelengths, and the SmC_a* Bragg reflection wavelength was largely temperature insensitive in the sample polymerized in the SmA* phase. One should also have in mind that the Bragg peak is not only governed by the helix pitch. It also scales with the average effective

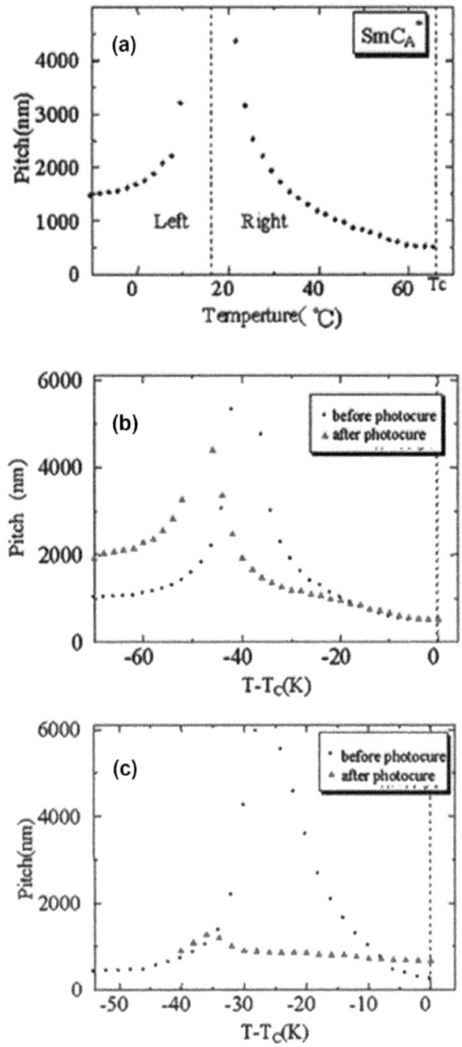

Figure 11.19 Polymer stabilization of the pitch in the AFLC mixture CS-4001. (a) Pure CS-4001. (b–c) with 2%, and 5% monomer, respectively, before and after photocure. In the case of 5% polymer (d) the pitch is stabilized at the value exhibited by the system at the photocuring temperature and the helix inversion is suppressed.
Reproduced from ref. 59 with permission from Taylor & Francis.

refractive index, which in turn is set by the average molecular tilt in SmC_a^* and SmC* and the orientational order. Therefore, in principle, small variations in Bragg reflection with temperature, especially close to phase transitions, are expected to occur even if the (network-stabilized) pitch is not changing.

11.6 Discussion

As shown in ref. 60 the anisotropic network of polymer strands formed in the nematic, SmA and SmC phases with rod-like molecules orients along the director field present during cross-linking. Once the network is formed, the LC director aligns along the network strands even after heating the liquid crystal to the isotropic phase and cooling back to the LC phase. Hence, the network works like a distributed aligning surface in which the elastic interactions between this network surface and the liquid crystal stabilize the quiescent director field.[9,10] Accordingly, the observed (non-linear) reduction in apparent molecular tilt and polarization from switching experiments can then be explained by considering the molecules in contact with the network being firmly attached to the network and not taking part in the switching, as described in Section 11.5.

There is no reason to believe that the situation would be very different in AFLC materials, provided that the cross-linking is carried out at elevated temperatures, *e.g.* in the orthogonal SmA or the synclinic SmC phase. However, when the network is formed in the anticlinic phase, (SmC_a* phase) the situation is more complex as the director then makes a zigzag structure. A rather naive picture would be that the strands formed in an anticlinic SmC_a* phase follow the local director and "kink" at the layer interfaces, preserving a tendency for anticlinic order, as in the case of mixing monomers with anticlinic polymers.[42] However, SEM studies on polymer strands formed in FLC materials have shown that the diameter of the strands can on average be about 0.1 μm[60] *i.e.* about two or three orders of magnitude larger than the length of the LC molecules and the smectic layer thickness, respectively. It is therefore unlikely that such relatively thick strands themselves can follow the zigzag director field. If the network indeed preserves anticlinic order on the smectic layer scale, the microscopic features of the network itself must be different when formed in the anticlinic state compared to the synclinic state.

One could also speculate that for elastic energy reasons (thick) strands formed in the anticlinic state might want to be *as parallel as possible* to the zigzag-shaped director field. This would correspond to the strands being formed *parallel to the molecular tilt plane*, as schematically illustrated in Figure 11.20. In the case of $\theta = 0°$ (SmA) (uniaxial symmetry) the strands would form along the layer normal **z**. For finite tilt values the network would in principle have biaxial order and for $\theta \approx 45°$ the strands would have the strongest confinement to the tilt plane but little or no preference for orienting along the layer normal. Hence, at $\theta \approx 45°$ the network of strands would then be isotropic in the tilt plane. Once the network has been created, the tilt plane always wants to be parallel to the planes of strands. This *ad hoc* model would be consistent with the results shown in Figure 11.6 where the surface-stabilized (helix free) AFLC structure is immediately reformed after heating to the isotropic phase and cooling down to the SmC_a* phase again. In addition, the apparent lack of birefringence of the cell when the OAFLC is

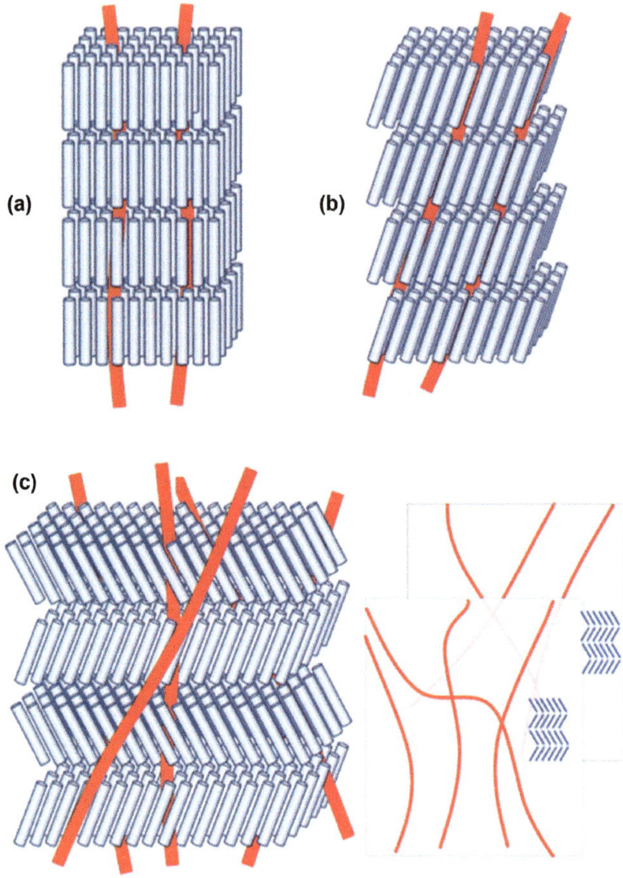

Figure 11.20 Speculation of network (schematic) formed in (a) SmA phase, (b) SmC phase, and (c) SmC_a^* phase. When formed in the SmA and SmC the network strands tend to form along the director field. When formed in the anticlinic SmC_a^* structure one might envision that the strands form as parallel to the average director as possible, *i.e.* foremost in the tilt plane which contains the director.

in the isotropic phase would further support the idea that the strands are randomly oriented in the plane of the cell. When the cross-linking is done in the SmC* phase or in the field-induced synclinic ferroelectric *F* states, the strands form along the homogeneous director, as is consistent with the stabilization of the synclinic state observed *e.g.* by Glossmann *et al.*,[38] Artal *et al.*,[11] Rudquist *et al.*[35] and demonstrated in Section 11.4.3.

On the other hand, the formation of strands is based on a polymerization-induced phase-separation between the AFLC host and the monomer. Most studies involve nematic reactive monomers and it would be interesting to further explore systems where the monomer itself has the SmC_a^* phase, with identical cores and tail-lengths to the AFLC host molecules. Maybe the phase-segregation could then occur on much smaller length scales, even down to smectic or molecular scales. Artal *et al.* suggested that the gel polymerized in the SmC_a* phase in their study[11] adopted the herringbone structure typical for this phase. This is not consistent with the formation of 100 nm wide strands. But in their study, the cross-linker in fact had an anticlinic smectic phase itself, with the identical "MHPOBC-core" as the AFLC host, and the details of the network are not known. Maybe careful tuning of monomers to exactly match the host molecular structure could be a most fruitful strategy for forming network structures on much smaller scales for further development in polymer-stabilized AFLC devices.

When considering the suppression of metastable *F*-states and the faster relaxation from *F* to *AF* in thin device cells, one must also take into account that the formed polymer structure could simply give rise to a very large number of microscopic disturbances in the local smectic structure and hereby also provide a similar amount of nucleation sites for the *F* to *AF* relaxation. Furthermore, we have not discussed the possibilities of local concentration variations of the formed network. The UV intensity might vary along the cell normal, and migration of monomers and host molecules might occur when the polymerization proceeds. One might also envisage that the network, if concentrated at the cell surfaces, could somehow even alter the phase behaviour and/or preference for tilt-order at the surfaces. Despite being known for more than 20 years, the concept of polymer-stabilized AFLCs is still, to a very large extent, to be explored and further developed from a scientific as well as from an application point of view.

References

1. A. D. L. Chandani, E. Gorecka, Y. Ouchi, H. Takezoe and A. Fukuda, Antiferroelectric chiral smectic phases responsible for the tristable switching in MHPOBC, *Jpn. J. Appl. Phys.*, 1989, **28**, L1265–L1268.
2. A. Fukuda, Y. Takanishi, T. Isozaki, K. Ishikawa and H. Takezoe, Antiferroelectric chiral smectic liquid crystals, *J. Mater. Chem.*, 1994, **4**, 997–1016.
3. Y. Yamada, N. Yamamoto, K. Mori, K. Nakamura, T. Hagiwara, Y. Suzuki, I. Kawamura, H. Orihara and Y. Ishibashi, Ferroelectric liquid crystal display using tristable switching, *Jpn. J. Appl. Phys., Part 1*, 1990, **29**, 1757–1764.
4. N. Yamamoto, N. Koshoubu, K. Mori, K. Nakamura and Y. Hamada, Full-color antiferroelectric liquid crystal display, *Ferroelectrics*, 1993, **149**, 295–304.
5. S. T. Lagerwall, *Ferroelectric and Antiferroelectric Liquid Crystals*, 2nd edn., Wiley, 2011.
6. T. Niori, T. Sekine, J. Watanabe and H. Takezoe, Distinct ferroelectric smectic liquid crystals consisting of banana shaped achiral molecules, *J. Mater. Chem.*, 1996, **6**, 1231–1233.

7. D. R. Link, G. Natale, R. Shao, J. E. Maclennan, N. A. Clark, E. Körblova and D. M. Walba, Spontaneous formation of macroscopic chiral domains in a fluid smectic phase of achiral molecules, *Science*, 1997, **278**, 1924–1927.
8. A. Eremin and A. Jakli, Polar bent-shaped liquid crystals – from molecular bend to layer splay and chirality, *Soft Matter*, 2013, **9**, 615–637.
9. J. Strauss and H.-S. Kitzerow, Gray-scale in polymer-stabilized antiferroelectric liquid crystal displays, *Appl. Phys. Lett.*, 1996, **69**, 725–727.
10. J. Strauss and H.-S. Kitzerow, Antiferroelectric gels, *Rev. Bunsenges. Phys. Chem.*, 1998, **102**, 1609–1614.
11. M. C. Artal, M. B. Ros, J. L. Serrano, M. R. de la Fuente and M. A. Pérez-Jubindo, Antiferroelectric liquid-crystal gels, *Chem. Mater.*, 2001, **13**, 2056–2067.
12. B. Atorf, A. Hoiachen, M. D. Ros, N. Gimeno, C. Tchierske, G. Dantlgraber and H. Kitzerow, Switching performance of a polymer-stabilized antiferroelectric liquid crystal based on bent-core molecules, *Appl. Phys. Lett.*, 2012, **100**, 223301.
13. G. P. Crawford and S. Zumer, *Liquid Crystals in Complex Geometries Formed by Polymer and Porous Networks*, Taylor & Francis, London, 1996.
14. P. Lakshmi Madhuri, S. Krishna Prasar and G. G. Nair, Influence of polymer stabilization on the dielectric relaxations of an antiferroelectric liquid crystal, *RSC Adv.*, 2014, **4**, 3121.
15. U. Singh and S. Bradshaw, Modifying optical Bragg reflections from an antiferroelectric liquid crystal with photopolymer networks, *Phys. Rev. E*, 2003, **87**, 042507.
16. U. Singh and H. Gleeson, The effect of polymer stabilization on phase transitions in a series of antiferroelectric heterocyclic esters, *Mol. Cryst. Liq. Cryst.*, 2005, **439**, 135–145.
17. U. Singh and H. Gleeson, Optical Bragg scattering from a polymer-stabilized anti-ferroelectric liquid crystal, *Ferroelectrics*, 2005, **321**, 53–61.
18. H. de Vries, Rotatory power and other optical properties of certain liquid crystals, *Acta Crystallogr.*, 1951, **4**, 219–226.
19. J. P. F. Lagerwall, P. Rudquist, S. T. Lagerwall and F. Giesselmann, On the phase sequence of antiferroelectric liquid crystals and its relation to orientational and translational order, *Liq. Cryst.*, 2003, **30**, 399–414.
20. S. T. Lagerwall, K. D´havé and P. Rudquist, Future antiferroelectric liquid crystal displays, *SID 01 DIGEST*, 2001, 120–123.
21. R. A. M. Hikmet, From liquid crystalline molecules to anisotropic gels, *Mol. Cryst. Liq. Cryst.*, 1991, **198**, 357–370.
22. R. A. M. Hikmet, Anisotropic gels in liquid crystal devices, *Adv. Mater.*, 1992, **4**, 679–683.
23. S. M. Kelly, Anisotropic networks, elastomers and gels, *Liq. Cryst.*, 1998, **24**, 71–82.
24. I. Dierking, Polymer network-stabilized liquid crystals, *Adv. Mater.*, 2000, **12**, 167–239.

25. J. Ohyama, M. Ishikawa, Y. Tanaka and H. Hatoh, Influence of mechanical stress on molecular alignment and electro-optical properties in antiferroelectric LCDs, *Asia Display*, 1995, **95**, 447.
26. S. Zhang, B. Wen, S. S. Kest, M. E. Neubert, P. L. Taylor and C. Rosenblatt, Freedericksz transition in an anticlinic liquid crystal, *Phys. Rev. Lett.*, 2000, **84**, 4140–4143.
27. P. Rudquist, J. P. F. Lagerwall, J. G. Meier, K. D'havé and S. T. Lagerwall, On the tilt plane orientation and the origin of the pretransitional effect, *Phys. Rev. E*, 2002, **66**, 061708.
28. J.-S. Yu, J.-G. Yoo, D.-J. Jeong, S.-C. Park, Y.-J. Chang and H.-G. Yang, Suppression of the pretransition by polymer in antiferroelectric liquid crystal, *Ferroelectrics*, 2010, **276**, 301–313.
29. K. D'havé, P. Rudquist, S. T. Lagerwall, H. Pauwels, W. Drzewinski and R. Dabrowski, Solution of the dark state problem in antiferroelectric liquid crystal displays,, *Appl. Phys. Lett.*, 2000, **76**, 3528–3530.
30. S. T. Lagerwall, A. Dahlgren, P. Jägemalm, P. Rudquist, K. D'havé, H. Pauwels, R. Dabrowski and W. Drzewinski, Unique electrooptical properties of liquid crystals designed for molecular optics, *Adv. Funct. Mater.*, 2001, **11**, 87–94.
31. K. D'havé, A. Dahlgren, P. Rudquist, J. P. F. Lagerwall, G. Andersson, M. Matuszczyk, S. T. Lagerwall, R. Dabrowski and W. Drzewinski, Antiferroelectric liquid crystals with 45° tilt – a new class of promising electro-optic materials, *Ferroelectrics*, 2000, **244**, 115–128.
32. P. Rudquist, Orthoconic antiferroelectric liquid crystals, *Liq. Cryst.*, 2013, **40**(12), 1678–1697.
33. H. Okada, T. Sakurai, T. Katoh, M. Watanabe, H. Onnagawa, N. Nakatani and K. Miyashita, Electrooptic responses of antiferroelectric liquid crystals with very short reverse pulse voltages, *Jpn. J. Appl. Phys.*, 1993, **32**, 4339–4343.
34. P. Nayek, S. Ghosh, S. Kundu, S. K. Roy, T. P. Majumder, N. Bennis, J. M. Oton and R. Dabrowski, Electrooptical and dielectric properties of a high-tilt antiferroelectric liquid crystal mixture W-193B, *J. Phys. D: Appl. Phys.*, 2009, **42**, 225504.
35. P. Rudquist, D. Elfström, S. T. Lagerwall and R. Dabrowski, Polymer-stabilized orthoconic antiferroelectric liquid crystals, *Ferroelectrics*, 2006, **344**(1), 177–188.
36. B. Caillaud, L. Dupont, P. Gautier, J.-L. de Bougrenet de la Tocnaye, P. Rudquist, P. Jägemalm and S. T. Lagerwall, Safe and ergonomic welding masks by fast liquid crystals, *Mol. Cryst. Liq. Cryst.*, 2008, **494**, 195–204.
37. B. Caillaud, L. Dupont, P. Gautier, J.-L. de Bougrenet de la Tocnaye, P. Rudquist, D. Engström, P. Jägemalm and S. T. Lagerwal, A new applications of ferroelectric liquid crystals – Fast electrooptic helmet visors for pulsed welding applications, *Ferroelectrics*, 2008, **364**, 66–71.
38. J. Glossmann, A. Hoischen, T. Roder and H.-S. Kitzerow, Asymmetric switching and storage effects in ferroelectric and antiferroelectric gels and polymers, *Ferroelectrics*, 2000, **243**, 95–106.

39. D. J. Broer, J. Boven, G. N. Mol and G. Challa, In situ photo-polymerization of oriented liquid-crystalline acrylates 3. Oriented polymer networks from a mesogenic diacrylate, *Makromol. Chem.*, 1989, **190**, 2255.
40. D. J. Broer, In radiation curing in polymer science and technology, in *Polymerization Mechanisms*, ed. J. P. Fouassier and J. F. Rabek, Elsevier Science Publishers Ltd., London, 1993, Vol. III, ch. 12.
41. D. S. Hermann, A. Hult, L. Komitov, S. T. Lagerwall and M. Lindgren, Pyroelectric polymers for nonlinear optics, *Ferroelectrics*, 1998, **213**, 11–21.
42. I. Nishiyama and J. W. Goodby, Effect of polymerization on the stability of antiferroelectric liquid-crystalline phases: Liquid-crystalline properties of some chiral acrylates and their corresponding polyacrylates, *J. Mater. Chem.*, 1993, **3**, 169–182.
43. E. A. Soto-Bustamente, S. V. Yablonski, B. I. Ostrovskii, L. A. Beresnev, L. M. Blinov and W. Haase, Antiferroelectric achiral mesogenic polymer, *Chem. Phys. Chem.*, 1996, **260**, 447–452.
44. E. A. Soto-Bustamente, S. V. Yablonski, B. I. Ostrovskii, L. A. Beresnev, L. M. Blinov and W. Haase, Antiferroelectric behavior of achiral mesogenic polymer mixtures, *Liq. Cryst.*, 1996, **21**, 829–839.
45. D. R. Link, N. A. Clark, B. I. Ostrovskii and E. A. Soto-Bustamente, Bilayer-by-bilayer antiferroelectric ordering in freely suspended films of an achiral polymer-monomer liquid crystal mixture, *Phys. Rev. E*, 2000, **61**, R37–R40.
46. M. Ozaki, H. Moritake, K. Nakayama and K. Yoshino, Smectic layer rotation in antiferroelectric liquid crystal, *Jpn. J. Appl. Phys., Part 2*, 1994, **33**, L1620–L1623.
47. N. A. Clark and S. T. Lagerwall, Submicrosecond bistable electrooptic switching in liquid crystals, *Appl. Phys, Lett.*, 1980, **36**, 899–901.
48. L. A. Parry-Jones, Field-driven helix distortion and switching in antiferroelectric liquid crystals, PhD Thesis, University of Oxford, United Kingdom, 2001.
49. P. Rudquist, Orthoconic antiferroelectric liquid crystals, in *Chiral Liquid Crystals*, ed. W. Kuczynski, Polish Academy of Sciences, Institute of Molecular Physics, Poznan, 2005, ISBN 83-922875-1.
50. P. Rudquist, D. Krüerke, S. Lagerwall and D. M. Walba, The hysteretic behaviour of V-shaped switching materials, *Ferroelectrics*, 2000, **242**, 21–33.
51. X. Quintana, P. L. Castillo, J. M. Oton, N. Bennis, A. LAra, V. Urruchi and R. Dabroqski, Video-rate multiplexed driving scheme for passive antiferroelectric liquid crystal displays, *Opto-Electron. Rev.*, 2004, **12**, 291–297.
52. P. Rudquist, J. P. F. Lagerwall, M. Buivydas, F. Gouda, S. T. Lagerwall, N. A. Clark, J. E. Maclennan, R. Shao, D. A. Coleman, S. Bardon, T. Bellini, D. R. Link, G. Natale, M. A. Glaser, M. D. Wand and X.-H. Chen, *J. Mater. Chem.*, 1999, **6**, 1257–1261.
53. M. Wessling, D. Elfström, D. Engström, S. T. Lagerwall and P. Rudquist, 34.3: Fast orthoconic antiferroelectric liquid crystals for field-sequential color applications, SID Symposium Digest of Technical Papers, 2012, 43.

54. D. Engström, P. Rudquist, J. Bengtsson, K. D'havé and S. Galt, Three-level phase modulator based on orthoconic antiferroelectric liquid crystals, *Opt. Lett.*, 2006, **31**, 3158–3160.
55. N. Bennis, I. Merta, A. Karlbarczyk, M. Maciejewski, P. Marc, A. Spadlo and L. R. Jaroszewicz, Real time phase modulation measurements in liquid crystals, *Opto-Electron. Rev.*, 2017, **25**, 69–73.
56. K. Hiraoka, H. Takezoe and A. Fukuda, Dielectric relaxation modes in the antiferroelectric smectic CA* phase, *Ferroelectrics*, 1993, **147**, 13–25.
57. M. Buivydas, F. Gouda, S. T. Lagerwall and B. Stebler, The molecular aspect of the double absorption peak in the dielectric spectrum of the antiferroelectric liquid crystal phase, *Liq. Cryst.*, 1995, **18**, 879–886.
58. P. Lakshmi Madhuri, S. K. Prasad and G. G. Nair, Influence of polymer stabilization on the dielectric relaxations of an antiferroelectric liquid crystal, *RCS Adv.*, 2014, **4**, 3121–3130.
59. H. Furue, E. Umeno and J. Hatano, Temperature dependence of helical structure of polymer-stabilized antiferroelectric liquid crystals, *Mol. Cryst. Liq. Cryst.*, 2006, **437**, 203/[1447]–209/[1453].
60. I. Dierking, A review of polymer-stabilized ferroelectric liquid crystals, *Materials*, 2014, 7, 3568–3587.

CHAPTER 12

Polymer-stabilized Frustrated Phases

INGO DIERKING

School of Physics and Astronomy, University of Manchester, Oxford Road, Manchester M13 9PL, UK
Email: ingo.dierking@manchester.ac.uk

12.1 Introduction

In practice, and without any external field applied, frustration results from the competition between helical orientational ordering due to chirality and geometric constraints. This is most prominently manifested in the formation of Blue Phases,[1–3] thermodynamically located between the isotropic liquid and the cholesteric phase, as well as the Twist Grain Boundary phases,[4–6] which occur between the cholesteric and the respective fluid smectic phase (SmA* or SmC*). Frustration is generally relieved *via* the formation of regular structures of defects, which in turn require a penalty to be paid in the form of elastic energy. The latter is the reason why frustrated phases are in most cases only exhibited over narrow temperature intervals, as thermodynamics quickly tends to win over chirality. The schematic structures of Blue Phases (BP) and Twist Grain Boundary phases (TGB), as well as their texture appearance in polarizing microscopy are depicted in Figure 1.9 of Chapter 1.

12.2 Polymer-stabilized Blue Phases (PSBPs)

Given that the temperature interval of these frustrated phases is generally very limited, their suitability for applications has long been thought to be

Soft Matter Series No. 8
Polymer-modified Liquid Crystals
Edited by Ingo Dierking

Published by the Royal Society of Chemistry, www.rsc.org

minimal, and their existence of mere academic interest.[7] Only in more recent years have systems been discovered, often based on dimers, that exhibit frustrated phases of appreciable width of several tens of degrees. This has been the case for both Blue Phases[8–10] and TGB phases.[11–13] A significant breakthrough in the widening of the Blue Phase regime came through the pioneering work of Kikuchi *et al.*,[14] who reported on a special system of polymer-stabilized Blue Phases. Normally, the stabilizing effect of a polymer network is dependent on the monomer concentration, and the stability regime increase for increasing concentration is of a few degrees, as depicted in Figure 12.1, which is taken from a study investigating a variety of different methods to widen Blue Phase existence.[15]

In contrast, the work by Kikuchi suggested that a very broad temperature range of stability >60 K, can be achieved when specific mixtures of polymers at certain concentrations are employed. These were demonstrated and it was suggested that the polymer is formed within the disclination lines of the Blue Phase, thus lowering the free energy of the system and therefore increasing its width,[14] as it is schematically illustrated in Figure 12.2.

It should be noted that this investigation is repeatable, also with independently synthesised materials,[16] nevertheless, there seems to be no true consensus to date on why this specific combination of materials is so successful in stabilizing the Blue Phase over a very wide temperature range. In general, the width of the BP, as well as the electrooptic properties, appears

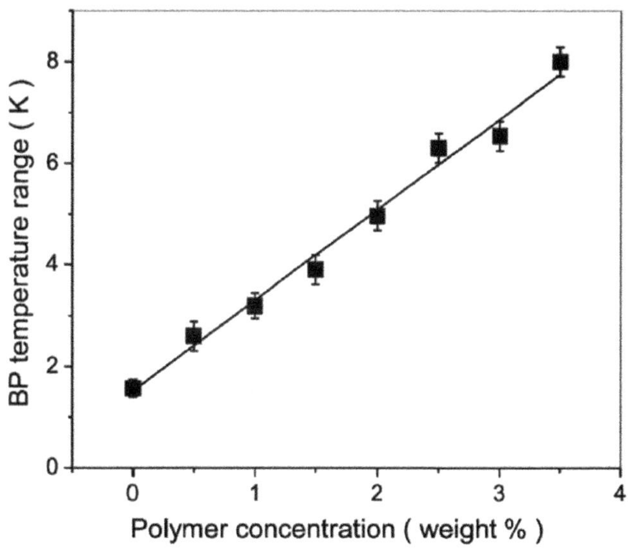

Figure 12.1 Blue Phase temperature range as a function of polymer concentration for a non-optimized system. The temperature range is often increased from about one degree to several degrees for increasing polymer concentration.
Reproduced from ref. 15 with permission from the Royal Society of Chemistry.

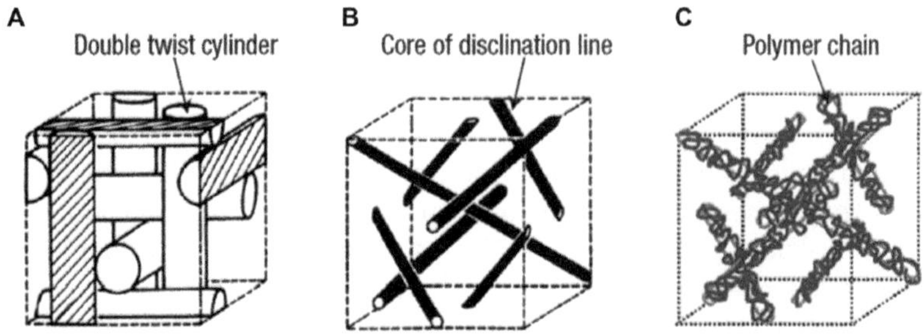

Figure 12.2 (a) Structural model of the Blue Phase constructed from double twist cylinders. (b) Network of disclination line defects, which by polymer stabilization is filled with a polymer network, as shown in (c).
Reproduced from ref. 14 with permission from Springer Nature, Copyright 2002.

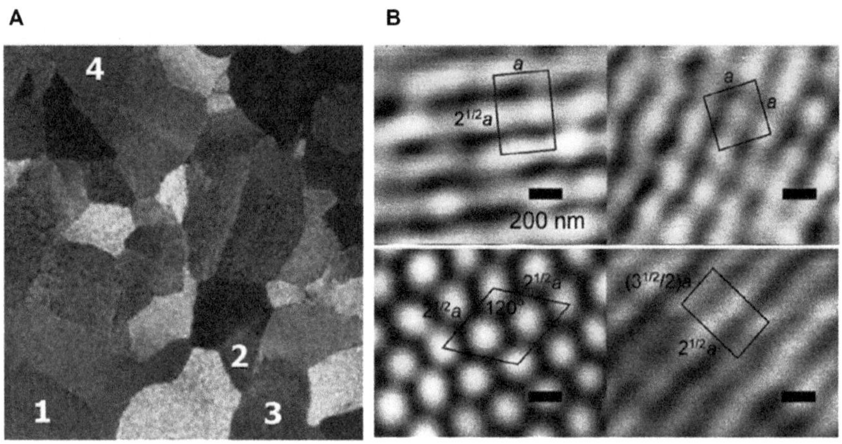

Figure 12.3 (a) Platelet texture of the Blue Phase with different lattice plane directions, 1–4. (b) These were shown by Confocal Laser Scanning Microscopy to represent the [110], [100], [111], and [211] planes of the cubic lattice, respectively.
Reproduced from ref. 25 with permission from American Chemical Society, Copyright 2008.

to be strongly dependent on host material properties, chemical composition of the monomers,[17–21] polymerization kinetics influencing the polymer morphology[22] and monofunctional monomers as diluters.[23,24]

Evidence for the hypothesis that the polymer network is formed in the disclination line defects of the Blue Phase was presented *via* Confocal Laser Scanning Microscopy (CLSM),[25] where different orientation planes were imaged, displaying polymer structures which were expected for the cubic unit cell of Blue Phases. The results of this demonstration are depicted in Figure 12.3.

It is believed that the polymer network forms in the defect regions of the Blue Phase due to a lowering of the free energy. The creation of defects in frustrated phases in general occurs at the expense of an energy addition, which accounts for the delicate balance between helical structures and geometric constraints. The cores of the disclination line defects are of isotropic order, and the whole phase structure is less disturbed by the formation of random coil polymer structures in these isotropic regions, absorbing some of the energy cost. Reducing the free energy by addition of the polymer thus widens and stabilizes the defect structure, which is also illustrated by the fact that the selective reflection spectrum of the BP is not significantly disturbed. The overall BP structure is maintained during the formation of the stabilizing polymer network. A similar argument holds for the addition of nanoparticles to Blue Phases, which locate at the defects and stabilize the structure. Diffraction experiments have provided further evidence for this interpretation,[26] as shown in Figure 12.4.

The interesting feature of polymer-stabilized Blue Phases (PSBPs) is the fact that they can be employed to modulate light *via* a relatively large Kerr effect,[27,28] which gives rise to an induced birefringence:

$$\Delta n_{\text{ind}} = \lambda K E^2 \tag{12.1}$$

where λ is the wavelength, K the Kerr constant and E the applied electric field amplitude. Eqn (12.1) holds in the limit of small applied electric fields, and the induced birefringence approaches saturation for larger field amplitudes.[29] The birefringence increases quadratically with the applied electric field, and the Kerr constant is temperature dependent, which is shown in Figure 12.5 together with measurements of the individual refractive indices.

Between crossed polarizers the induced birefringence gives rise to an electro-optic effect, which can be used for light modulation and possibly new display technologies. It has a profound advantage over conventional displays, as it does not require orientation layers, which reduces the number of rejects and saves costs in the production process. Furthermore, the electro-optic effect occurs in the plane of the substrate, and thus provides good contrast and viewing angle characteristics. The optical modulation is fast, which allows much higher refresh rates than conventional nematic displays. Response times are in the microsecond range, depending on temperature *via* the viscosity of the PSBP and decaying exponentially with increasing temperature. They additionally depend on polymer concentration, decreasing with increasing concentration, as is commonly observed for all electro-optic effects involving polymer stabilization, due to an increased elastic interaction (Figure 12.6a). Furthermore, the observed threshold voltage to drive the Blue Phase out of its stabilizing template increases with increasing polymer concentration, while switching saturation is also achieved at larger voltages for increasing polymer content. The switching process itself exhibits a hysteresis effect which becomes more pronounced at smaller polymer concentrations[31] (Figure 12.6b). Due to the importance of PSBPs for potential display applications, the next part, Chapter 13, provides a much more detailed account of their applicational aspects.

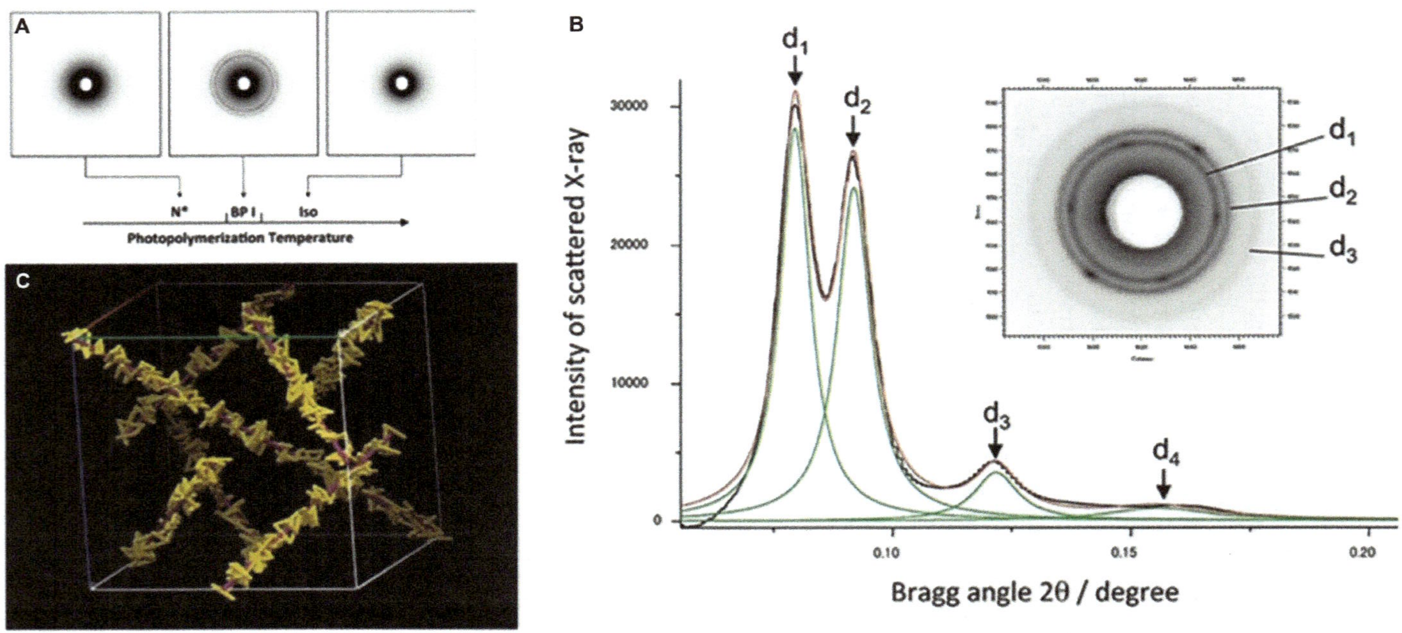

Figure 12.4 (a) X-ray diffraction patterns of the isotropic, BP and cholesteric N* phases. (b) Extremely Small Angle X-ray Diffraction revealing the cubic structure of the polymer stabilized BP, and (c) corresponding model of the PSBP. Reproduced from ref. 26 with permission from the Royal Society of Chemistry.

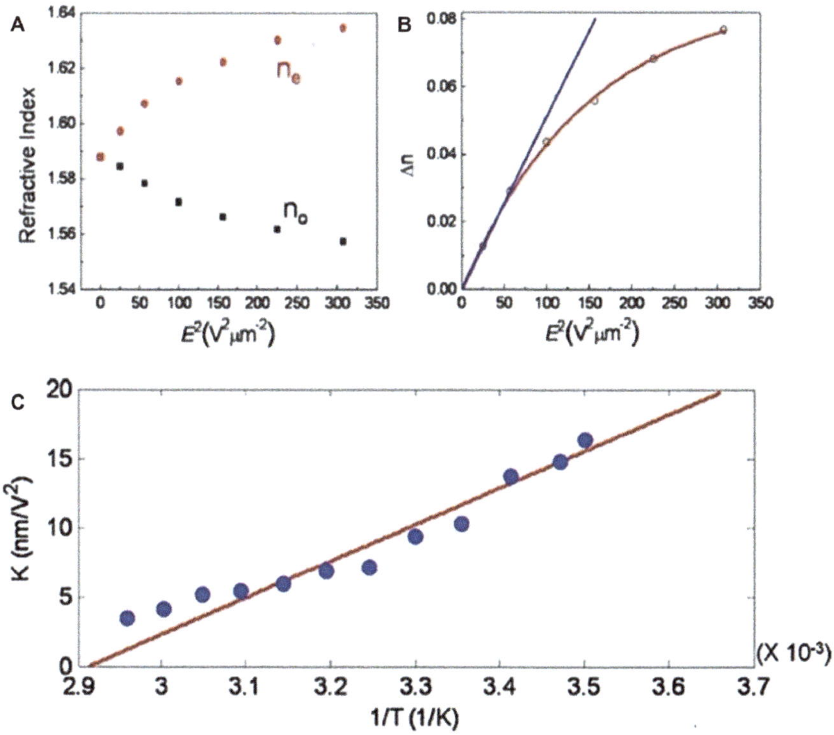

Figure 12.5 (a) Measurements of the ordinary and extraordinary refractive index due to the Kerr effect. (b) Linear dependence of the Kerr induced birefringence at small applied electric field, and saturation at larger field amplitudes. Reproduced from ref. 29 with permission from AIP Publishing. (c) Temperature dependence of the Kerr constant, *K*. Reproduced from ref. 30 with permission from The Optical Society.

It should further be noted that the applications of polymer-stabilized Blue Phases are not limited to display devices, despite the fact that they most likely have their biggest potential in that area. As with cholesteric liquid crystals, which can act as helical photonic band gap materials, Blue Phases can also be used for example to create switchable gratings, which can be of the one-dimensional or the two-dimensional type,[32] as shown in Figure 12.7a. Also, lasing has been demonstrated for PSBPs[33] (Figure 12.7b).

So far, we have only discussed conventional systems of Blue Phases, stabilized by a polymer network which is formed through UV illumination and photo-polymerization of the BP state. More recently another possibility has been suggested, which relies on simply mixing a polymer into an existing Blue Phase.[34] Polystyrene of varying degrees of polymerization, thus molecular weight, was investigated, and it was found that while the transition temperatures were strongly reduced with increasing polystyrene volume fraction, the width of the Blue Phase temperature range increased considerably with decreasing polystyrene molecular weight (see Figure 12.8).

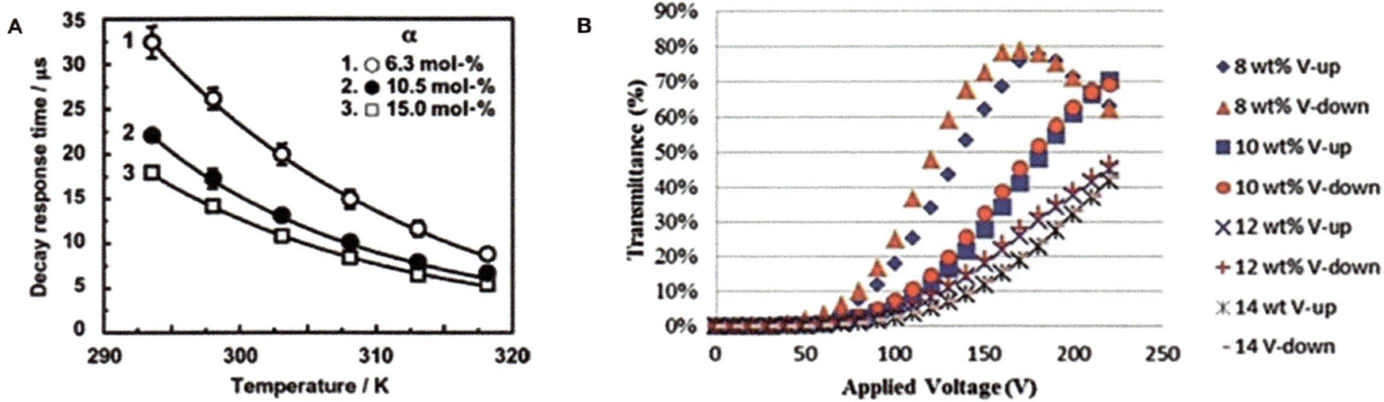

Figure 12.6 (a) Temperature dependence of the electro-optic decay time for several polymer concentrations of a PSBP. Reproduced from ref. 27 with permission from John Wiley and Sons, Copyright © 2005 WILEY-VCH Verlag GmbH & Co. KGaA, Weinheim. (b) Transmittance and hysteresis as a function of applied voltage for a PSBP at different polymer concentrations. Reproduced from ref. 31 with permission from Taylor & Francis.

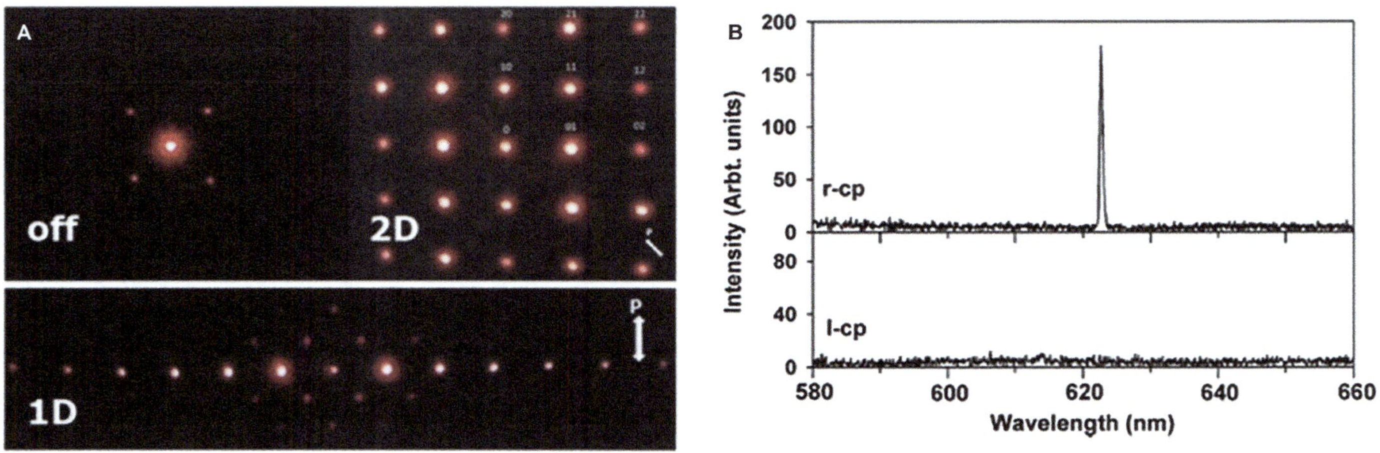

Figure 12.7 Examples of non-display applications of polymer-stabilized Blue Phases. (a) One- and two-dimensional switchable diffraction gratings and (b) lasing from PSBPs.
Part (a) reproduced from ref. 32 with permission from AIP Publishing. Part (b) reproduced from ref. 33 with permission from John Wiley and Sons, Copyright © 2006 WILEY-VCH Verlag GmbH & Co. KGaA, Weinheim.

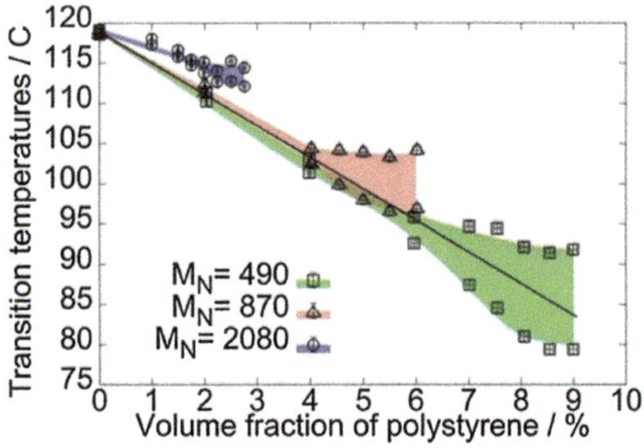

Figure 12.8 Concentration dependence of the stabilization of the Blue Phase by simple addition of polystyrene of different molecular weight, M_N. The smaller the degree of polymerization, the larger is the stabilization effect.
Reproduced from ref. 34 with permission from the Royal Society of Chemistry.

In fact, if the BP temperature regime is extrapolated to values of small degrees of polymerization, for example in the range of oligomers, one could expect very wide Blue Phases, similar to those observed for dimers of mesogens. This naturally leads to an approach which was reported recently, using specially designed bifunctional bent-core mesogens.[35] It is well known that bent-core dopants can induce Blue Phases in chiral hosts[36] or stabilize existing Blue Phases.[15] This is due to the chiral matrix acting as a bias of highly chiral conformations of the bent-core dopant, as demonstrated by computer simulations.[37] Yang *et al.*[35] have designed a bifunctional, photo-polymerizable, achiral bent-core mesogen, which readily induced a Blue Phase of considerable temperature stability. Subsequent photo-polymerization further increased this BP regime to close to approximately 60°, including room temperature (Figure 12.9).

12.3 Polymer-stabilized Twist Grain Boundary Phases (PSTGB)

In contrast to Blue Phases, Twist Grain Boundary phases[4–6] seem to have been studied much less with respect to polymer stabilization. This is probably due to the fact that they appear optically very similar to the much easier to handle cholesteric phases, and thus do not hold any particular properties which make them interesting for specific applications. TGB phases are frustrated phases, resulting from a competition between helicity and smectic layer formation, and were introduced in Chapter 1, Figure 1.9. They generally only occur over very narrow temperature intervals, and can exhibit SmA* as well as SmC* behaviour, depending on the local director structure within individual smectic blocks. There have been some reports of relatively wide

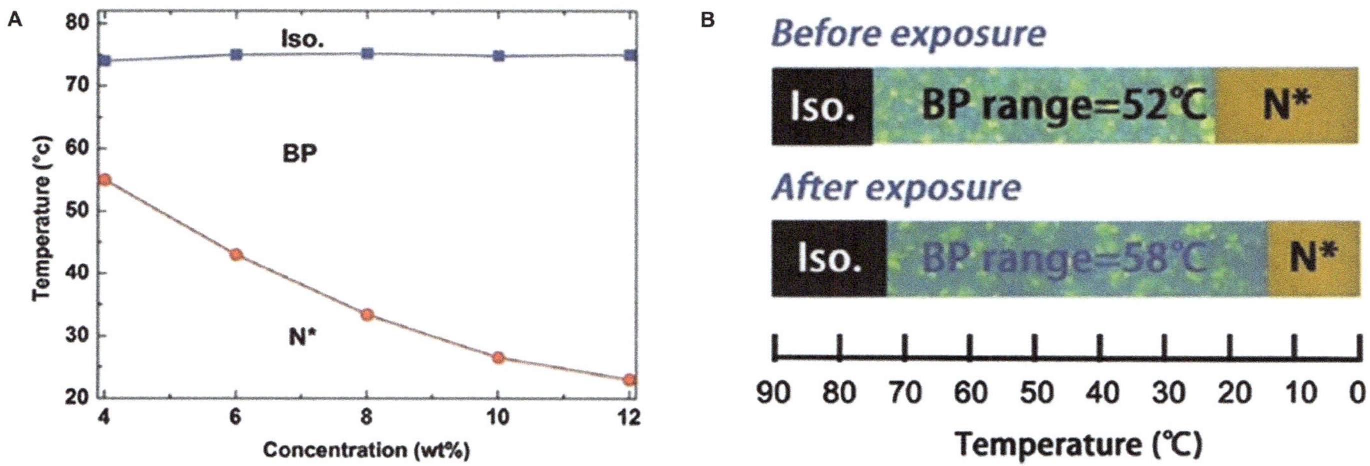

Figure 12.9 Wide temperature range BP stabilization by specially designed photo-reactive bent-core monomers. Reproduced from ref. 35 with permission from the Royal Society of Chemistry.

TGB phases,[11–13,38,39] which interestingly often consist of asymmetric dimers involving a cholesteryl group. Other reports were published for chiral liquid crystalline side-group polymers.[40–42]

As demonstrated in Figure 6.7 of Chapter 6, the polymer network formed by UV irradiation of photoreactive monomers in a TGB phase exhibits a discontinuously helical superstructure, where the polymer appears to be formed in the smectic blocks.[43] During the polymerization, the polymer is thus expelled from the highly twisted regions of the twist dislocations into the lesser twisted regions. A similar effect had already been observed in the cholesteric phase, where the network was expelled from the oily streaks defects, or in the Blue Phase, where it is expelled from the double twist cylinders into the isotropic regions of the defect cores. The addition of the monomer to a chiral host phase can already induce a Twist Grain Boundary state prior to polymerization. This occurs at the expense of the non-twisted smectic phase, as shown for increasing monomer concentration in Figure 12.10. For somewhat larger cooling rates not only the smectic A* phase is transformed into a TGBA* phase, but also the smectic C* to a TGBC* phase.[43]

The effects of polymerization of the monomers in the twisted SmA* or TGBA* state can best be visualized by comparison to the equivalent non-helical state, *i.e.* polymerization in SmA*. This is demonstrated in Figure 12.11 where four different phases are shown, while the polymerization was carried out in the temperature region of the equilibrium between smectic A* and TGBA*.

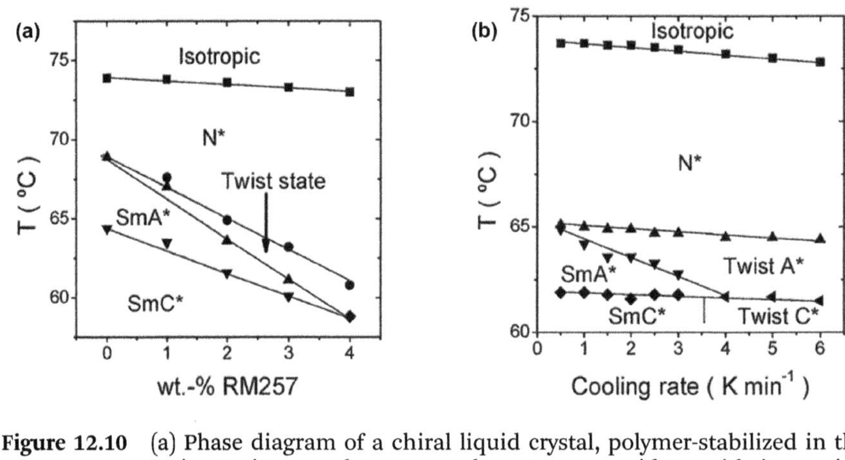

Figure 12.10 (a) Phase diagram of a chiral liquid crystal, polymer-stabilized in the Twist Grain Boundary state. The TGB state widens with increasing polymer concentration at the expense of the SmA* phase, which eventually disappears completely. (b) The stabilizing effect becomes more pronounced for larger cooling rates, where a TGBC* state can also be observed.
Reproduced from ref. 43 with permission from the Royal Society of Chemistry.

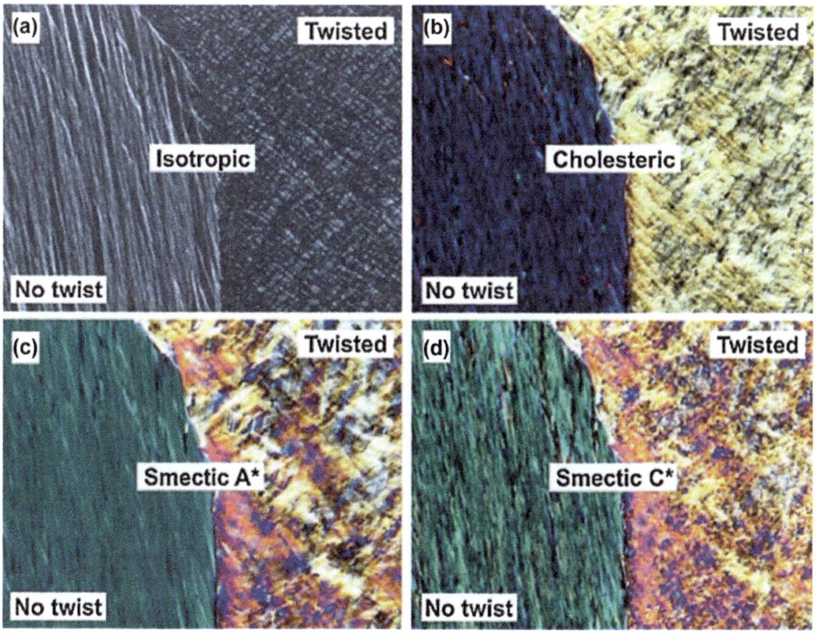

Figure 12.11 Shown are the coexistence regions between twisted and non-twisted states. After polymerization in the temperature regime of the SmA*/TGBA* phase and heating above the clearing temperature, the polymer network can be made visible in the isotropic phase through the residual birefringence (a). In the cholesteric phase the pitch is unwound to a nematic director configuration in former SmA* regions (b). (c) Shows the SmA*/TGBA* region, while (d) depicts SmC* and an induced TGBC* state.
Reproduced by permission from ref. 43 with permission from the Royal Society of Chemistry.

Part (a) shows the sample in the isotropic phase, where the polymer network is visible between crossed polarizers due to the residual birefringence. The left side of the figure was polymer-stabilized in the non-twisted smectic A* phase, and thus depicts a birefringent pattern which shows the expected unidirectional orientational order of the SmA* phase. The right side of the figure, on the other hand, was polymerized in the TGBA* state and shows a helical structure with helix axis out of the plane of the paper. Lowering the temperature into the cholesteric phase in part (b), one can see that the polymer network on the left side of the figure supresses the helical superstructure, exhibiting a nematic director field, while the right side is twisted as in the cholesteric phase. The helix axis is again out of the plane of the paper. Part (c) shows the smectic A* phase on the left and the TGBA* phase on the right. This is the original orientation in which the polymer network was formed, and thus the left side is non-helical and the right side has a helix with axis out of the paper plane. Continued cooling into the smectic C* phase temperature range (d) shows SmC* on the left and TGBC* on the right.

One can thus conclude that polymer stabilization in the Twist Grain Boundary state of a liquid crystal produces a discretely helical polymer network structure in contrast to a continuously helical structure formed in the cholesteric phase. In terms of the stabilizing effect of the polymer network, TGB networks are similar to those of cholesteric phases.

References

1. P. P. Crooker, in *Chirality in Liquid Crystals*, ed. H.-S. Kitzerow and C. Bahr, Springer, Berlin, 2001, ch. 7.
2. D. C. Wright and N. D. Mermin, *Rev. Mod. Phys.*, 1989, **61**, 385.
3. B. Pansu, *J. Phys. II*, 1995, **5**, 573.
4. S. R. Renn and T. C. Lubensky, *Phys. Rev. A*, 1988, **38**, 132.
5. J. W. Goodby, M. A. Waugh, S. M. Stein, E. Chin, R. Pindak and J. S. Patel, *Nature*, 1989, **337**, 449.
6. H.-S. Kitzerow, in *Chirality in Liquid Crystals*, ed. H.-S. Kitzerow and C. Bahr, Springer, Berlin, 2001, ch. 10.
7. See the quotation of Sir Charles Frank, in H.-S. Kitzerow, *Proc. SPIE 7232, Emerging Liquid Crystal Technologies IV*, 2009, **1**, DOI: 10.1117/12.813372.
8. H. J. Coles and M. N. Pivnenko, *Nature*, 2005, **436**, 997.
9. C. V. Yelamaggad, I. S. Shashikala, G. Liao, D. S. Shankar Rao, S. Krishna Prasad, Q. Li and A. Jakli, *Chem. Mater.*, 2006, **18**, 6100.
10. A. Yoshizawa, M. Sato and J. Rokunohe, *J. Mater. Chem.*, 2005, **15**, 3285.
11. M. B. Pandey, R. Dhar, A. S. Achalkumar and C. V. Yelamaggad, *J. Phys.: Condens. Matter*, 2007, **19**, 436219.
12. W.-K. Lee, K.-N. Kim, M. F. Achard and J.-I. Jin, *J. Mater. Chem.*, 2006, **16**, 2289.
13. C. V. Yelamaggad, G. Shankar, U. S. Hiremath and S. K. Prasad, *J. Mater. Chem.*, 2008, **18**, 2927.
14. H. Kikuchi, M. Yokota, Y. Hisakado, H. Yang and T. Kajiyama, *Nat. Mater.*, 2002, **1**, 64.
15. I. Dierking, W. Blenkhorn, E. Credland, W. Drake, R. Kociuruba, B. Kayser and T. Michael, *Soft Matter*, 2012, **8**, 4355.
16. Dabrowski group, private communication.
17. T. N. Oo, T. Mizunuma, Y. Nagano, H. Ma, Y. Ogawa, Y. Haseba, H. Higuchi, Y. Okumura and H. Kikuchi, *Opt. Mater. Express*, 2011, **1**, 1502.
18. T. Mizunuma, T. N. Oo, Y. Nagano, H. Ma, Y. Haseba, H. Higuchi, Y. Okumura and H. Kikuchi, *Opt. Mater. Express*, 2011, **1**, 1561.
19. J.-L. Zhu, S.-B. Ni, C. P. Chen, X.-L. Song, C.-Y. Chen, J.-G. Lu and Y. Su, *Liq. Cryst.*, 2014, **41**, 891.
20. J. Yan and S.-T. Wu, *J. Displ. Technol.*, 2011, 7, 490.
21. R. Kizhakidathazhath, H. Higuchi, Y. Okumura and H. Kikuchi, *ChemistrySelect*, 2017, **2**, 6728.

22. T. Iwata, K. Suzuki, N. Amaya, H. Higuchi, H. Masunaga, S. Sasaki and H. Kikuchi, *Macromolecules*, 2009, **42**, 2002.
23. Y. Chen, J. Yan, M. Schadt, S.-H. Liu, K.-L. Cheng, J.-W. Shiu and S.-T. Wu, *J. Displ. Technol.*, 2013, **9**, 592.
24. Y. Chien and S.-T. Wu, *J. Appl. Polym. Sci.*, 2014, DOI: 10.1002/APP.40556.
25. K. Higashiguchi, K. Yasui and H. Kikuchi, *J. Am. Chem. Soc.*, 2008, **130**, 6326.
26. H. Kikuchi, S. Izena, H. Higuchi, Y. Okumura and K. Higashiguchi, *Soft Matter*, 2015, **11**, 4572.
27. Y. Hisakado, H. Kikuchi, T. Nagamura and T. Kajiyama, *Adv. Mater.*, 2005, **17**, 96.
28. J. Yan and S.-T. Wu, *Opt. Mater. Express*, 2011, **1**, 1527.
29. L. Rao, J. Yan, S.-T. Wu, S. Yamamoto and Y. Haseba, *Appl. Phys. Lett.*, 2011, **98**, 081109.
30. J. Yan, M. Jiao, L. Rao and S.-T. Wu, *Opt. Express*, 2010, **18**, 11450.
31. G. Lim, J.-H. Hwang, H. Kikuchi and S.-K. Hong, *Mol. Cryst. Liq. Cryst.*, 2015, **609**, 54.
32. J.-L. Zhu, J.-G. Lu, J. Qiang, E.-W. Zhong, Z.-C. Ye, Z. He, X. Guo, C.-Y. Dong, Y. Su and H.-P. D. Shieh, *J. Appl. Phys.*, 2012, **111**, 033101.
33. S. Yokoyama, S. Mashiko, H. Kikuchi, K. Uchida and T. Nagamura, *Adv. Mater.*, 2006, **18**, 48.
34. N. Kasch, I. Dierking and M. Turner, *Soft Matter*, 2013, **9**, 4789.
35. W.-Q. Yang, G.-Q. Cai, Z. Liu, X.-Q. Wang, W. Feng, Y. Feng, D. Shen and Z.-G. Zheng, *J. Mater. Chem. C*, 2017, **5**, 690.
36. M. Nakata, Y. Takanishi, J. Watanabe and H. Takezoe, *Phys. Rev. E*, 2003, **68**, 041710.
37. D. J. Earl, M. A. Osipov, H. Takezoe, Y. Takanishi and M. R. Wilson, *Phys. Rev. E*, 2005, **71**, 021706.
38. A. S. Pandey, R. Dhar, M. B. Pandey, A. S. Achalkumar and C. V. Yelamaggad, *Liq. Cryst.*, 2009, **36**, 13.
39. C. V. Yelamaggad and M. Mathews, *Liq. Cryst.*, 2003, **30**, 125.
40. G. H. Hsiue and J. H. Chen, *Macromolecules*, 1995, **28**, 4366.
41. C.-S. Hsu and C.-H. Tsai, *Liq. Cryst.*, 1997, **22**, 669.
42. J. H. Chen, G. H. Hsiue and C. P. Hwang, *Chem. Mater*, 1997, **9**, 51.
43. P. Archer and I. Dierking, *Soft Matter*, 2009, **5**, 835.

CHAPTER 13

Polymer-stabilized Blue Phase Liquid Crystal Displays

Y. LI

National Engineering Lab for TFT-LCD Materials and Technologies, Department of Electronic Engineering, Shanghai Jiao Tong University, Shanghai 200240, China
Email: yan.li@sjtu.edu.cn

13.1 Introduction

Blue phases (BPs)[1–3] are liquid crystal (LC) mesophases that exist within a narrow temperature range between chiral nematic and isotropic phases. When the first of these compounds was discovered,[4] it was blueish in appearance, hence their name. On a microscopic level, BPs are composed of double-twist cylinders (DTCs). Inside each DTC, LC directors are twisted from -45° to $+45^\circ$ around any radius of the cylinder. There are three types of BPs: BPI, BPII, and BPIII, in order of ascending temperature. They are categorized according to the arrangements of their DTCs. BPI and BPII are both lattice structures; the former is body-centered cubic and the latter is simple cubic.[2] BPIII, however, is an amorphous structure. The cubic structures of BPI and BPII are shown in Figure 13.1(a).

In such structures, it is impossible to make the LC directors match everywhere. Thus, defects occur at the points where DTCs are in contact. These defects tend to destabilize the BP structures. Thus, the temperature

Soft Matter Series No. 8
Polymer-modified Liquid Crystals
Edited by Ingo Dierking

Published by the Royal Society of Chemistry, www.rsc.org

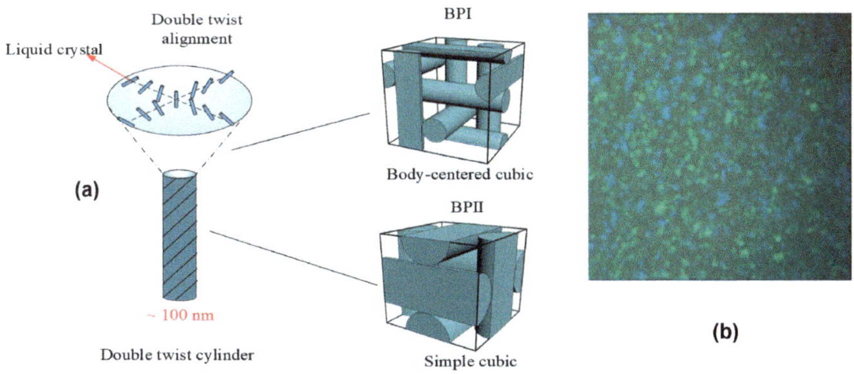

Figure 13.1 (a) DTCs in the cubic structures of BPI and BPII. (b) Platelet texture of a PS-BPLC composite under a polarizing microscope.

range over which a BP exists is usually very narrow, hindering the use of blue phase liquid crystals (BPLCs) in practical applications.

Many approaches,[5–17] such as polymer stabilization, polymer templates, and nanoparticle doping, have been attempted to widen the temperature range of BPs. Among them, the most commonly used method is the polymer stabilization method proposed by Kikuchi *et al.* in 2002.[14] They added a small fraction of a monomer and photo initiator to a BP system, which consisted of an LC host and a chiral dopant. When the mixture was exposed to ultraviolet (UV) light at BP temperature, a cross-linked polymer network was selectively concentrated along the defect lines, and thus the DTC structure was stabilized over a wide temperature range (>60 K). Recently, Li *et al.* realized BP stabilization using visible laser light.[18–20]

Polymer-stabilized blue phase liquid crystals (PS-BPLCs) are attractive for display[21] and photonic[22] applications due to several features: a sub-millisecond gray-to-gray response time;[23] natural self-assembly, which eliminates the need for alignment layers; a quasi-isotropic voltage-off state; and a reasonably wide temperature range. However, there are several challenges to overcome before PS-BPLCs can be widely implemented: high operating voltage, relatively low optical efficiency, and hysteresis.[24] The most critical problem for polymer-stabilized blue phase liquid crystal displays (PS-BPLCDs) is the need to reduce operating voltage to less than 10 V for a-Si thin-film transistor (TFT) driving. In recent decades, extensive research work has been performed to improve the performance of PS-BPLCDs in both material and device aspects. PS-BPLCs hold great potential for applications in transmissive, reflective, and transflective displays.

In this chapter, we start by introducing BPLC materials and their physical properties, and then delve into some typical examples of transmissive, reflective, and transflective PS-BPLCDs.

13.2 Physical Properties of PS-BPLCs

13.2.1 Optical Properties of PS-BPLCs Without Electric Field

In the lattice structures of BPs, the local refractive index variation results in selective Bragg reflections. The maximum Bragg reflection wavelength, λ_B, can be determined by[25]

$$\lambda_B = \frac{2na}{\sqrt{h_{\mathrm{m}}^2 + k_{\mathrm{m}}^2 + l_{\mathrm{m}}^2}}, \tag{13.1}$$

where n and a denote the average refractive index and lattice constant of BPs, respectively; and h_{m}, k_{m}, and l_{m} are the Miller indices of the crystal plane. In BPI, the lattice constant a corresponds to one pitch length, whereas in BPII, a corresponds to one-half pitch length. Figure 13.1(b) shows the characteristic platelet texture of a PS-BPLC under a polarizing microscope. There are multiple Bragg reflections due to different crystal plane orientations. For display applications, the Bragg reflection wavelengths are usually tuned to the UV region so that the BPLCs are actually transparent in the visible spectrum.

For wavelengths that are far from the Bragg reflection region, BPs appear quasi-optically isotropic. As light passes through a PS-BPLC cell, it experiences an isotropic refractive index and a weak optical rotatory effect.[26] The latter is caused by the average twisting power of DTCs over multiple randomly oriented domains.

13.2.2 Electric Field Effects

In BPLCs, there are three main types of electric field effects:[27–32] local reorientation, lattice distortion, and phase transition. At different field strengths, different effects dominate the process, resulting in various electro-optical properties.

When a relatively low electric field ($E < E_c$) is applied to a BPLC cell, local reorientation dominates.[30] LC molecules inside the DTCs tend to reorientate parallel to the electric field for a positive $\Delta\varepsilon$ LC (or perpendicular to the electric field for a negative $\Delta\varepsilon$ LC).[33] Here, E_c, the critical electric field, is a material parameter determined by the stiffness of the polymer network.[30,34,35] Thanks to the small diameter of the DTCs (usually several hundred nanometers), the response time tends to be in the submillisecond range, which is much faster than the Fréedericksz transition[36] in nematic LC devices. The relaxation time (with the same applied voltage) of local reorientation, τ_1, is insensitive to cell gap variation, but mainly determined by the rotational viscosity (γ_1), elastic constant k, and pitch length P as[32]

$$\tau_1 \approx \gamma_1 P^2 / [k(2\pi)^2]. \tag{13.2}$$

From mean field theory,[37] the elastic constant k is related to the nematic order parameter as $k \sim S^2$, where $S \sim (1 - T/T_c)^{\beta}$ is the approximated expression when the temperature is not too close to T_c (e.g., $T_c - T > 1$ °C) using Haller's semi-empirical equation.[38] Here, T is the absolute temperature, T_c is the clearing temperature of a PS-BPLC composite, and β is a material constant. According to the modified Arrhenius model,[39,40] the rotational viscosity is $\gamma_1 \sim S\exp(E_a/k_BT)$, where E_a is the activation energy of molecular rotation and k_B is the Boltzmann constant. Therefore, the temperature dependence of τ_1 is obtained as[41]

$$\tau_1 \approx \frac{\gamma_1 P^2}{k(2\pi)^2} \sim \frac{S\exp(E_a/k_BT)P^2}{S^2(2\pi)^2} \approx B\frac{\exp(E_a/k_BT)}{(1 - T/T_c)^{\beta}}. \tag{13.3}$$

PS-BPLCs usually have relatively low clearing temperatures compared to nematic LCs, and thus their response time is much more sensitive to temperature variation.

When an electric field higher than E_c is applied, the electrostriction effect starts to manifest.[30] Under such a strong field, the polymer network and lattice structure of BPs become deformed. According to eqn (13.1), the electric-field-induced change in the lattice constant a would also lead to a shift in the Bragg reflection wavelength.[42] Lattice distortion usually involves a large number of LC molecules, and the response time is in the range of several milliseconds or longer. The total relaxation time, τ, is determined by both location reorientation and lattice distortion. Thus, double relaxation can be observed in some relaxation–time curves.[30] In addition, the electrostriction effect leads to a discrepancy in the electro-optical properties of PS-BPLCs when they are driven by ascending and descending voltages. This phenomenon, referred to as hysteresis, is commonly observed in chiral and polymer network systems.[24] Hysteresis affects the accuracy of gray-level control and should therefore be minimized.

An irreversible process, a phase transition, can occur when the electric field is sufficiently high. A BP may transfer to another BP or to a chiral nematic phase, and then ultimately to a nematic phase.[2,43] The response time of such a phase transition is usually several seconds or longer.

Therefore, PS-BPLCs should be operated with electric fields of strength less than E_c to ensure fast response and negligible hysteresis. A high polymer concentration is helpful for increasing E_c as well as suppressing electrostriction and phase transition.[30]

Macroscopically, PS-BPLCs can be regarded as a Kerr medium. Along the E field direction, the refractive index is increased to $n_e(E)$, whereas in the orthogonal directions, it decreases to $n_o(E)$. The induced birefringence can be expressed by the Kerr effect in the low field region as[44]

$$\Delta n_{\text{ind}} = n_e(E) - n_o(E) = \lambda K E^2, \tag{13.4}$$

where K is the Kerr constant and λ is the wavelength.

The induced birefringence should gradually saturate in a finite material system. Yan *et al.* proposed an extended Kerr model to describe the saturation phenomenon in the high electric field region:[45]

$$\Delta n_{\mathrm{ind}}(E) = \Delta n_s[1 - \exp(-(E/E_s)^2)], \tag{13.5}$$

where Δn_s is the saturation birefringence and E_s the saturation electric field. The Kerr constant can then be expressed as

$$K = \Delta n_s/(\lambda E_s^{\,2}). \tag{13.6}$$

According to Gerber's model, the Kerr constant is jointly determined by several parameters:[32]

$$K \approx \frac{\Delta n \Delta \varepsilon}{k}\frac{\varepsilon_0 P^2}{\lambda(2\pi)^2}, \tag{13.7}$$

where Δn and $\Delta\varepsilon$ are the birefringence and dielectric anisotropy of the LC host, respectively; k is the average elastic constant; and P is the pitch length of a BPLC composite.

Because of the $\Delta n/\lambda$ term, the Kerr constant is optical frequency dependent or wavelength dependent.[46] In the visible spectrum, according to the single band model, Δn is proportional to $\frac{\lambda^2 \lambda^{*2}}{\lambda^2 - \lambda^{*2}}$, where λ^* is the resonant wavelength of the LC composite.[47] Thus, the optical dispersion relation of the Kerr constant can be obtained as[46]

$$K = A\frac{\lambda \lambda^{*2}}{\lambda^2 - \lambda^{*2}}, \tag{13.8}$$

where is A is a proportional constant.

Because of the $\Delta\varepsilon$ term, K also varies with the frequency of the alternating current electric field.[48–50] The dielectric dispersion property of the Kerr constant follows the extended Cole–Cole equation[49]

$$K(f) = K_\infty + (K_s - K_\infty)\frac{1 + \left(\frac{f}{f_r}\right)^{1-\alpha}\sin\frac{1}{2}\alpha\pi}{1 + 2\left(\frac{f}{f_r}\right)^{1-\alpha}\sin\frac{1}{2}\alpha\pi + \left(\frac{f}{f_r}\right)^{2(1-\alpha)}}, \tag{13.9}$$

where f_r is the relaxation frequency; K_s and K_∞ are the Kerr constant at static and high frequency, respectively; and α is a material constant between 0 and 1. Moreover, depending on the polarity of $\Delta\varepsilon$, the Kerr constant of a PS-BPLC can be positive or negative. Negative Kerr constants are usually very small because of the small $|\Delta\varepsilon|$ of negative LC hosts.[33] For most display applications, PS-BPLC materials with a large positive Kerr constant are used to obtain a low operating voltage.

The parameters Δn, $\Delta\varepsilon$, and k are also related to the nematic order parameter S and temperature T.[37,47] Assuming pitch P does not change with temperature, the Kerr constant can be approximated as:

$$K \approx \frac{\Delta n \cdot \Delta\varepsilon}{k} \frac{\varepsilon_0 P^2}{\lambda(2\pi)^2} \sim \Delta n_0 S \frac{S}{T} \frac{1}{S^2} \frac{\varepsilon_0 P^2}{\lambda(2\pi)^2} \sim \frac{1}{T}, \tag{13.10}$$

where ε_0 is the permittivity in vacuum and Δn_0 is the birefringence when $T=0$. To satisfy the boundary condition that at the clearing point, K should vanish, eqn (13.10) is modified to[41]

$$K \sim \frac{1}{T} - \frac{1}{T_c}. \tag{13.11}$$

The Kerr constant is a crucial parameter of PS-BPLCs. The operating voltage V_o is approximately proportional to $\frac{1}{\sqrt{K}}$ according to the Kerr effect.[41] From the material perspective, according to eqn (13.10), the Kerr constant can be increased by employing large Δn, $\Delta\varepsilon$ host LCs,[51–56] shifting the pitch length P to the infrared region,[57,58] and reducing k.[59–61]

According to eqn (13.2) and (13.7), by increasing the elastic constant k (*e.g.*, using a stronger polymer network), relaxation process could be made faster, but the tradeoff is a decreased Kerr constant and increased voltage. By increasing the pitch length P, the Kerr constant can be enhanced; however, the response slows as a result. Enlarging the dielectric anisotropy of the LC host, $\Delta\varepsilon$, also enhances the Kerr constant, but its associated high viscosity, γ_1, dramatically increases the response time. Therefore, improving the Kerr constant may result in an inferior response time, and *vice versa*. To comprehensively evaluate the electro-optical performance of PS-BPLCs, a figure of merit is defined as[62]

$$FoM = \frac{K}{\tau}. \tag{13.12}$$

A high figure of merit requires a large Kerr constant, K, and a fast response time, τ. As both K and τ are temperature dependent, for a given material and device structure, the optimal temperature can be identified when the figure of merit is maximized.

13.3 Modeling Physics of BPLCDs

For most PS-BPLCDs, the brightness modulation of a pixel is realized by a phase retardation change, which is a result of field-induced birefringence. Phase retardation, Γ , is the difference in the optical phases of the o and e waves when light propagates in an anisotropic medium. For a uniform medium, the phase retardation for normal incidence can be expressed as

$$\Gamma = \frac{d_{opt}(n_{eff} - n_o)2\pi}{\lambda}, \tag{13.13}$$

where d_{opt} is the light travelling distance, λ is the wavelength, n_o is the ordinary refractive index experienced by the o wave, and n_{eff} is the effective refractive index experienced by the e wave. n_{eff} can be determined by the angle between the light propagation direction and optical axis when the refractive indices n_o and n_e are known. For a transmissive PS-BPLCD, the BPLC cell is sandwiched between crossed polarizers. Transmittance, the ratio of output light intensity (with Γ phase retardation) under crossed polarizers to that under parallel polarizers (with zero phase retardation), can be modulated by Γ[63,64]

$$T_{tr} = \sin^2\frac{\Gamma}{2}. \tag{13.14}$$

Here we assume that the polarizers completely absorb the light polarized along the absorption axes and ignore the optical loss of the polarizers and BPLC cells, as well as interface reflections. When voltage is applied, each small BP unit is optically isotropic, so no phase retardation occurs, leading to a good dark state. When the phase retardation Γ is π at an appropriate voltage, the transmittance T_{tr} is maximized and a bright state is obtained. Induced birefringence, phase retardation, and transmittance can be tuned by varying the voltage. In most PS-BPLCDs, both the intensity and direction of the electric field and induced birefringence vary from position to position. Thus, it is necessary to use numerical modeling to calculate the electro-optical properties of PS-BPLCDs.

To accurately calculate the LC response to an external field in PS-BPLC devices, both Landau free energy and electric energy should be considered, which is an extremely complicated process.[65,66] Macroscopically, the electric field effect can be simplified into induced birefringence, which is the Kerr or extended Kerr effect. Then the anisotropic refractive index distribution in equilibrium under a constant field distribution can thus be obtained.[67]

Figure 13.2(a) shows the flow chart for the simplified PS-BPLCD modeling under a constant field distribution. The first step is to compute the potential distribution inside a PS-BPLCD by solving the Laplace equation and then determining the electric field distribution in the medium. The most common numerical means of obtaining the potential distribution are the finite difference and finite element methods. Commercial software such as Techwiz, Dimos, and Comsol can be used to determine the potential distribution. The second step is to calculate the induced birefringence distribution based on the Kerr or extended Kerr effect, and assign the local optic axis direction of each unit along the electric direction. If the Kerr model is used in this step, the calculated birefringence should be confined to less than the intrinsic birefringence of the BPLC system.[67] Next, the extended Jones matrix[68] method is employed to compute the related electro-optical properties, such as the voltage-dependent transmittance (VT) curve, viewing angle, and color shift.

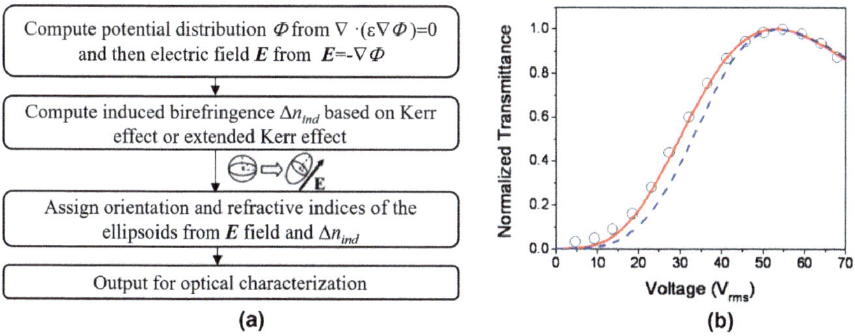

Figure 13.2 (a) Flowchart of BPLCD modeling. (b) Model fittings with experimental data of an in-plane-switching (IPS) cell using the truncated Kerr model (dashed blue line) and extended Kerr effect (solid red line). Circles are data obtained from ref. 69. (b) is reproduced from ref. 45 with permission from AIP Publishing, Copyright 2010.

In step 3, although the Kerr effect can be used to calculate the refractive index difference $\Delta n_{\text{ind}} = n_e - n_o$ given the Kerr constant and operation wavelength, the absolute values of n_o and n_e are not clearly revealed. To obtain n_o and n_e, deductions are performed based on a few approximate assumptions that are reasonable and supported by experimental results.[45,70] The detailed procedures of the deduction are explained as follows:

When $n_e - n_o$ is small, the average refractive index $n_{\text{average}} = \sqrt{(2n_e^2 + n_o^2)/3}$ can be approximated by $(2n_e + n_o)/3$.[45] Assuming that this n_{average} does not vary with electric field, then $n_{\text{average}} = n_{\text{iso}}$, where n_{iso} is the isotropic refractive index in the voltage-off state. Hence, the ordinary refractive index change under an electric field can be expressed as[45]

$$\begin{aligned}\delta n = n_{\text{iso}} - n_o(E) &= \frac{2n_e(E) + n_o(E)}{3} - n_o(E) \\ &= \frac{n_e(E) - n_o(E)}{3} = \frac{\Delta n_{\text{ind}}(E)}{3}.\end{aligned} \tag{13.15}$$

Thus, we can obtain

$$n_o(E) = n_{\text{iso}} - \frac{\Delta n_{\text{ind}}(E)}{3} \tag{13.16}$$

and

$$n_e(E) = n_{\text{iso}} + \frac{2\Delta n_{\text{ind}}(E)}{3}. \tag{13.17}$$

The value of n_{iso} can be directly measured using the Abbe refractometer in the absence of an electric field.[70]

Figure 13.2(b) displays the normalized voltage-dependent transmittance (normalized to the peak transmittance) of an in-plane-switching (IPS) cell. The circles denote experimental data. The blue curve is simulated using the Kerr model, and the red curve is computed from the extended Kerr model. Close agreement is achieved between the simulated and experimental results. The extended Kerr model is slightly more accurate than the Kerr model in this case.

When the electrode dimension is small (*e.g.*, less than 5 μm), the transmittance in some PS-BPLC cells is significantly higher than that predicted by the Kerr or extended Kerr models.[71] Xu *et al.* proposed a new model that accounts for the refraction effect to explain this phenomenon.[72] When exposed to a nonuniform electric field, BPLCs exhibit different n_o and n_e in different positions (layers). Thus, refraction occurs at the interfaces between subdivided layers, bending light and increasing its optical path length. As a result, phase retardation increases in a portion of the light, and thus the average transmittance increases.

The Kerr and extended Kerr models are the most commonly used models for BPLCDs. The Kerr model is available in some commercial software such as Techwiz and Dimos. The numerical modelling methods provide a useful tool for understanding the underlying physics of PS-BPLCDs, and for improving their performance by optimizing display structures.

13.4 Transmissive PS-BPLCDs

Transmissive PS-BPLCDs, which require backlight illumination, can achieve excellent performance indoors, with high contrast ratios and vivid color. However, their performance is substantially decreased under bright ambient lighting such as that found outdoors. The fast response of BPLCs not only reduces motion blurring of images but also enables color-sequential operation without color breakup.[73] The latter in particular can triple both the optical efficiency and resolution density by removing color filters.

13.4.1 IPS Mode

The IPS mode is one of the oldest PS-BPLCD operation modes[67] and is commonly used in labs for measuring the electro-optical properties of PS-BPLC materials. The device configuration of a typical IPS cell is depicted in Figure 13.3(a). The interdigital strip electrodes (pixel and common electrodes) are formed on the bottom substrate. The strip orientation is set at 45° with respect to the transmission axes of the crossed polarizers. The electrode width is w, the gap between adjacent electrodes is l, and the cell gap is d.

In the voltage-off state, a PS-BPLC is quasi-optically isotropic, so no phase retardation is induced in the light path. With no change in the incident polarization, a dark state is achieved between crossed polarizers.

When a voltage is applied between the pixel and common electrodes, a nonuniform electric field distribution is generated and birefringence is

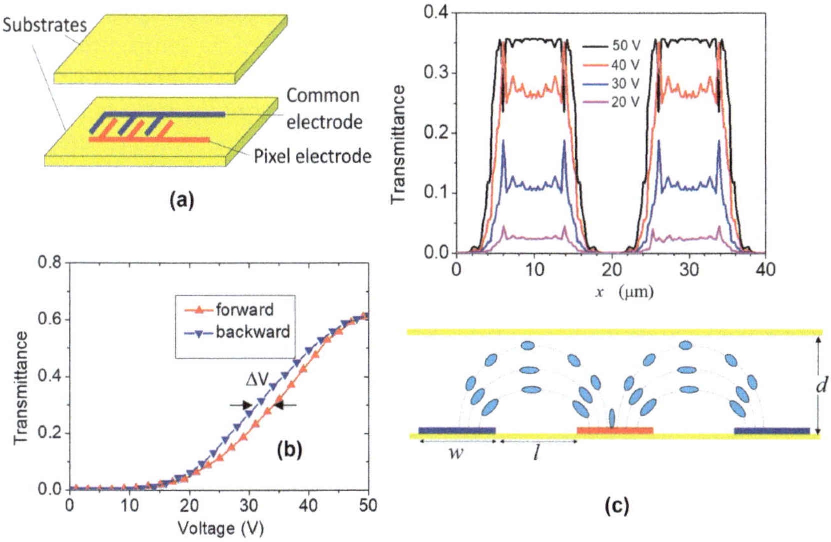

Figure 13.3 (a) Device configuration of an IPS cell. Measured VT curves (b) and simulated transmittance distribution curves at different voltages (c) of an IPS PS-BPLC cell with $w = 10$ μm and $l = 10$ μm at 633 nm. The Kerr constant for the simulation is $K = 13$ nm V^{-2}.
(b) is adapted from ref. 74 with permission from AIP Publishing, Copyright 2011.

induced according to the extended Kerr effect. The electric field has its strongest horizontal component near the patterned stripe electrode edges, leading to high transmittance as shown in Figure 13.3(c). However, right on top of the electrodes, where the electric field is mostly vertical, the optical axis of the refractive index ellipsoid is almost parallel to the direction of light propagation, resulting in low phase retardation. Thus, the transmittance is almost zero. Such regions are often referred to as dead zones.

According to the Laplace equation $\nabla^2\Phi = 0$, where Φ is the electric potential, the dissipation distance of the electric field from the bottom LC cell surface in the longitudinal z-direction is approximately proportional to $w + l$.[21] For an IPS cell with $w = 5$ μm and $l = 10$ μm, the penetration depth is ~3 μm, whereas for one with $w = 2$ μm and $l = 4$ μm, it is ~2 μm.[67] Therefore, the induced birefringence is mainly confined near the bottom substrate. Because of the limited penetration depth, a strong electric field is needed to accumulate sufficient phase retardation and transmittance. This explains why the operating voltage is usually high in IPS modes. For an exemplary configuration with $w = 10$ μm and $l = 10$ μm, peak transmittance is achieved at ~50 V, as shown in Figure 13.3(b). But, on the positive side, once the cell gap exceeds the penetration depth, its electro-optical performance is insensitive to cell gap variation. This cell gap insensitivity is particularly attractive for large panel manufacturing because it loosens the requirement for cell gap uniformity.

The existence of dead zones makes the average transmittance of the IPS PS-BPLC cell relatively low. The peak transmittance of the configuration with $w = 10$ μm and $l = 10$ μm is approximately 61%, as shown in the VT curve in Figure 13.3(b). To achieve high transmittance, a large l/w ratio is favorable, but a large l in turn increases the operating voltage. For the same configuration, the operating voltage is more or less inversely proportional to the square root of the Kerr constant.[21] Thus, employing a BPLC material with a large Kerr constant is essential for achieving low operating voltage.

In an IPS PS-BPLC cell, hysteresis is usually observable. For display applications, hysteresis is defined as the ratio of half-maximum transmittance voltage between forward and backward voltage scans (ΔV) to the maximum transmittance voltage (V_p), as shown in Figure 13.3(b). The ratio for the IPS cell ($w = 10$ μm and $l = 10$ μm) using a BPLC material Chisso JC-BP01M is measured as 5.8%;[75] the ratio should be less than 5% for practical applications.[24] Hysteresis in BPs is mainly caused by electrostriction.[30] When the electric field exceeds E_c, the DTC structure starts to unwind and the LC relaxation process does not follow the same route as the rising path. From the material perspective, polymer network strength plays a crucial role in determining the extent of hysteresis. A stronger polymer network is favorable for reducing hysteresis, but the usual tradeoff is increased operating voltage. From the device perspective, it is critical to keep the peak electric field below E_c to suppress electrostriction. The electric field distribution is highly non-uniform in the IPS mode. Although the electric field is below E_c in most LC regions, the maximum electric field, which occurs near the electrode edges, is much higher than the critical field and causes noticeable hysteresis. Therefore, efforts have been devoted to improving electric field uniformity by optimizing curing conditions[76–78] and device configurations.[34,74,75]

When no voltage is applied, the optical behavior of BPLCs resembles that of a quasi-isotropic material. In simulations, assuming no scattering or optical rotatory effects occur, the off-state light leakage at off-axis incidence is merely derived from the effective angle deviation between two crossed linear polarizers (*i.e.*, two crossed linear polarizers at normal incidence are no longer perpendicular at an oblique incidence). When voltage is applied, the induced birefringence also exhibits a two-domain-like profile from the electric field distribution in the IPS mode. The iso-brightness contour of the IPS cell is thus more symmetric and the viewing angle is relatively large.

Figure 13.4(a) shows the simulated isocontrast plot of a simple IPS PS-BPLC cell with strip electrode width $w = 10$ μm and spacing $l = 10$ μm ($\lambda = 633$ nm, $K = 13$ nm V^{-2}) without any compensation film, whereas Figure 13.4(b) is the isocontrast plot of the same cell with a biaxial compensation film ($n_x = 1.5110$, $n_y = 1.5095$, and $N_z = 0.5$).[21] A contrast ratio more than 200 : 1 can be easily expanded to an angle range of more than 75°, which is comparable to that of a conventional four-domain IPS nematic LCD.

In reality, there is inevitably some scattering[79] in the visible range, even when the Bragg reflection is shifted to the UV region. Moreover, the optical

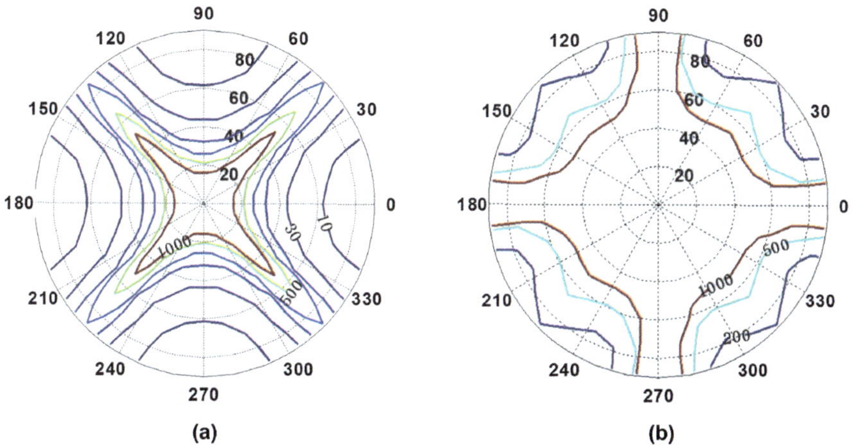

Figure 13.4 Simulated isocontrast plots of the IPS PS-BPLC cell (a) without any compensation film and (b) with a biaxial film, with $N_z = 0.5$ and $(n_x - n_y) \times d_{\text{film}} = \lambda/2$. IPS cell: $w = 10$ μm, $l = 10$ μm, $d = 10$ μm, $K = 13$ nm V^{-2}, and $\lambda = 633$ nm.

rotatory effect affects polarization as light transverses the PS-BPLC and further leads to light leakage in the dark state.[26] When optimal adjustments are made to the polarizer angle and Bragg reflection wavelength, the contrast ratio of a PS-BPLC cell is estimated to be approximately 3000 : 1 for white light, consisting of 60% green (514 nm), 30% red (633 nm), and 10% blue (457 nm) spectral content at normal incidence.[26]

The IPS mode of PS-BPLCDs has a simple planar structure comparable to that of conventionally fabricated LCDs.[80] Its electro-optical properties do not depend on cell gap variation, rendering it favorable for large-size panel manufacturing. However, several concerns have hindered the widespread application of IPS PS-BPLCDs: high operating voltage, relatively low transmittance, noticeable hysteresis, and slow charging time.[81] To overcome these problems, new device configurations[82–90] such as etched substrates,[91] double-sided in-plane switching,[92,93] protrusion electrodes,[94–97] corrugated electrodes,[98] and vertical field switching (VFS) structures[74,75,99] have been proposed.

13.4.1.1 Etched Substrate

Rao *et al.* proposed an IPS structure with an etched substrate for PS-BPLCDs.[91] Through this etching, the single-sided IPS structure can achieve approximately 30% lower operating voltage than the IPS mode without etching. As shown in Figure 13.5(a), the glass substrate between the strip electrodes is etched at a depth h that enables the electric field to penetrate both sides of the substrate. Consequently, phase retardation also accumulates from BPLC molecules above and within the etched part of the substrate. With a double penetration depth, the required induced birefringence

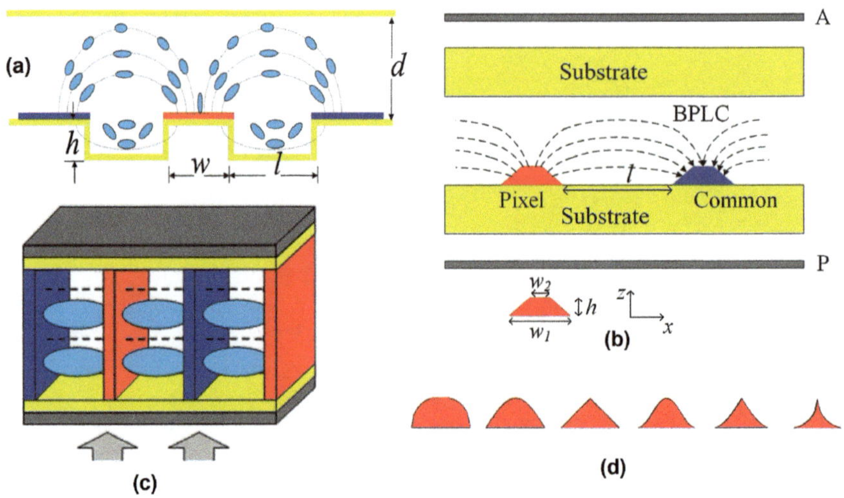

Figure 13.5 Schematic structures of PS-BPLCDs with an etched substrate (a), trapezoidal protrusion electrodes (b), and wall electrodes (c). (d) Variation in the protrusion electrode shape.

is smaller and the voltage is lowered. A larger etching depth h indicates more BPLC molecules in the etched region contributing to transmittance, and thus a lower operating voltage is required. However, once h exceeds the downward penetration depth of the electric field, the voltage reduction ceases. According to the simulation of Rao *et al.*, the downward penetration depth for a configuration with $w = 5$ μm and $l = 10$ μm is approximately 2 μm, and that for a configuration with $w = 2$ μm and $l = 4$ μm is approximately 1 μm. Such a shallow etching depth can be easily realized using the wet etching process.

13.4.1.2 Protrusion Electrode

Employing protrusion electrodes is a relatively effective approach for reducing the operating voltage of PS-BPLCDs.[94,95] When Indium-Tin-Oxide (ITO) electrodes are coated on top of transparent protrusions, which are usually a few micrometers high, a much stronger horizontal electric field is generated between the protrusion electrodes and the field penetrates much deeper into the BPLC bulk. For a trapezoidal protrusion as shown in Figure 13.5(b) with $w_1 = 2$ μm, $w_2 = 1$ μm, $h = 2$ μm, $l = 4$ μm, and a Kerr constant of $K = 10$ nm V^{-2}, the simulated peak transmittance is approximately 71% at 17 V. Compared with the conventional IPS structure with electrode width $w = 2$ μm, electrode gap $l = 4$ μm, peak transmittance voltage V_o =38 V, and peak transmittance = 66.5%, the protrusion electrode structure exhibits a significantly lower driving voltage and slightly higher transmittance. In such structures, directly on top of the trapezoidal electrodes, the electric field is still mostly vertical, barely introducing phase retardation or transmittance. The trapezoidal shape of the protrusions renders the dead-zone ratio much smaller than that in the conventional IPS mode. Consequently, the average transmittance is higher.

The dimensions and shape of protrusions play critical roles in determining the electro-optical performances of PS-BPLCDs. As the thickness of trapezoidal electrodes (with a specific taper angle) increases, the operating voltage continues to decrease because more BPLC molecules are driven by the horizontal electric field and help to increase phase retardation. When the thickness of electrodes becomes substantial, the structure is usually referred to as a "wall-shaped electrode" structure. In the special case depicted in Figure 13.5(c), when the electrode thickness reaches the maximum (the gap of the BPLC cell) $h = d$, and electrode width $w = w_1 = w_2$, the horizontal electric field uniformly penetrates the entire BPLC bulk, resulting in fairly low voltage and uniform transmittance distribution. However, due to the dead zones created by the electrodes themselves, the average transmittance is limited to the ratio $l/(w + l)$.

On the other hand, if the protrusion thickness is fixed, a larger taper angle (steeper protrusions) leads to a lower operating voltage. According to their simulation result,[95] the display has the highest average transmittance for a 45° taper angle. Moreover, among various types of pyramid structures, such as ellipse, sine-curved, linear, S-shaped, reversed sine-curved, and reversed-ellipse structures [Figure 13.5(d)], the reversed-ellipse structure achieves the highest transmittance.[95]

Protrusion structures with a typical thickness of roughly 2 μm have been frequently used to obtain wide viewing angles in multidomain vertical alignment LCDs. In 2013, AU Optronics employed trapezoidal protrusion electrodes in a 10-inch active matrix PS-BPLCD panel at the Display Week of the Society for Information Display.[96]

13.4.1.3 Corrugated Electrode

In 2010, Jiao *et al.* proposed a corrugated electrode structure,[98] which is in principle different from the aforementioned designs, to simultaneously lower voltage and boost transmittance. The device structure is depicted in Figure 13.6(a). Both the top and bottom substrates are fabricated with large-period corrugated structures. The pixels and common electrodes are coated on the inner surfaces of the top and bottom substrates, respectively. The minimum distance between the pixel and common electrodes is defined as the cell gap d. In such a structure, although the optic axes of the induced refractive index ellipsoids are not completely oriented in the desired horizontal direction, the horizontal component is reasonably large. Furthermore, although the induced birefringence is small at low voltage, it is uniformly distributed inside the whole cell except at the turning edge areas. In this way, the phase retardation can be accumulated along the entire traveling distance inside the BPLC layer. Consequently, the on-state voltage is substantially reduced. Transmission dead zones can be found in the

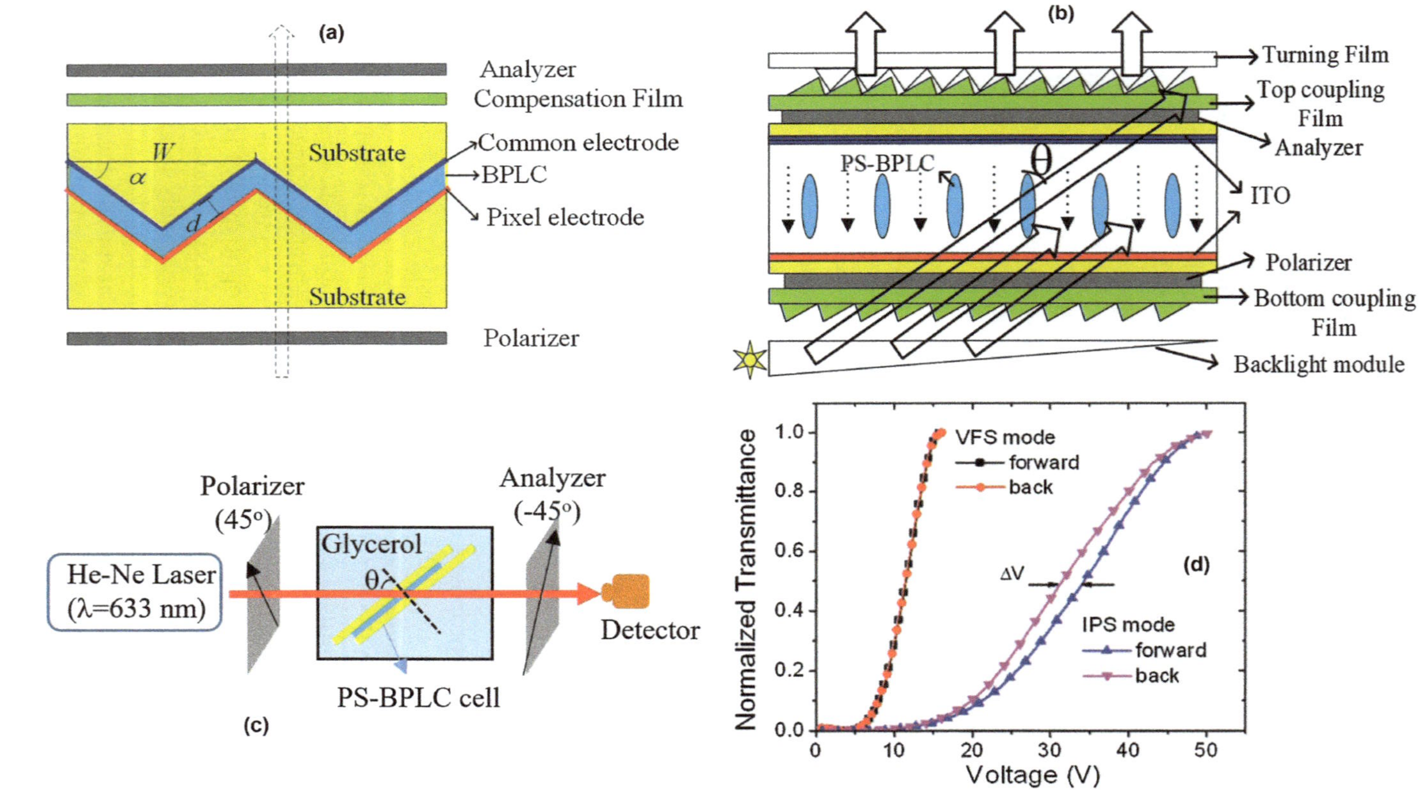

Figure 13.6 Schematic structures of PS-BPLCDs with corrugated electrodes (a) and in VFS mode (b). (c) Experimental setup for characterizing the VFS cell. (d) Measured VT curves and hysteresis of IPS and VFS cells. $\lambda = 633$ nm. Part (d) is reproduced from ref. 74 with permission from AIP Publishing, Copyright 2011.

turning edge areas because the induced birefringence mostly occurs along the vertical direction and makes no contribution to phase retardation. A larger period is correlated with a smaller dead-zone ratio and higher average transmittance. According to the simulations by Jiao *et al.*, which employed a PS-BPLC with a Kerr constant = 12.7 nm V^{-2}, low voltage of 9.9 V and high transmittance of 85.6% can be achieved when the period of the corrugation $W = 40$ μm, inclination angle $\alpha = 60°$, and cell gap $d = 3.5$ μm.

A larger inclination angle for the corrugated electrodes tends to reduce the operation voltage effectively. First, a larger inclination angle, α, means that a larger portion of the induced birefringence is in the desired horizontal direction, contributing to transmittance. Second, with a larger inclination angle, the effective path length of incident light inside the LC medium ($d/\cos\alpha$) is increased, so the requirement for induced birefringence decreases, lowering the operating voltage. A thinner cell gap is also helpful for this purpose. As the cell gap decreases, although the effective path length $d/\cos\alpha$ is decreased, the induced birefringence Δn_{ind} increases in a quadratic manner due to the stronger electric field. So, overall, a lower voltage can be realized with a thinner cell gap. Moreover, with a thinner cell gap, the dead zones become narrower if all the other parameters are kept the same, so transmittance increases.

The dimensions of the corrugated substrates are similar to those of backlight films such as turning films, and these substrates can be fabricated relatively easily using mold-pressing or printing methods. Although in Figure 13.6(a) the corrugations are triangular in the interest of simplicity, in practice, all the sharp edges should be round and smooth. The performance of the PS-BPLCD will not be strongly affected because these edges are dead zones with almost no transmittance. However, highly precise periodicity is required to maintain a uniform cell gap and avoid short circuiting between electrodes.

13.4.1.4 *Vertical Field Switching*

The previously discussed PS-BPLCD structures usually require complicated or unconventional electrode structures. Due to the existence of dead zones and the nonuniformity of induced birefringence, their overall transmittance is not high. In 2011, Cheng *et al.* proposed a VFS mode that significantly reduced operating voltage, achieving almost 100% transmittance, fast response, and negligible hysteresis with simple planar thin ITO electrodes.[74]

Its device configuration is depicted in Figure 13.6(b). The BPLC cell consists of two planar ITO glass substrates without an alignment layer. A uniform vertical electric field is generated within the BPLC bulk. For normally incident light, no phase retardation or transmittance is induced between cross polarizers. For oblique incidence, however, light undergoes uniform phase retardation. A coupling film is attached to the bottom substrate to couple the oblique input light and retain a large angle inside the BPLC cell. Otherwise, the refraction angle in the BPLC layer is dramatically reduced due to Snell's law. Coupling and turning films are placed above the top substrate and used to steer the output light toward the viewer in the normal direction.

In such a structure, a larger oblique angle, θ, is more favorable for achieving low operating voltage. When θ is larger, not only does the effective refractive index difference between *o* and *e* waves increase, but the light path length inside the PS-BPLC layer also lengthens. Consequently, a smaller induced birefringence and a lower voltage are required for realizing the same phase retardation. Because the induced birefringence is uniformly distributed within the cell, the transmittance at different lateral locations can reach the maximum simultaneously. Thus, the average transmittance of the BPLC cell can approach 100%.

To verify the effectiveness of the VFS structure, a simple experiment was performed to simulate the beam propagation path in the VFS mode, as shown in Figure 13.6(c). Such an experimental setup is also commonly adopted for measuring the electro-optical properties of PS-BPLC materials. A PS-BPLC cell is immersed in glycerol ($n = 1.47$ at $\lambda = 633$ nm). The index match of glass and glycerol enables light to pass through the BPLC at a very large oblique angle. Using a BPLC mixture Chisso JC-BP01M with a large Kerr constant of $K = 13$ nm V^{-2} (obtained by fitting a VT curve in a IPS cell) at 23 °C and $\lambda = 633$ nm, with a 70° incident angle and 5.74 μm cell gap, roughly 100% transmittance (normalized to parallel polarizers) is achieved at 16 V, as shown in Figure 13.6(d). The low operating voltage is roughly 3.2 times lower than that in an IPS cell ($w = 10$ μm, $l = 10$ μm, and $d = 7.5$ μm) filled with the same BPLC material. However, the Kerr constant obtained from the VT curve of the VFS cell is approximately 7.5 nm V^{-2}, roughly 70% smaller than that of the IPS cell. A possible explanation is that because the electric field varies greatly from the electrode edges to the center of the electrode gap in an IPS cell, the Kerr constant extracted through curve fitting has a larger uncertainty.

There is some optical loss due to the Fresnel reflection of the coupling and turning films, but the optical efficiency for the added films is more than 85% according to the simulation.[74] Therefore, the VFS mode should have comparable or higher optical efficiency than the IPS mode.

Moreover, the VFS mode exhibits almost zero hysteresis, whereas the IPS mode has a ΔV/Vp of approximately 5.8%. The VFS mode shows a twofold faster decay time than the IPS mode when released from peak-transmittance voltages. Both the hysteresis-free and fast-decay properties of the VFS mode can be attributed to how, with a uniform electric field well below the critical field, local reorientation dominates the switching process.

However, transmittance in such a configuration is highly sensitive to variation in the incident angle. A smaller angle, θ, leads to a less accumulation of phase retardation and lower transmittance, whereas a larger angle causes a greater accumulation of phase retardation and higher transmittance. A small divergence angle of backlight would result in considerable variation in transmittance. This means that without a compensation film, the VFS mode inherently has a narrow viewing angle. For instance, according to the simulation, when the incident angle is 70°, the contrast ratio quickly decreases to less than 10 : 1 when the oblique angle exceeds 10°. To enlarge the viewing angle, Cheng *et al.* proposed three methods:[99] adding compensation films, employing a wire-grid polarizer, and using an E-type analyzer.[100] Among them, the E-type analyzer method has exhibited the best performance. An alternative means of realizing a wide viewing angle for the VFS mode is to use a front diffusor or curved turning film combined with a well-collimated backlight.

The VFS mode exhibits excellent electro-optical performance aside from the narrow viewing angle. Therefore, its optimal application is in projectors, because viewing angle is not a problem with such equipment and more space is available for generating collimated light. He *et al.* proposed a color-sequential transmissive projector based on a VFS PS-BPLCD.[101]

13.5 Reflective PS-BPLCDs

Reflective PS-BPLCDs can be subdivided into reflective projectors and ambient-light-lit displays. For the former, the phase retardation required for peak brightness is halved because a backlight passes through a reflective panel twice. This achieves substantially lower operating voltage than that of transmissive displays. For the latter, ambient light is modulated by the PS-BPLC cell because the Bragg reflection intensity is continuously tuned by the voltage. Low power consumption and sunlight readability can be realized under these circumstances, but the displays become unreadable under dim ambient light conditions.

13.5.1 Reflective Projectors

Rao *et al.* proposed a reflective PS-BPLC projector display[102] based on liquid-crystal-on-silicon (LCOS) technology. As shown in Figure 13.7(a), an IPS BPLC panel is topped by a polarizing beam splitter (PBS). In the voltage-off state, the PS-BPLC is almost isotropic, causing no changes to the incidents polarization. So, the light is simply reflected back, leading to a dark state. By contrast, in the voltage-on state, the horizontal fringe field generated by the IPS electrodes introduces phase retardation to the s-polarized light. Hence, the reflected p-polarized light can pass through the PBS and reach the projection screen. The grayscale can be controlled by varying the applied voltage. The operating voltage in the reflective mode is approximately two-thirds that in the transmissive mode when the same electrode configuration (IPS $w = 10$ μm and $l = 10$ μm) and same material are employed. However, the voltage achieved using planar electrodes is still too high for LCOS displays. To reduce the voltage, protrusion electrodes were proposed in a previous study, which decreased the voltage to approximately 10 V.[102]

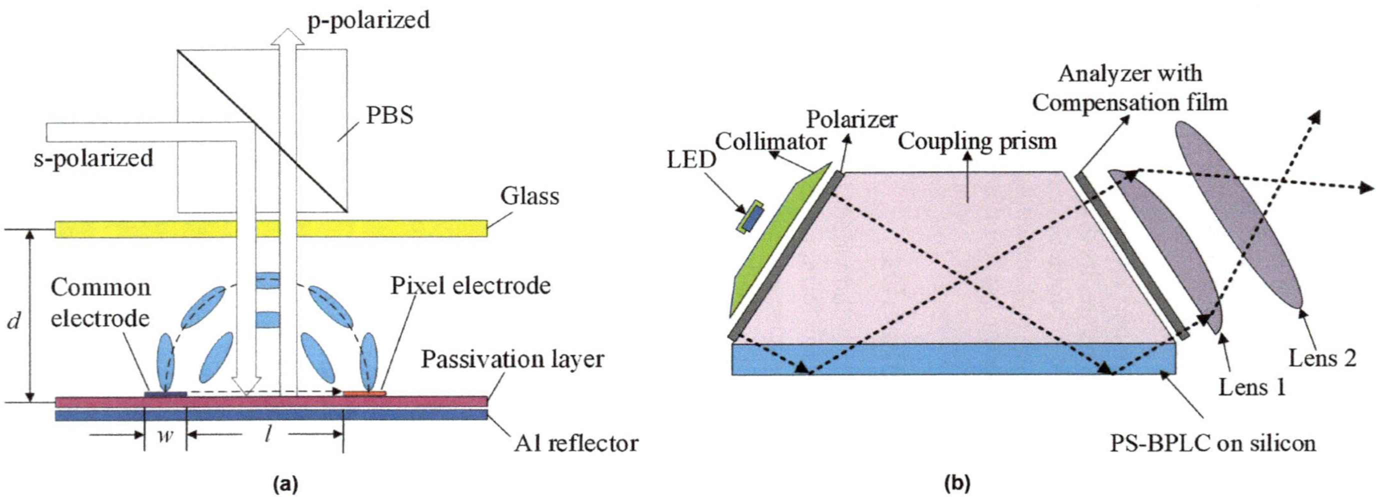

Figure 13.7 Basic device configurations of the reflective BPLC projection displays with IPS electrodes (a) and in VFS mode (b).

He *et al.* proposed a reflective-type projector based on a VFS PS-BPLCD.[101] Figure 13.7(b) demonstrates that large oblique incidence eliminates the need for a PBS in this design, distinguishing it from conventional LCOS projectors. The image degradation due to turning films can be eliminated using a large trapezoidal coupling prism. Moreover, the polarizer and analyzer can be placed perpendicular to the incident light, ensuring a favorable contrast ratio using conventional polarizers. A compensation film is laminated in front of the analyzer to compensate for the change in polarization due to reflection. When a PS-BPLC with a Kerr constant of $K = 13.7\ \mathrm{nm\,V^{-2}}$ and a cell gap of 4.87 μm is used, the operating voltage is lowered to 8 V. If a 1.5 μm cell is used, the voltage can reach 4.8 V, enabling TFT driving.

13.5.2 Direct-view Reflective PS-BPLCD Based on Bragg Reflection

Yan *et al.* proposed a direct-view reflective PS-BPLCD that uses ambient light to display images.[103] The working principle of this display is based on Bragg reflection, which is drastically different from the phase retardation effect in conventional LC displays. This display offers high sunlight readability and low power consumption and does not require a polarizer.

In conventional PS-BPLCDs, the Bragg reflection band is shifted to the UV region, which makes BPLCs appear almost optically transparent in the visible region. In a reflective PS-BPLCD, the pitch lengths are intentionally tuned to reflect vivid red, green, or blue through control of the chiral dopant concentration. The basic working principle is illustrated in Figure 13.8(a). Bragg reflection is strongest in the voltage-off state. As voltage increases, the LC molecules tend to align parallel to the electric field and perpendicular to the substrates. Hence, the local refractive index difference decreases and so does the reflectance. The voltage-dependent reflectances of the three primary color pixels are plotted in Figure 13.8(b,c,d). A painted absorption layer is added to the substrate to absorb the transmitted light, achieving a dark state. As the reflectance of each primary color can be continuously tuned by the electric field, a full color display can be realized by either laterally arranging the subpixels or stacking them. The submillisecond response time of the reflective PS-BPLCD allows for the display of dynamic videos without image blurring.

Although the lattice in pure BPLCs tends to become deformed when an electric field is applied, resulting in a shift in the reflection color,[27] the cubic structures in PS-BPLCs are stabilized by the polymer network so that lattice deformation only occurs in relatively strong fields.[42] Thus, this display is operated in the low field region, so that LC reorientation mainly occurs without noticeable lattice deformation, as shown in Figure 13.8.

Surface alignment has also been found to play a crucial role in generating a uniform domain and saturated colors in the voltage-off state. Both 90°-twisted and homogeneous cells exhibit stronger reflection and narrower

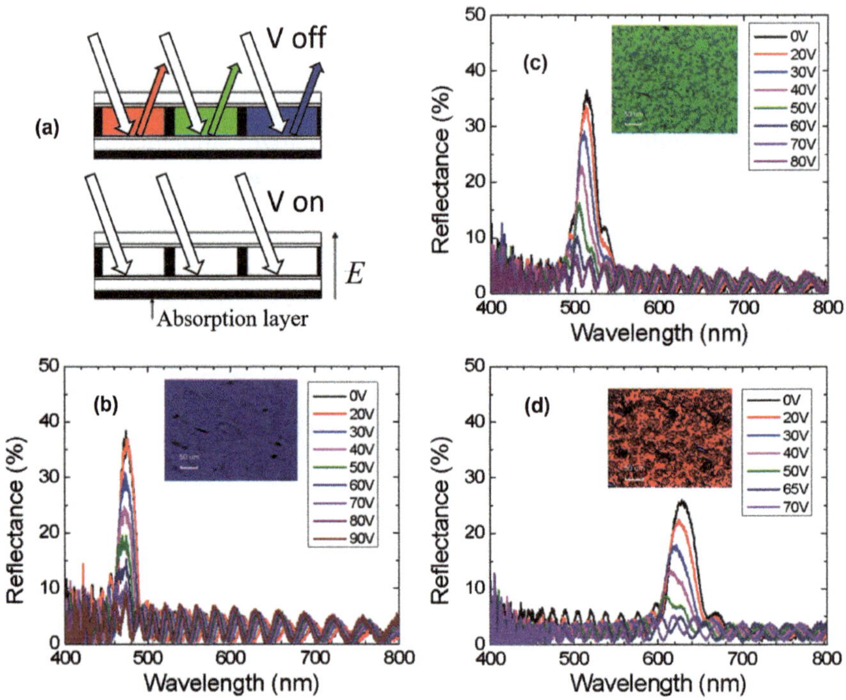

Figure 13.8 (a) Working principle of the reflective display based on Bragg reflection. Voltage-dependent reflection spectra of (b) blue cell, peak wavelength $\lambda = 477$ nm; (c) green cell, $\lambda = 514$ nm; and (d) red cell, $\lambda = 634$ nm. Inset: reflective microscope photographs with crossed polarizers. Parts (b), (c) and (d) are reproduced from ref. 103 with permission from AIP Publishing, Copyright 2013.

bandwidth than cells containing no polyimide if the cell gap is much larger than the lattice constant of the BP. Later, Chen *et al.* found that applying a vertical electric field before curing could also result in monodomain texture and a narrower reflection band.[104]

13.6 Transflective PS-BPLCDs

Transflective PS-BPLCDs integrate the features of transmissive and reflective displays. In the transmissive mode, backlight passes through the PS-BPLC cell once, whereas in the reflective mode, ambient light transverses it twice. The two modes can be operated simultaneously and independently. They are operational over a large dynamic range of ambient illumination conditions, from darkness to bright sunlight, at the cost of structural complexity. Because of the quasi-optical isotropy of a PS-BPLC in the off state, a wide viewing angle can be achieved in both transmissive and reflective modes in transflective PS-BPLCDs through a simple compensation scheme. In transflective PS-BPLCDs, there are two major challenges: (1) how to realize low operating voltage and (2) how to match reflectance and transmittance to enable single gamma curve driving.

A transflective-type PS-BPLCD employing protrusion electrodes was proposed by Li *et al.* in 2010.[105] The PS-BPLCD panel is sandwiched between crossed circular polarizers, as shown in Figure 13.9(a). Each pixel is divided into a transmissive (T) region and a reflective (R) region. The backlight passes through the T region once from the bottom, and ambient light is reflected by the bumpy reflector and transverses the BPLC cell twice in the R region. In both regions, trapezoidal protrusion electrodes are formed on the bottom substrate to generate a strong fringing field. The electrode gap in the T region is made smaller than that in the R region so that a stronger electric field and larger induced birefringence are generated in the T region. As a

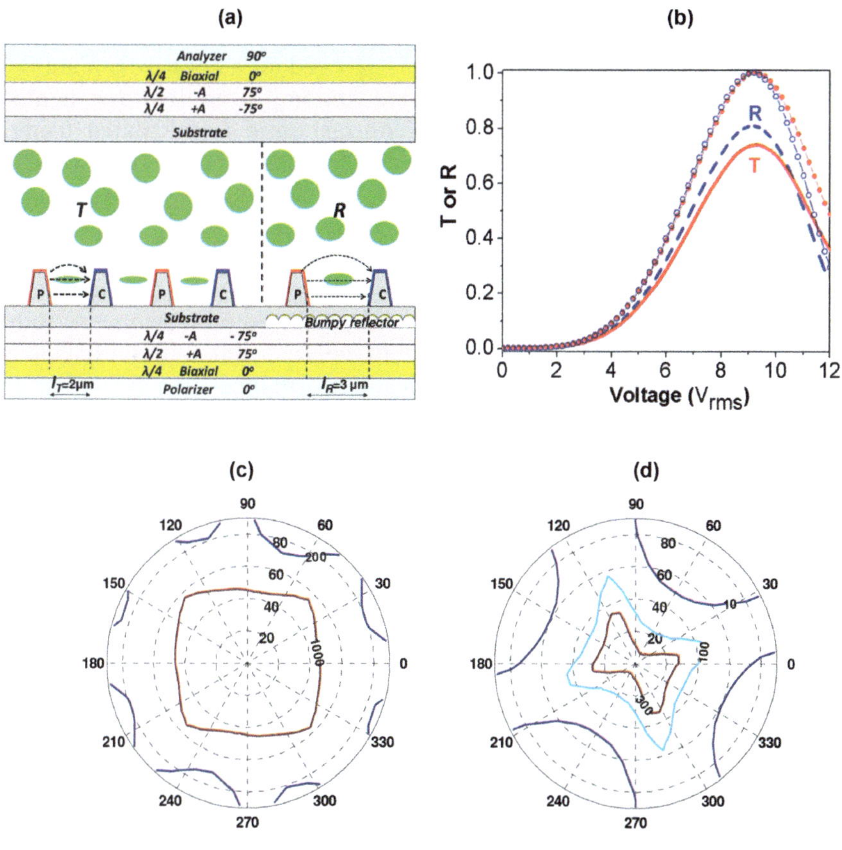

Figure 13.9 (a) Schematic structure of the transflective PS-BPLCD. (b) Simulated VT and VR curves for the proposed transflective BPLCD. The solid red and dashed blue lines represent simulated VT and VR curves, and the closed and open circles represent normalized transmittance and reflectance, respectively. Simulated isocontrast contour plots for the (c) T and (d) R modes of the proposed transflective LCD.
Reproduced from ref. 105, with permission from OSA, Copyright 2011.

result, the double passage of the ambient light through the R region can be thoroughly compensated, and reflectance and transmittance can be matched.

Simulation was performed of a transflective PS-BPLCD with the following parameters: $w_1 = 0.5$ μm, $w_2 = 1$ μm, $h = 2$ μm, $l_T = 2$ μm, and $l_R = 3$ μm. The simulated VT and voltage-dependent reflectance (VR) curves for incident light with $\lambda = 550$ nm are depicted in Figure 13.9(b). The red and blue curves represent the transmittance and reflectance, respectively. The normalized transmittance and reflectance (closed and open circles) match closely, which indicates single gamma curve driving could be enabled.

To obtain the normally black mode for both the T and R regions, two broadband and wide-view circular polarizers[106] were used. To achieve a more favorable contrast ratio over wide viewing angles, two biaxial $\lambda/4$ plates can be placed below the top linear polarizer and above the bottom linear polarizer, respectively, as shown in Figure 13.9(a). The quasi-optically isotropic off state of a PS-BPLC means that there is no need for a negative C-plate[107] or an in-cell phase retarder,[108] which are generally required for nematic transflective-LCDs.[109,110] The average contrast ratio is calculated using a simulation that assumes white light is 60% green ($\lambda = 550$ nm), 30% red ($\lambda = 650$ nm), and 10% blue ($\lambda = 450$ nm). As shown in Figure 13.9(c), the contrast ratio in the T region is higher than 1000 : 1 within a 45° viewing cone and higher than 100 : 1 over the entire viewing cone. In the R region, contrast ratio is higher than 10 : 1 over a 50° viewing cone. In more recent studies of transflective PS-BPLCDs[82,111–119] enhanced protrusion electrodes, double penetration electrodes, corrugated electrodes and so on, have been employed to improve their electro-optical performances.

13.7 Conclusion

In conclusion, we have introduced the physical properties of PS-BPLC materials, discussed the modeling physics of PS-BPLCD devices, and reviewed recent developments in transmissive, reflective, and transflective PS-BPLCDs. The most critical concern regarding PS-BPLCDs is the realization of operating voltages less than 10 V to enable amorphous-silicon TFT driving. This requires both material advances and novel device design.

Microscopically, a BP lattice structure composed of DTCs leads to Bragg reflection. Macroscopically, the refractive index of PS-BPLCs can vary from isotropic to anisotropic, following the Kerr effect. Gerber's model explains the dependence of the Kerr constant on wavelength, electric frequency, and temperature. The Kerr constant is essential for realizing low operating voltage. Recently, a record high Kerr constant of 33.1 nm V^{-2} was obtained using a large $\Delta\varepsilon$ and Δn host LC.

Numerical modeling methods for PS-BPLCDs, including the Kerr and extended Kerr models, were illustrated. A refraction model, which is more accurate when the electrode dimension is small, was also introduced.

Numerical modeling is valuable for understanding the electro-optical properties of various PS-BPLCD devices. Moreover, it provides a useful tool for novel PS-BPLCD device design and optimization.

For transmissive PS-BPLCDs, a number of electrode structure designs, including IPS structures with etched substrates, corrugated electrodes, protrusion electrodes, and VFS mode were proposed to lower the operating voltage and improve transmittance. With a large-Kerr-constant BPLC material an operating voltage of less than 10 V, high transmittance, submillisecond response time, and negligible hysteresis have been realized.

Reflective PS-BPLC projectors employing IPS and VFS modes were presented. Because light transverses a PS-BPLC cell twice, the operating voltage is further reduced. A direct-view PS-BPLCD based on Bragg reflection was also demonstrated. The polarizer- and color-filter-free display provides vivid color, fast response, and good sunlight readability.

Transflective PS-BPLCDs with a single cell gap, reasonably high optical efficiency, matched transmittance and reflectance, relatively low operating voltage, and wide viewing angles were discussed. Both sunlight readability and favorable indoor performance were demonstrated to be achievable.

With promising progress in PS-BPLC materials and display devices, alignment-free and fast-response PS-BPLCDs hold great potential for various display applications.

Acknowledgements

Support was provided by the National Natural Science Foundation of China (61727808).

References

1. H. Kikuchi, in *Liquid Crystalline Functional Assemblies and Their Supramolecular Structures*, ed. T. Kato, Springer, Berlin, Heidelberg, Germany, 2008, p. 99.
2. P. P. Crooker, in *Chirality in Liquid Crystals*, ed. H.-S. Kitzerow and C. Bahr, Springer, New York, NY, 2001, p. 186.
3. D. Xu, F. Peng and S. T. Wu, in *Handbook of Visual Display Technology*, ed. J. Chen, W. Cranton and M. Fihn, Springer, Berlin, Heidelberg, 2015, p. 1.
4. F. Reinitzer, *Monatshefte für Chemie und verwandte Teile anderer Wissenschaften*, 1888, **9**, 421.
5. W. He, G. Pan, Z. Yang, D. Zhao, G. Niu, W. Huang, X. Yuan, J. Guo, H. Cao and H. Yang, *Adv. Mater.*, 2009, **21**, 2050.
6. A. Yoshizawa, Y. Kogawa, K. Kobayashi, Y. Takanishi and J. Yamamoto, *J. Mater. Chem.*, 2009, **19**, 5759.
7. L. Wang, W. He, X. Xiao, Q. Yang, B. Li, P. Yang and H. Yang, *J. Mater. Chem.*, 2012, **22**, 2383.

8. L. Wang, W. He, Q. Wang, M. Yu, X. Xiao, Y. Zhang, M. Ellahi, D. Zhao, H. Yang and L. Guo, *J. Mater. Chem. C*, 2013, **1**, 6526.
9. L. Wang, W. He, M. Wang, M. Wei, J. Sun, X. Chen and H. Yang, *Liq. Cryst.*, 2013, **40**, 354.
10. B. Li, W. He, L. Wang, X. Xiao and H. Yang, *Soft Matter*, 2013, **9**, 1172.
11. G. Zhu, X. W. Lin, W. Hu, Z. G. Zheng, H. F. Wang, H. Q. Cui, D. Shen and Y. Q. Lu, *Opt. Mater. Express*, 2011, **1**, 1478.
12. H. Liu, D. Shen, X. Wang, Z. Zheng and S. Li, *Opt. Mater. Express*, 2016, **6**, 436.
13. H. S. Kitzerow, H. Schmid, A. Ranft, G. Heppke, R. A. M. Hikmet and J. Lub, *Liq. Cryst.*, 1993, **14**, 911.
14. H. Kikuchi, M. Yokota, Y. Hisakado, H. Yang and T. Kajiyama, *Nat. Mater.*, 2002, **1**, 64.
15. F. Castles, F. V. Day, S. M. Morris, D. H. Ko, D. J. Gardiner, M. M. Qasim, S. Nosheen, P. J. W. Hands, S. S. Choi and R. H. Friend, *Nat. Mater.*, 2012, **11**, 599.
16. H. Yoshida, Y. Tanaka, K. Kawamoto, H. Kubo, T. Tsuda, A. Fujii, S. Kuwabata, H. Kikuchi and M. Ozaki, *Appl. Phys. Express*, 2009, **2**, 121501.
17. E. Karatairi, B. Rozic, Z. Kutnjak, V. Tzitzios, G. Nounesis, G. Cordoyiannis, J. Thoen, C. Glorieux and S. Kralj, *Phys. Rev. E: Stat., Nonlinear, Soft Matter Phys.*, 2010, **81**, 041703.
18. Y. Yuan, L. Yan, C. C. Ping, R. Na, W. Li, L. Xiao, H. Wei, P. Zhou, J. Lu and Y. Su, *SID Symp. Dig. Tech. Pap.*, 2015, **46**, 549.
19. Y. Yuan, Y. Li, C. P. Chen, S. Liu, N. Rong, W. Li, X. Li, P. Zhou, J. Lu and R. Liu, *Opt. Express*, 2015, **23**, 20007.
20. N. Rong, Y. Li, X. Li, P. Zhou, S. Liu, J. Lu and Y. Su, *J. Disp. Technol.*, 2016, **12**, 1008.
21. Z. Ge, S. Gauza, M. Jiao, H. Xianyu and S. T. Wu, *Appl. Phys. Lett.*, 2009, **94**, 101104.
22. Y. Li, S. Huang, P. Zhou, S. Liu, J. Lu, X. Li and Y. Su, *Adv. Mater. Technol.*, 2016, **1**, 1600102.
23. K. M. Chen, S. Gauza, H. Xianyu and S. T. Wu, *J. Disp. Technol.*, 2010, **6**, 49.
24. K. M. Chen, S. Gauza, H. Xianyu and S. T. Wu, *J. Disp. Technol.*, 2010, **6**, 318.
25. R. J. Miller and H. F. Gleeson, *Phys. Rev. E: Stat. Phys., Plasmas, Fluids, Relat. Interdiscip. Top.*, 1995, **52**, 5011.
26. Y. Liu, Y. F. Lan, H. Zhang, R. Zhu, D. Xu, C. Y. Tsai, J. K. Lu, N. Sugiura, Y. C. Lin and S. T. Wu, *Appl. Phys. Lett.*, 2013, **102**, 131102.
27. H. S. Kitzerow, *Mol. Cryst. Liq. Cryst.*, 1991, **202**, 51.
28. G. Heppke, M. Krumrey and F. Oestreicher, *Mol. Cryst. Liq. Cryst.*, 1983, **99**, 99.
29. H. S. Kitzerow, P. P. Crooker, S. L. Kwok, J. Xu and G. Heppke, *Phys. Rev. A: At., Mol., Opt. Phys.*, 1990, **42**, 3442.

30. D. Xu, J. Yan, J. Yuan, F. Peng, Y. Chen and S. T. Wu, *Appl. Phys. Lett.*, 2014, **105**, 011119.
31. P. Pierański, P. E. Cladis, T. Garel and R. Barbet-Massin, *J. Phys.*, 1986, **47**, 139.
32. P. R. Gerber, *Mol. Cryst. Liq. Cryst.*, 1985, **116**, 197.
33. Y. Li, Y. Chen, J. Yan, Y. Liu, J. Cui, Q. Wang and S. T. Wu, *Opt. Mater. Express*, 2012, **2**, 1135.
34. L. Rao, J. Yan, S. T. Wu, Y. C. Lai, Y. H. Chiu, H. Y. Chen, C. C. Liang, C. M. Wu, P. J. Hsieh, S. H. Liu and K. L. Cheng, *J. Disp. Technol.*, 2011, **7**, 627.
35. J. Yan, Y. Chen, S. T. Wu, S. H. Liu, K. L. Cheng and J. W. Shiu, *J. Appl. Phys.*, 2012, **111**, 063103.
36. D. K. Yang and S. T. Wu, *Fundamentals of Liquid Crystal Devices*, Wiley, Hoboken, NJ, USA, 2006.
37. W. Maier and A. Saupe, *Zeitschrift für Naturforschung A*, 1960, **15**, 287.
38. I. Haller, *Prog. Solid State Chem.*, 1975, **10**, 103.
39. S. T. Wu and C. S. Wu, *Phys. Rev. A: At., Mol., Opt. Phys.*, 1990, **42**, 2219.
40. S. T. Wu and C. S. Wu, *Liq. Cryst.*, 1990, **8**, 171.
41. L. Rao, J. Yan and S. T. Wu, *J. Soc. Inf. Disp.*, 2010, **18**, 954.
42. S. Y. Lu and L. C. Chien, *Opt. Lett.*, 2010, **35**, 562.
43. F. Porsch and H. Stegemeyer, *Liq. Cryst.*, 1987, **2**, 395.
44. J. Kerr, *Philos. Mag.*, 1875, **50**, 337.
45. J. Yan, H. C. Cheng, S. Gauza, Y. Li, M. Jiao, L. Rao and S. T. Wu, *Appl. Phys. Lett.*, 2010, **96**, 071105.
46. M. Jiao, J. Yan and S. T. Wu, *Phys. Rev. E: Stat., Nonlinear, Soft Matter Phys.*, 2011, **83**, 041706.
47. S. T. Wu, *Phys. Rev. A: At., Mol., Opt. Phys.*, 1986, **33**, 1270.
48. H. Xianyu, S. T. Wu and C. L. Lin, *Liq. Cryst.*, 2009, **36**, 717.
49. Y. Li, Y. Chen, J. Sun, S. T. Wu, S. H. Liu, P. J. Hsieh, K. L. Cheng and J. W. Shiu, *Appl. Phys. Lett.*, 2011, **99**, 181126.
50. S. Yabu, H. Yoshida, G. Lim, K. Kaneko, Y. Okumura, N. Uehara, H. Kikuchi and M. Ozaki, *Opt. Mater. Express*, 2011, **1**, 1577.
51. L. Rao, J. Yan, S. T. Wu, S. I. Yamamoto and Y. Haseba, *Appl. Phys. Lett.*, 2011, **98**, 081109.
52. Y. Chen, D. Xu, S. T. Wu, S. I. Yamamoto and Y. Haseba, *Appl. Phys. Lett.*, 2013, **102**, 141116.
53. T. Ishitani, Y. Niikura, M. Ikenaga, M. Kobayashi, M. Kato, T. Nagi, Y. Oe, M. Nakano, S. Seo and Y. Hirakata, *SID Symp. Dig. Tech. Pap.*, 2012, **43**, 18.
54. Y. Haseba, S.-I. Yamamoto, K. Sago, A. Takata and H. Tobata, *SID Symp. Dig. Tech. Pap.*, 2013, 254.
55. M. Wittek, N. Tanaka, D. Wilkes, M. Bremer, D. Pauluth, J. Canisius, A. Yeh, R. Yan, K. Skjonnemand and M. Klasen-Memmer, *SID Symp. Dig. Tech. Pap.*, 2012, **43**, 25.
56. M. Wittek, N. Tanaka, M. Bremer, D. Pauluth, K. Tarumi, M. C. Wu, D. M. Song and E. S. Lee, *SID Symp. Dig. Tech. Pap.*, 2011, **42**, 292.

57. H. Choi, H. Higuchi and H. Kikuchi, *Soft Matter*, 2011, 7, 4252.
58. J. Yan, Z. Luo, S. T. Wu, J. W. Shiu, Y. C. Lai, K. L. Cheng, S. H. Liu, P. J. Hsieh and Y. C. Tsai, *Appl. Phys. Lett.*, 2013, **102**, 011113.
59. J. L. Zhu, S. B. Ni, Y. Song, E. W. Zhong, Y. J. Wang, C. P. Chen, Z. Ye, G. He, D. Q. Wu, X. L. Song, J. G. Lu and Y. Su, *Appl. Phys. Lett.*, 2013, **102**, 071104.
60. J. L. Zhu, S. B. Ni, C. P. Chen, X. L. Song, C. Y. Chen, J. G. Lu and Y. Su, *Liq. Cryst.*, 2014, **41**, 891.
61. J. Yan and S. T. Wu, *J. Disp. Technol.*, 2011, **7**, 490.
62. J. Yan, Y. Chen, S. T. Wu and X. L. Song, *J. Disp. Technol.*, 2013, **9**, 24.
63. S. T. Wu, U. Efron and L. D. Hess, *Appl. Opt.*, 1984, **23**, 3911.
64. S. T. Wu, U. Efron and L. D. Hess, *Appl. Phys. Lett.*, 1984, **44**, 1033.
65. B. P. D. Gennes and J. Prost, *The Physics of Liquid Crystals*, 2nd edn, Clarendon Press, Oxford, UK, 1995.
66. S. Meiboom, J. P. Sethna, P. W. Anderson and W. F. Brinkman, *Phys. Rev. Lett.*, 1981, **46**, 1216.
67. Z. Ge, L. Rao, S. Gauza and S. T. Wu, *J. Disp. Technol.*, 2009, **5**, 250.
68. P. Yeh, *J. Opt. Soc. Am.*, 1982, **72**, 507.
69. H. Kikuchi, Y. Haseba, S. I. Yamamoto, T. Iwata and H. Higuchi, *SID Symp. Dig. Tech. Pap.*, 2009, **40**, 578.
70. J. Yan, M. Jiao, L. Rao and S. T. Wu, *Opt. Express*, 2010, **18**, 11450.
71. K. M. Chen, J. Yan, S. T. Wu, Y. P. Chang, C. C. Tsai and J. W. Shiu, *J. Disp. Technol.*, 2011, 7, 362.
72. D. Xu, Y. Chen, Y. Liu and S. T. Wu, *Opt. Express*, 2013, **21**, 24721.
73. P. Bos, T. Buzak and R. Vatne, *SID Symp. Dig. Tech. Pap.*, 1985, **26**, 157.
74. H. C. Cheng, J. Yan, T. Ishinabe and S. T. Wu, *Appl. Phys. Lett.*, 2011, **98**, 261102.
75. H. C. Cheng, J. Yan, T. Ishinabe, N. Sugiura, C. Y. Liu, T. H. Huang, C. Y. Tsai, C. H. Lin and S. T. Wu, *J. Disp. Technol.*, 2012, **8**, 98.
76. Y. Liu, S. Xu, D. Xu, J. Yan, Y. Gao and S. T. Wu, *Liq. Cryst.*, 2014, **41**, 1339.
77. C. Y. Fan, H. C. Jau, T. H. Lin, F. C. Yu, T. H. Huang, C. Liu and N. Sugiura, *J. Disp. Technol.*, 2011, 7, 615.
78. D. Xu, J. Yuan, M. Schadt and S. T. Wu, *Appl. Phys. Lett.*, 2014, **105**, 081114.
79. C. W. Chen, H. C. Jau, C. T. Wang, C. H. Lee, I. C. Khoo and T. H. Lin, *Opt. Express*, 2012, **20**, 23978.
80. R. Kiefer, B. Weber, F. Windscheid and G. Baur, *Proc. Jpn. Display*, 1992, **92**, 547.
81. C. D. Tu, C. L. Lin, J. Yan, Y. Chen, P. C. Lai and S. T. Wu, *J. Disp. Technol.*, 2013, **9**, 3.
82. Y. Li and S. T. Wu, *J. Disp. Technol.*, 2011, 7, 359.
83. K. S. Bae, Y. J. Lee, C. J. Yu, H. Lee, J. H. Park and J. H. Kim, *Jpn. J. Appl. Phys.*, 2012, **51**, 072201.
84. M. K. Kim, M. S. Kim, B. G. Kang, G. H. Yang, A. Mukherjee, S. W. Kang, H. Lee, S. T. Shin and S. H. Lee, *Mol. Cryst. Liq. Cryst.*, 2011, **543**, 194/[960].

85. Z. Su, Y. Chen, J. Lu and X. Tang, *Liq. Cryst.*, 2013, **40**, 976.
86. J. L. Mao, J. Wang, H. X. Fan and Q. H. Wang, *Liq. Cryst.*, 2016, **43**, 535.
87. J. F. Algorri, V. Urruchi and J. M. Sánchez-Pena, *Liq. Cryst.*, 2015, **42**, 941.
88. H. Fan, J. Cui and Q.-H. Wang, *Liq. Cryst.*, 2015, **42**, 481.
89. H. Fan, Q. H. Wang, J. Cui and P. Tang, *Liq. Cryst.*, 2015, **42**, 41.
90. Y. Sun, Y. Zhao, Y. Li, P. Li and H. Ma, *J. Disp. Technol.*, 2014, **10**, 797.
91. L. Rao, H. C. Cheng and S. T. Wu, *J. Disp. Technol.*, 2010, **6**, 287.
92. Y. Chen, Y. Sun and G. Yang, *Liq. Cryst.*, 2011, **38**, 555.
93. H. C. Jau, P. H. Liao, H. W. Li, H. K. Hsu, C. H. Chen, C. C. Wang and T. H. Lin, *J. Soc. Inf. Disp.*, 2012, **20**, 351.
94. L. Rao, Z. Ge, S. T. Wu and S. H. Lee, *Appl. Phys. Lett.*, 2009, **95**, 231101.
95. S. Yoon, M. Kim, M. S. Kim, B. G. Kang, M. K. Kim, A. K. Srivastava, S. H. Lee, Z. Ge, L. Rao, S. Gauza and S. T. Wu, *Liq. Cryst.*, 2010, **37**, 201.
96. C. Y. Tsai, F. C. Yu, Y. F. Lan, P. J. Huang, S. Y. Lin, Y. T. Chen, T. I. Tsao, C. T. Hsieh, B. S. Tseng and C. W. Kuo, *SID Symp. Dig. Tech. Pap.*, 2015, **46**, 542.
97. M. Kim, M. S. Kim, B. G. Kang, M.-K. Kim, S. Yoon, S. H. Lee, Z. Ge, L. Rao, S. Gauza and S. T. Wu, *J. Phys. D: Appl. Phys.*, 2009, **42**, 235502.
98. M. Jiao, Y. Li and S. T. Wu, *Appl. Phys. Lett.*, 2010, **96**, 011102.
99. H. C. Cheng, J. Yan, T. Ishinabe, C. H. Lin, K. H. Liu and S. T. Wu, *J. Disp. Technol.*, 2012, **8**, 627.
100. P. I. Lazarev and M. V. Paukshto, *J. Soc. Inf. Disp.*, 2001, **9**, 101.
101. S. He, J. H. Lee, H. C. Cheng, J. Yan and S. T. Wu, *J. Disp. Technol.*, 2012, **8**, 352.
102. L. Rao, S. He and S. T. Wu, *J. Disp. Technol.*, 2012, **8**, 555.
103. J. Yan, S. T. Wu, K. L. Cheng and J. W. Shiu, *Appl. Phys. Lett.*, 2013, **102**, 081102.
104. Y. Chen and S. T. Wu, *Appl. Phys. Lett.*, 2013, **102**, 171110.
105. Y. Li, M. Jiao and S. T. Wu, *Opt. Express*, 2010, **18**, 16486.
106. Z. Ge, M. Jiao, R. Lu, T. X. Wu, S. T. Wu, W. Y. Li and C. K. Wei, *J. Disp. Technol.*, 2008, **4**, 129.
107. C. H. Lin, Y. R. Chen, S. C. Hsu, C. Y. Chen, C. M. Chang and A. Lien, *J. Disp. Technol.*, 2008, **4**, 123.
108. C. Doornkamp, B. M. I. van der Zande, S. J. Roosendaal, L. W. G. Stofmeel, J. J. van Glabbeek, J. T. M. Osenga and J. A. M. Steenbakkers, *J. Soc. Inf. Disp.*, 2004, **12**, 233.
109. X. Zhu, Z. Ge, T. X. Wu and S. T. Wu, *J. Disp. Technol.*, 2005, **1**, 15.
110. Z. Ge and S. T. Wu, *Transflective Liquid Crystal Displays*, John Wiley & Sons, Chichester, West Sussex, UK, 2010.
111. J. Wang, J. L. Mao, H. X. Fan and Q. H. Wang, *Chin. Phys. B*, 2016, **25**, 4.
112. Y. Li, S. Huang, N. Rong, J. G. Lu, C. P. Chen, X. Li, P. Zhou, Y. Yuan, S. Liu, G. He and Y. Su, *J. Disp. Technol.*, 2016, **12**, 122.
113. P. Li, Y. Sun and Q. Wang, *Liq. Cryst.*, 2013, **40**, 1024.
114. D. Wu, Q. H. Wang, F. Zhou, J.-P. Cui and C. Q. Song, *J. Disp. Technol.*, 2011, 7, 459.

115. Z. Feng, C. Jian Peng, W. Qiong Hua, L. Da Hai and W. Di, *J. Disp. Technol.*, 2011, 7, 170.
116. J. P. Cui, F. Zhou, Q. H. Wang, D. Wu and D. H. Li, *J. Disp. Technol.*, 2011, 7, 398.
117. J. P. Cui, Q. H. Wang and F. Zhou, *J. Soc. Inf. Disp.*, 2011, **19**, 709.
118. H. Q. Cui, Z. C. Ye, W. Hu, X. Lin, T. C. Chung, T. S. Jen and Y. Q. Lu, *J. Inf. Disp.*, 2011, **12**, 115.
119. C. Song, Q. H. Wang, J. P. Cui, F. Zhou, L. Qi and D. H. Li, *J. Disp. Technol.*, 2011, 7, 250.

CHAPTER 14

Polymer Dissolved Liquid Crystals

INGO DIERKING

School of Physics and Astronomy, University of Manchester, Oxford Road, Manchester M13 9PL, UK
Email: ingo.dierking@manchester.ac.uk

14.1 Introduction

When modifying liquid crystals by polymers, one of the most obvious ways is the simple dissolution of polymers in the liquid host. There are surprisingly few studies that seem to cover this aspect, which is most likely due to the fact that phase separation occurs as the polymer content increases. Much of the general work has been summarized and reviewed by Mucha.[1] It appears that the general consensus is that different applications or uses of composite systems involving liquid crystals and polymers can be met by different classes of material. When the main goal is to fabricate a material of fibres in the most general sense, main-chain polymers seem to be the materials of choice. If polymer properties are desirable while maintaining some of the liquid crystalline advantages, such as electro-optic switching for instance, it is better to utilize the side-group polymers, as introduced by Finkelmann.[2] For modern electro-optic applications and displays, polymer dispersed liquid crystals (PDLC)[3] and polymer-stabilized liquid crystals (PSLC),[4] as discussed in detail earlier, are exploited. The first is mainly an advantage for self-supporting applications, such as addressable foils, because the liquid crystal is confined and encapsulated within a continuous polymer

Soft Matter Series No. 8
Polymer-modified Liquid Crystals
Edited by Ingo Dierking

Published by the Royal Society of Chemistry, www.rsc.org

matrix, while the second introduces stabilization of structures and director fields with only small negative effects on electro-optic parameters, such as response times and threshold voltages. Both are at polymer concentrations that induce phase separation, the first at high polymer concentrations with a continuous polymer matrix and dispersed liquid crystal droplets, while the second, at low polymer concentrations, represents a bi-continuous system.

14.2 Phase Diagrams

Investigations of polymer dissolved liquid crystals were in the past mainly carried out for the determination of phase separation curves, and the respective phase diagrams.[5–14] Most of the time, side-group polymers have been used as components, added to the low molar weight liquid crystal, but branched polymers have also been reported.[15] A typical phase diagram observed is depicted in Figure 14.1. In the scenario of phase separation, the kinetics are, of course, also of importance, as studied by Lin *et al.*[16]

A thermodynamic description was obtained by combining the Flory–Huggins[17,18] theory with the Maier–Saupe theory,[19,20] thus describing normal isotropic polymer mixing with the occurrence of nematic phases.

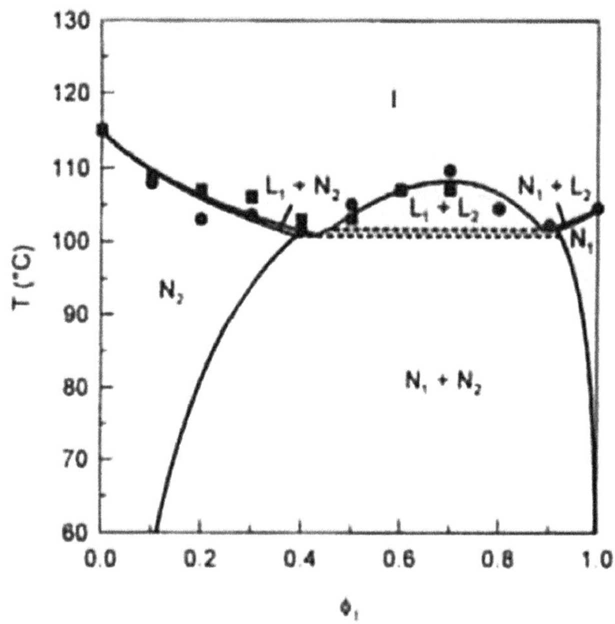

Figure 14.1 Typical phase diagram of a polymer – liquid crystal mixture. ϕ_1 is the fraction of liquid crystal molecules. Two-phase regions of a few degrees are indicated. The solid lines represent the theoretically calculated phase diagram.
Reproduced from ref. 9 with permission from Taylor and Francis.

This approach was put forward by Kelkar and Manohar,[21] and Shen and Kyu[22] to give the free energy as:

$$g = \phi \ln \phi + \frac{1-\phi}{r} \ln(1-\phi) + \chi\phi(1-\phi) - \phi\Sigma(S) - \frac{1}{2}\nu\phi^2 S^2 \qquad (14.1)$$

where ϕ is the volume fraction of the liquid crystal, r the number of lattice sites occupied by a polymer chain, S the nematic order parameter, $\chi = \mathrm{C} + \mathrm{D/T}$, a factor describing the liquid crystal polymer interaction, where C and D are constants and T is the temperature, and ν is a nematic interaction parameter, which depends on the induced dipole-induced dipole interactions, thus the polarizability of the liquid crystal molecules. Minimization of the free energy from this combined Flory–Huggins–Maier–Saupe model provides the phases observed as a function of concentration and temperature, thus the phase diagram. The model largely describes experimentally obtained phase diagrams as, for example, presented in reference.[16]

14.3 Rheology and Viscoelasticity

Besides the determination of phase diagrams, there are of course more practical aspects of polymer dissolved liquid crystals. One can be found at large polymer concentration with the liquid crystal doped into the polymer, which is often a side-group polymer. In these liquid crystal–polymer blends,[23,24] the liquid crystal acts very much like a softener in plastics and swelling of the polymer can be observed.[25,26] The effect has also been compared between isotropic solvents and liquid crystalline solvents.[27] This has initiated interest in liquid crystals as lubricants, *i.e.* in the science of tribology, which was reviewed by Carrion *et al.*[28]

A topic which has largely been researched is the effect of liquid crystals on the viscoelastic and rheological properties of polymers and *vice versa*, the effect of polymers dissolved in liquid crystals, as these are very pronounced and of fundamental importance for applications. A summary of results so far has been provided by Jamieson *et al.*[29] A clear reduction of the melt viscosity is observed,[30] and the rheological properties have been elucidated *via* dynamic light scattering,[31–33] as well as by electro-rheology for dissolved main-chain polymers[34] (see Figure 14.2) and side-group polymers[35] in liquid crystal hosts.

14.4 Photorefractivity

Another well-studied aspect is the photorefractive effects[36–38] of polymer dissolved liquid crystals. Switchable Bragg gratings (Figure 14.3) are of particular interest for high and low molar mass liquid crystal mixtures[39,40] and have also been demonstrated for materials involving fullerenes.[41] In general, the diffraction efficiency is found to be increased with increasing polymer content of the mixtures. Other related applications may be found in

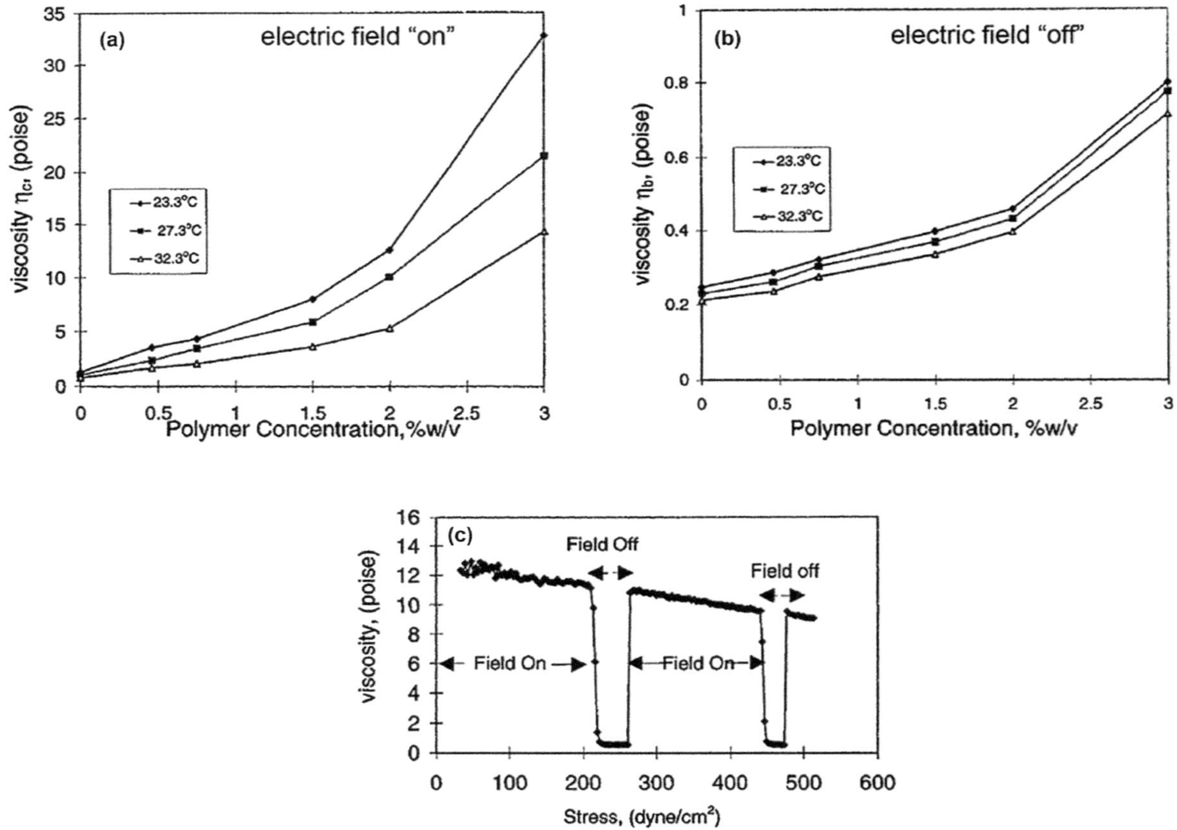

Figure 14.2 Electro-rheological viscosity with (a) electric field on and (b) electric field off. The viscosity under electric field application strongly increases by a factor larger than an order of magnitude (note the different scales on plots (a) and (b)). (c) The dynamics of this effect can be used to switch the viscosity between high and low values. In general the viscosity is found to decrease with increasing temperature, as expected, and increases with increasing polymer concentration.
Reproduced from ref. 34 with permission from American Chemical Society, Copyright 1997.

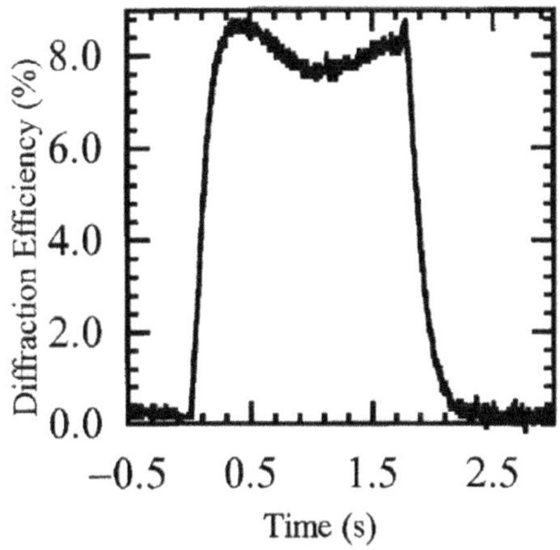

Figure 14.3 Demonstration of a switchable diffraction grating. Reproduced from ref. 39 with permission from AIP Publishing.

real-time holography.[42] Further, polymeric dyes have been added to low molar mass liquid crystals, to increase the lasing efficiency of tuneable LC lasers.[43,44]

Conjugated polymers in liquid crystal solutions were studied with respect to organic electronics and one-dimensional photoconductors[45] and also for hydrogen bonded systems.[46] It was also shown that dissolved conjugated polymers can stabilize the nematic mesophase through organizational coupling of the liquid crystal molecules to the polymer chains.[47] This means that dissolution of a polymer can broaden the range of mesogenic phase existence. We will return to this latter aspect below.

14.5 Polymer Dissolved Ferroelectric Liquid Crystals

So far, all of the discussions have been restricted to the nematic or chiral nematic (cholesteric) phase, thus phases with only orientational but no positional order. Of much interest for research and applications are ferroelectric SmC* liquid crystals (FLC). This is due to their very rapid response times, which are about two orders of magnitude faster for low molar mass FLCs than for nematics, at bi-stable (digital) switching, and even faster when the electroclinic effect is exploited. The latter effect is linear in electric fields and can produce sub-microsecond response times, although the modulation depth is rather limited, and the temperature regime narrowly localized around the SmA*–SmC* transition. Ruth *et al.*[48] investigated the interesting case where the ferroelectric liquid crystal was composed of the polysiloxane liquid crystal polymer and its side-group mesogen. This was done to assure uniform and thorough mixing of the polymer and the low molar mass mesogenic components at all concentrations.

It was found that the temperature width of the SmC* phase increased with increasing polymer content (Figure 14.4(a)). The spontaneous polarization,

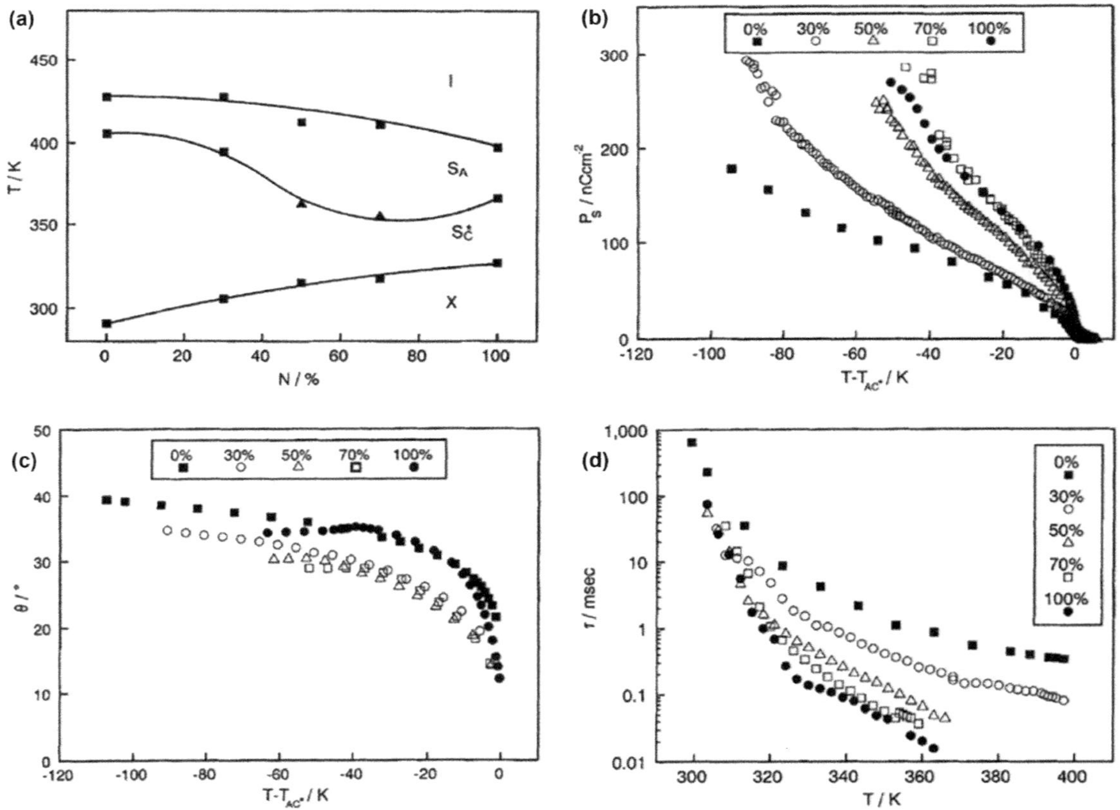

Figure 14.4 Electro-optic parameters of mixtures of a ferroelectric side-group polymer with its low molecular mass mesogenic group. (a) The ferroelectric SmC* phase is broadened. (b) The spontaneous polarization is reduced to approximately half the value of the low molecular ferroelectric when all the material is a side-group polymer, while (c) the tilt angle appears to go through a slight minimum, with both values for side-group polymer and low molar mass mesogen being practically equal. (d) The response time decreases with increasing temperature and increases for the polymer.
Reproduced from ref. 48 with permission from Taylor and Francis.

P_S, decreased for 100% side-group polymer to a value of about 2/3 of the low molar mass mesogen (Figure 14. 4(b)), while the tilt angle firstly only slightly decreased with increasing polymer concentration, and then increased again, eventually re-achieving its original value, when comparing all low molar mass FLC and all side-group polymers. The response times increased by about a factor of 100 when changing from concentrations of 100% low molar mass FLC to 100% side-group polymer (Figure 14.4(d)), but are still of the same order of magnitude as a fast switching nematic today, $\tau \approx 10$ ms. In a different system it has been shown that the electroclinic effect, which decreases to either side of the SmA*–SmC* transition, is maintained over a larger temperature range for increasing polymer concentration.[49] Also, these investigations indicate that simple dissolution of a polymer in a smectic liquid crystal can stabilize the phase.

14.6 Polymers Dissolved in Blue Phases and TGB Phases

As it has been pointed out several times above, Blue Phases have received a considerable amount of attention in recent years, due to the prospects of producing novel types of displays with excellent view characteristics and higher speeds than nematics, while having no need for alignment layers.[50] This is one of the production steps which causes many of the rejects, as often mechanical rubbing is involved, which may lead to dust particles. BP display prototypes were produced on the basis of polymer stabilization, which, under special conditions, enabled temperature ranges of about 60 K, including room temperature. This would be ideal for display production, but the conditions for which this is achieved are not yet transparent and depend on the polymer materials used, the exact mixture composition, and most likely also on polymerization conditions. Besides several other methods like the inclusion of chiral dopants, nano-particle dispersion, twist inversion compounds and bent-core dopants, Dierking *et al.*[51] also reported polymer stabilization of Blue Phases, with the moderate success of increasing the phase width from about 1 K to 8 K, *i.e.* by 800%. This is a relatively remarkable stabilization, considering that none of the components, neither the liquid crystal, nor the monomer, or the polymerization conditions were in any way optimized. But it is also surely not enough to produce a commercial display. Nevertheless, none of the alternative methods produced better stabilization than polymer networks. Also, combinations of different methods of stabilization did not increase the temperature range any further.

In the same group, Kasch *et al.*[52] then resorted to dissolving polystyrene in a (different material) Blue Phase. There was no crosslinking between the polymer chains, thus in fact real liquid crystal–polymer mixtures, not polymer stabilized materials, were used. For long chain polystyrene with high molecular weight (9500 $g\,mol^{-1}$, approximately 90 monomer repeat units), the effect of stabilization was again rather moderate, but as short chain

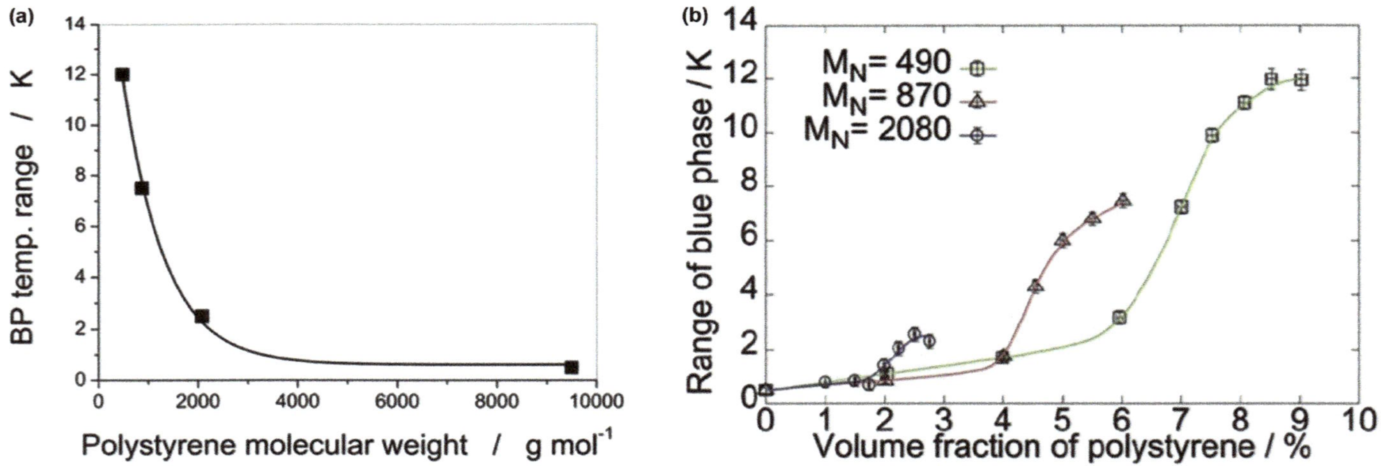

Figure 14.5 (a) With decreasing molecular weight of the liquid crystal dissolved polystyrene, the BP phase range is strongly enhanced. Extrapolation to a dimer leads to phase ranges of several tens of degrees. (b) The width of the Blue Phase also depends on concentration, rising sharply to saturation when all defects are filled with polystyrene. Due to the different molecular weights, thus chain lengths, investigated, this happens at different volume fractions.
Reproduced from ref. 52 with permission from the Royal Society of Chemistry.

polystyrene (5, 8, 20 monomer repeat units, corresponding to 490, 870, and 2080 g mol^{-1}, respectively) was mixed into the liquid crystal, a clear stabilization effect, in excess of any other method employed before, was observed (Figure 14.5(a)). In fact, the stabilization of the Blue Phase increased from 0.5 K to 12 K, *i.e.* by 2400%. The stabilizing effect of the low molecular weight polystyrene is in this case dependent on the concentration of the oligomer/polymer material within the liquid crystal (Figure 14.5(b)). It is plausible that this could also explain the observation that many of the Blue Phases of substantial broadness are observed with dimers,[53,54] which increase flexoelectricity and thereby stabilizes the BP structure.[55] Using chiral BP dimers would then be equivalent to extrapolating Figure 14.5(b) to 100% volume fraction, or to extrapolating Figure 14.5(a) to a molecular weight of 200 g mol^{-1}. The latter does in fact lead to a Blue Phase which is stable over 50 K as it was experimentally observed for dimers in reference.[53]

It should be mentioned that the above scenario can also be explained theoretically, by employing the Shen–Kyu model[22] with an appropriately constructed description of the Blue Phase free energy, based on the work by Fukuda.[56] The results of such a model calculation capture well the observed behaviour.[52] It may further be noted that the description can also be extended to describe the phase width and occurrence of twist grain boundary (TGB) phases in mixtures,[57] if one combines the Flory–Huggins[17,18] approach with the orientational ordering of Maier–Saupe[19,20] and the smectic layering of Kobayashi–McMillan.[58]

Finally, we can conclude that it is also of much interest to use polymers in liquid crystals to modify their properties, without crosslinking, thus without polymer stabilization. The observed effects on the liquid crystal may even be larger than for PSCTs.

References

1. M. Mucha, *Prog. Polym. Sci.*, 2003, **28**, 837.
2. Finkelmann and G. Rehage, *Makromol. Chem., Rapid Commun.*, 1980, **1**, 31.
3. J. W. Doane, N. A. Vaz, B. G. Wu and S. Zumer, *Appl. Phys. Lett.*, 1986, **48**, 269.
4. D.-K. Yang, L.-C. Chien and J. W. Doane, *Appl. Phys. Lett.*, 1992, **60**, 3102.
5. H. Finkelmann, H.-J. Kock and G. Rehage, *Mol. Cryst. Liq. Cryst.*, 1982, **89**, 23.
6. J. R. Dorgan and D. S. Soane, *Mol. Cryst. Liq. Cryst.*, 1990, **188**, 129.
7. W. Ahn, C. Y. Kim, H. Kim and S. C. Kim, *Macromolecules*, 1992, **25**, 5002.
8. H. W. Chiu, Z. L. Zhou, T. Kyu, L. G. Cada and L.-C. Chien, *Macromolecules*, 1996, **29**, 1051.
9. M.-C. Chang, H.-W. Chiu, T. Kyu, N. Leroux and L.-C. Chien, *Mol. Cryst. Liq. Cryst.*, 1997, **299**, 253.
10. M.-C. Chang, H.-W. Chiu, X. Y. Wang, T. Kyu, N. Leroux, S. Campbell and L.-C. Chien, *Liq. Cryst.*, 1998, **25**, 733.

11. T. Bouchaour, F. Benmouna, L. Leclercq, B. Ewen, X. Coqueret, M. Benmouna and U. Maschke, *Liq. Cryst.*, 2000, **27**, 413.
12. T. Bouchaour, F. Roussel, J.-M. Buisine, X. Coqueret, M. Benmouna and U. Maschke, *Liq. Cryst.*, 2003, **30**, 487.
13. N. Gogibus, F. Benmouna, B. Ewen, T. Pakula, X. Coqueret, M. Benmouna and U. Maschke, *J. Polym. Sci., Part B: Polym. Phys.*, 2003, **41**, 39.
14. S. K. Slimane, F. Roussel and U. Maschke, *J. Polym. Sci., Part B: Polym. Phys.*, 2007, **45**, 18.
15. D. K. Yoon and Y. C. Bae, *Liq. Cryst.*, 2001, **28**, 1539.
16. Z. Lin, H. Zhang and Y. Yang, *Macromol. Chem. Phys.*, 1999, **200**, 943.
17. P. J. Flory, *Principles of Polymer Chemistry*, Cornell University Press, Ithaca, 1953.
18. M. L. Huggins, *J. Am. Chem. Soc.*, 1942, **64**, 1712.
19. W. Maier and A. Saupe, *Z. Naturforsch.*, 1959, **14a**, 882.
20. W. Maier and A. Saupe, *Z. Naturforsch.*, 1960, **15a**, 287.
21. V. K. Kelkar and C. Manohar, *Mol. Cryst. Liq. Cryst.*, 1986, **133**, 267.
22. C. Shen and T. J. Kyu, *Chem. Phys.*, 1995, **102**, 556.
23. D. Dutta, H. Fruitwala, A. Kohli and R. A. Weiss, *Polym. Eng. Sci.*, 1990, **30**, 1005.
24. C. C. Riccardi, J. Borrajo, R. J. J. Williams, H. M. Siddiqi, M. Dumon and J. P. Pascault, *Macromolecules*, 1998, **31**, 1124.
25. F. V. Pereira, R. Borsali, A. A. Merlo and N. P. da Silveira, *Liq. Cryst.*, 2004, **31**, 655.
26. N. Bouchikhi, F. Semdani, L. A. Bedjaoui and U. Maschke, *Mol. Cryst. Liq. Cryst.*, 2012, **560**, 159.
27. B. D. Youcef, T. Bouchaour and U. Maschke, *Macromol. Symp.*, 2008, **273**, 66.
28. F.-J. Carrion, G. Martinez-Nicolas, P. Iglesias, J. Sanes and M.-D. Bermudez, *Int. J. Mol. Sci.*, 2009, **10**, 4102.
29. A. M. Jamieson, D. Gu, F. L. Chen and S. Smith, *Prog. Polym. Sci.*, 1996, **21m**, 981.
30. N. Motong, S. Thongyai and N. Clarke, *J. Appl. Polym. Sci.*, 2008, **107**, 1108.
31. D. Gu and A. M. Jamieson, *Mol. Cryst. Liq. Cryst.*, 1991, **209**, 147.
32. D. Gu, A. M. Jamieson and S.-Q. Wang, *J. Rheol.*, 1993, **37**, 985.
33. F. L. Chen, A. M. Jamieson, M. Kawasumi and V. Percec, *J. Polym. Sci., Part B: Polym. Phys.*, 1995, **33**, 1213.
34. Y.-C. Chiang, A. M. Jamieson, M. Kawasumi and V. Percec, *Macromolecules*, 1997, **30**, 1992.
35. Y. Zhao, S. Dong, A. M. Jamieson, X. Hu, J. Lal, S. Nazarenko and S. J. Rowan, *Macromolecules*, 2005, **38**, 5205.
36. I. C. Khoo, H. Li and Y. Liang, *Opt. Lett.*, 1994, **19**, 1723.
37. G. P. Wiederrecht, B. A. Yoon and M. R. Wasielewski, *Science*, 1995, **270**, 1794.
38. G. P. Wiederrecht, B. A. Yoon and M. R. Wasielewski, *Adv. Mater.*, 1996, **8**, 535.

39. H. Ono, I. Saito and N. Kawatsuki, *Appl. Phys. Lett.*, 1998, **72**, 1942.
40. H. Ono and N. Kawatsuki, *J. Appl. Phys.*, 1999, **85**, 2482.
41. H. Ono, R. Hasebe, T. Sasaki, K. Noda and N. Kawatsuki, *Opt. Commun.*, 2013, **300**, 210.
42. H. Ono, T. Kawamura, H. Norisada and N. Kawatsuki, *Proc. SPIE* **4107**, Liquid Crystals IV, 2000, 114.
43. F. Araoka, K.-C. Shin, Y. Takanishi, K. Ishikawa, H. Takezoe, Z. Zhu and T. M. Swager, *J. Appl. Phys.*, 2003, **94**, 279.
44. K.-C. Shin, F. Araoka, B. Park, Y. Takanishi, K. Ishikawa, Z. Zhu, T. M. Swager and H. Takezoe, *Jpn. J. Appl. Phys.*, 2004, **43**, 631.
45. Z. Zhu and T. M. Swager, *J. Am. Chem. Soc.*, 2002, **124**, 9670.
46. J. Hoogboom and T. M. Swager, *J. Am. Chem. Soc.*, 2006, **128**, 15058.
47. A. Lohr and T. M. Swager, *J. Mater. Chem.*, 2010, **20**, 8107.
48. J. Ruth, J. Naciri and R. Shashidhar, *Liq. Cryst.*, 1994, **16**, 883.
49. P. Tuli and H. J. Coles, *Liq. Cryst.*, 1993, **14**, 1087.
50. H. Kikuchi, M. Yokota, Y. Misakado, H. Yang and T. Kajiyama, *Nat. Mater.*, 2002, **1**, 64.
51. I. Dierking, W. Blenkhorn, E. Credland, W. Drake, R. Kociuruba, B. Kayser and T. Michael, *Soft Matter*, 2012, **8**, 4355.
52. N. Kasch, I. Dierking and M. Turner, *Soft Matter*, 2013, **9**, 4789.
53. H. J. Coles and M. N. Pivnenko, *Nature*, 2005, **436**, 997.
54. C. V. Yelamaggad, I. S. Shashikala, G. Liao, D. S. Shankar Rao, S. Krishna Prasad, Q. Li and A. Jakli, *Chem. Mater.*, 2006, **18**, 6100.
55. F. Castles, S. M. Morris, E. M. Terentjev and H. J. Coles, *Phys. Rev. Lett.*, 2010, **104**, 157801.
56. J.-I. Fukuda, *Phys. Rev. E*, 2010, **82**, 061702.
57. N. Kasch and I. Dierking, *J. Chem. Phys.*, 2015, **143**, 064907.
58. W. L. McMillan, *Phys. Rev. A*, 1971, **4**, 1238.

CHAPTER 15

Stabilization of Discotic Liquid Crystals

A. R. YUVARAJ AND SANDEEP KUMAR*

Raman Research Institute, C. V. Raman Avenue, Sadashivanagar, Bangalore 560080, India
*Email: skumar@rri.res.in

15.1 Introduction

Nowadays, people around the world are addicted to electronic devices in everyday life; these electronic gadgets are most commonly tablet/notebook computers, digital cameras, smartphones, flat panel TVs, MP3 players, *etc.* The electronics industry is desperately looking for new low-cost raw materials, easy fabrication steps, to be eco-friendly and produce non-hazardous products. In the development of display devices, calamitic liquid crystals have played a major and vital role.[1,2] However, their corresponding disc-shaped molecular counterparts have also been employed in displays[3–5] as well as in sensors.[5,6] Excellent progress in the preparation of novel discotic materials for modern research continues worldwide.[7–9] Stabilization of mesophases of discotic liquid crystals (DLCs) is really needed to improve the applicability of the research. Nevertheless, the structural and physical properties[9] of the mesomorphic compounds are used in potential research applications namely organic light emitting diodes, organic field effect transistors and organic photovoltaic devices.[10,11] As far as applicability is concerned, tuned or modified electro-optical characteristics of the LC compounds are required.

A composite of liquid crystal and polymer produces a major class of materials; these materials allow stabilized mesophases and tuneable

Soft Matter Series No. 8
Polymer-modified Liquid Crystals
Edited by Ingo Dierking

Published by the Royal Society of Chemistry, www.rsc.org

electro-optical properties to be obtained.[12–16] There are two categories in the polymer modification of materials, dependent on the concentration of the polymer matrices used namely (a) polymer-stabilized liquid crystal (PSLC) and (b) polymer dispersed liquid crystal (PDLC). In PDLC, a small quantity of LC droplets is present with the bulk continuous polymer matrix. In contrast, the major component of polymer stabilized liquid crystals (PSLC) is the liquid crystal. Further, stabilization occurs in the mesomorphism of the system, mainly by widening the mesophase range of the LC using the PSLC mechanism.

In the case of discotic liquid crystals, DLCs, there is no literature available regarding polymer stabilization of mesophases. However, discotic mesophases could be stabilized by two important methods namely charge transfer (CT) interaction and complimentary polytropic interaction (CPI). These aspects of mesophase stabilization of DLCs are briefly presented here.

15.2 Stabilization of Discotic Liquid Crystals by Charge Transfer Complexation

Charge transfer (CT) complexes usually enhance the stability and widen the mesophase range of discotic liquid crystals. In 1989, Ringsdorf *et al.* introduced the concept and mechanism of CT in DLCs.[17] They showed that mesomorphism can be induced in amorphous polymers by doping with electron acceptors. Due to the presence of low molecular weight electron acceptors, the disc-shaped counterparts arrange in a columnar fashion. In Figure 15.1, the typical columnar arrangements of discogens are given.

Praefcke *et al.* discovered a new type of nematic phase of DLCs using a binary non-polymeric system.[18] This system consists of electron donor disc-shaped molecules (alkyl pentakis[pheny1ethynyl]phenyl ethers) and small electron acceptor molecules (2,4,7-trinitrofluorenone) (TNF). This composite mixture is further studied in the presence of an electric field. An excellent bend elastic constant value 22×10^{-12} N is obtained for the composite, which is one order magnitude higher than the respective value of the normal discotic nematic phase.[19] Another approach in evaluation of CT complexes is by binding electron acceptors to the polymeric chain of electron donors.[20] For this study, the composite is fabricated using TNF and electron rich 2,3,6,7,10,11-hexapentyloxytriphenylene. This suggests the idea of producing homogeneous stable mesophase blends instead of physical mixing of the two-component system.[21] The stability and CT interactions of such homogeneous mixtures are evaluated by Ringsdorf and co-workers[21] using ^{2}H-NMR and dielectric studies. In order to broaden the study of homogeneous one-component systems, Andreu and his group synthesized tetrathiafulvalene-based DLCs associated with various alkyl spacers.[22] Among all these derivatives of tetrathiafulvalene, only a few compounds showed mesophases. From these studies, one could understand the importance of central core and alkyl spacer in obtaining the stabilized mesophases.

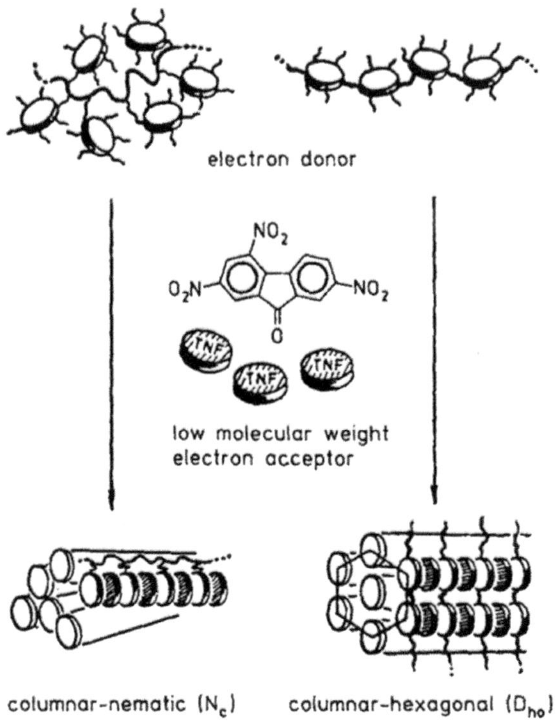

Figure 15.1 Arrangement of disc molecules influenced by CT mechanism. Reproduced from ref. 17 with permission from John Wiley and Sons, Copyright © 1989 by VCH Verlagsgesellschaft mbH, Germany.

Suzuki and Koide synthesized biphenylethynylbenzene derivatives and looked at the mesomorphic properties of their TNF complexes.[23] Nematic schlieren texture is found in the mesomorphic investigation. Interestingly, a stable focal conic fan texture is obtained when this synthesized disc-shaped LC is doped with TNF. Kumar and co-workers prepared a number of photoconducting and photorefractive carbazole-substituted TP derivatives which, on complexation with TNF, display columnar mesophases.[24–26] They also found that novel non-LC dibenzo[g,p]chrysene discotics exhibit columnar phases upon TNF doping.[27] Kouwer *et al.* described the induction of nematic discotic phases in non-liquid crystalline disc-shaped pentakis-(phenylethynyl)phenol compounds on doping with 2,4,7-trinitro-9-fluorenone.[28–30] The stability of the mesophase depends on the extent of complexation from CT interactions.[30] On the other hand, Hatsusaka and Ohta designed bis[octakis(3,4-dialkoxyphenoxy)-phthalocyaninato]lutetium-(iii) complexes to evaluate the intermolecular charge transfer system.[31] The stability of such co-ordination compounds and the mechanism of intermolecular CT are examined by electronic absorption parameters. The variation of CT strength is observed when temperature and polarity of the solvent are variables.

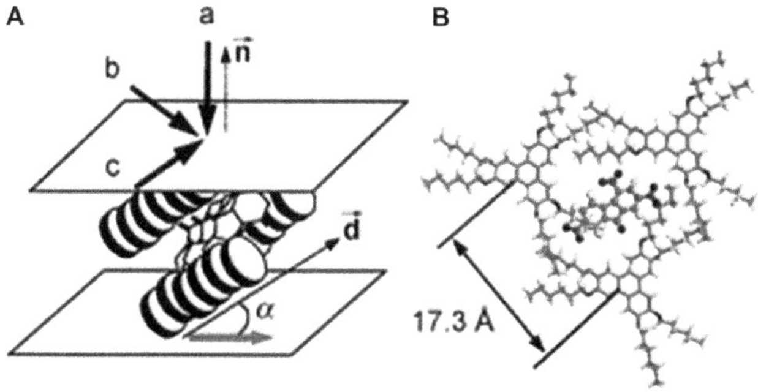

Figure 15.2 (A) The possible geometry for the given CT complex using experimental evidence; (B) Schematic representation of CT interaction between the molecules.
Reproduced from ref. 32*a* with permission from John Wiley and Sons, Copyright © 2007 WILEY-VCH Verlag GmbH & Co. KGaA, Weinheim.

In 2007, Kruglova and co-workers prepared a mixture of a triphenylene derivative and TNF and found enhanced stability of the mesophase.[32a] These results were obtained using neutron scattering, X-ray diffraction and dielectric relaxation spectroscopy. Possible orientation of discotic-TNF columns with respect to the film plane is presented (Figure 15.2).

The mesomorphic stability, resistance towards shear and the slower local dynamics of the CT complex are greater than that of corresponding triphenylene derivative alone.[32b] By analysing the geometry (Figure 15.2) of CT interaction, it is clear that TNF is not sandwiched between triphenylene molecules; instead, it is present in between the columnar arrangement of triphenylene cores. This effects stiffening of the system and also improves the charge mobility. Eventually, this arrangement increases the stability of the system. Further, Lakshminarayanan *et al.*[33] studied the electrical conductivity of a TP based LC system in the presence of electron deficient ferrocenium molecules. This donor–acceptor system increases the quasi-one-dimensional conductivity though the formation of CT complexes. The interdomain spacing between the discogens is occupied by dopants and enhances the stability of the mesophase and it is confirmed by SAXS.

Chen *et al.* explained the stability and CT interaction using density functional theory.[34a,b] Accordingly, TP derivatives are good for charge transport. Mono-substituted TPs are better than poly-substituted derivatives. Positive charge transport can be improved by amide or ester substitution. A theoretical study on intermolecular CT between homogeneous π-conjugated biphenazine derivatives was conducted by Zarate and Schott.[34c] Various conformations of biphenazine dimers are used to evaluate the physical properties such as rotation, intermolecular separation, constructive overlapping and stabilization.

In 2009, Kumar *et al.* studied the TP dimeric molecules with various spacers.[35a] These compounds are non-LCs in nature but showed columnar phases when TNF dopant was added.[35] It is noted that core-to-core interaction attributed to the TNF dopant is very important in the induction of columnar phases *via* CT interactions. Hirose *et al.*[36] constructed a CT complex using tetraalanoyloxypyrene derivatives with the electron acceptor 2,4,5,7-tetranitro-9-fluorenone. Stabilized mesophase is obtained after CT interaction between electron donor-acceptor molecules. Klivansky *et al.*[37] introduced a new donor–acceptor complementary disc-shaped TP derivative. This CT complex showed an association constant 2.6×10^4 M^{-1}. This value is remarkable in the case of electrostatic interaction between the donor–acceptor discogens. Thus, stabilization of the mesophase is high due to the strong electrostatic interactions.

Using DLC compounds, Haverkate *et al.*[38] explained an important morphology of CT complex *via* absorption of visible light. In the presence of visible light, a rapid CT effect could be observed in the system, which consists of a DLC and dopant TNF composite. So, this principle is suitable for photovoltaic investigations, as shown in Figure 15.3.

Pal *et al.*[39] designed dyad-based TP and pentaalknylbenzene room temperature mesogenic compounds. Along with the stability of the mesophase, excellent fluorescence properties are observed. Due to the appreciable CT in the system, these compounds are suitable for optoelectronic applications. Charge transfer (CT) interactions between the neighbouring molecules are most likely favoured by molecules of complementary nature.[40] In other words, a mixture of electron donor and electron acceptor DLC compounds interact with each other by electrostatic forces of attraction. Usually, CT influences the widening of the mesomorphic range near to room temperature.[41,42] The LC properties of discogens could be tuned by changing the proportions of blending materials. The CT complexes of donor/acceptor mixture exhibit excellent alignment properties often without any external stimuli.[42,43] These CT composites also exhibit several important properties, mainly an ambipolar CT[44] and ferroelectricity.[45] The self-assembled alternate face-to-face stacks of the donor–acceptor mixture are based on several factors. They are mainly due to the electron donating ability of donor

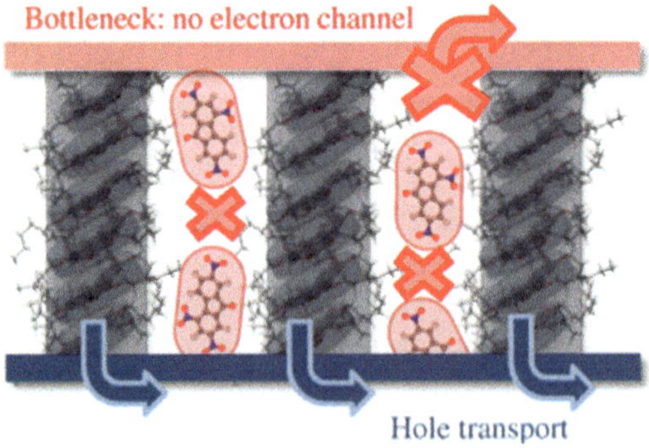

Figure 15.3 Model for the binary mixture of TP derivative and TNF self-assembled heterojunction in a photovoltaic device.
Reproduced from ref. 38 with permission from American Chemical Society, Copyright 2012.

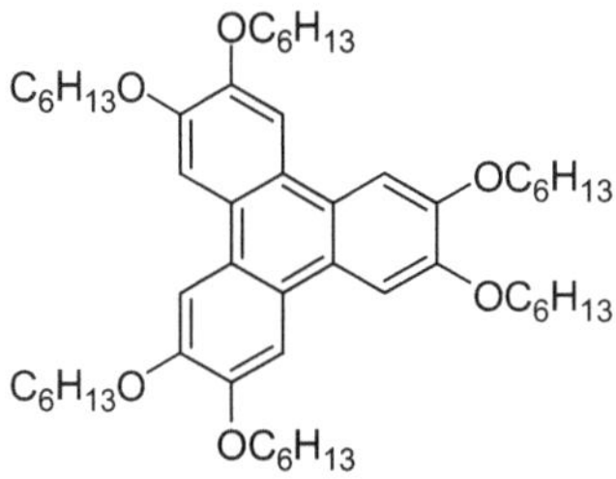

1

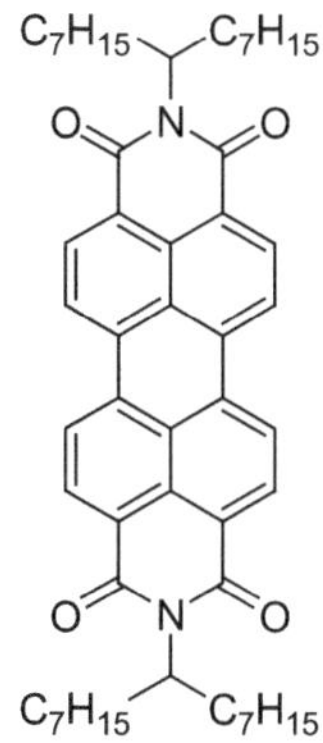

2

Figure 15.4 The structure of 2,3,6,7,10,11-hexakis(hexyloxy)triphenylene (**1**) with N,N′-bis(1-heptyloctyl)perylene-3,4,9,10-tetracarboxylic diimide (**2**).

molecules, their structural similarity and complementary π-surfaces.[46] Many different donor–acceptor systems are developed and reported in the literature[47–50] but attention has not been focussed on obtaining the proper microscopic shapes.[51,52]

Li *et al.*[53] blended discogens (Figure 15.4) to build alternate stacking columns through CT interactions of compound **1** with **2**. One could easily find the difference between these two compounds in symmetry and shape. Due to CT interaction, the composite showed a wide-range mesophase near room temperature as well as identical inter-columnar distances ~2.06 nm. Interestingly, homeotropic alignment of the substrates is observed on cooling from the isotropic temperature under normal conditions.

Recently, Pal *et al.*[54] reported an interesting phenomenon when a discotic compound interacts with TNF dopants. This process is governed by CT interaction between electron donor–acceptor molecules. Initially, the columnar rectangular phase is observed for the triphenylene-based oligomeric units connected to an azobenzene core. The columnar hexagonal phase is obtained when TNF is added to give an exactly 1 : 2 ratio (Figure 15.5). This CT matrix favours the stabilization of the mesophase as well as enhancing the charge mobility.

15.3 Stabilization of Columnar Phase by Complementary Polytropic Interaction

Complementary polytropic interaction (CPI) means the uniform stacking of large and small ring components of a mixture within the column without any electrostatic interaction or donor–acceptor mechanism. An atom-centred multipolar interaction and Van der Waals attraction between two complementary compounds arranged on an atom-by-atom basis, is known as CPI.[55,56]

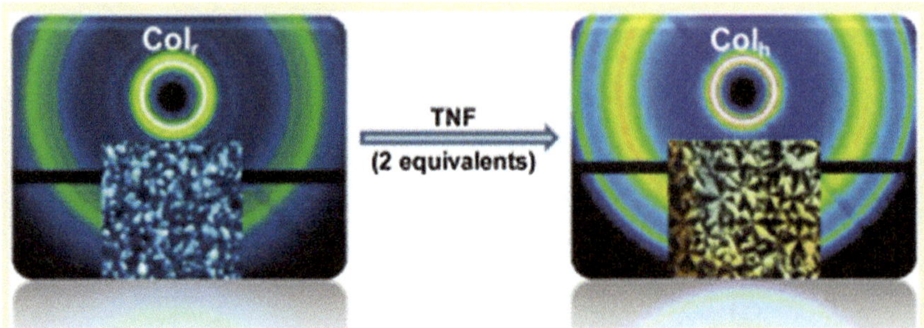

Figure 15.5 Transition from columnar rectangular to columnar hexagonal phase. Reproduced from ref. 54 with permission from American Chemical Society, Copyright 2017.

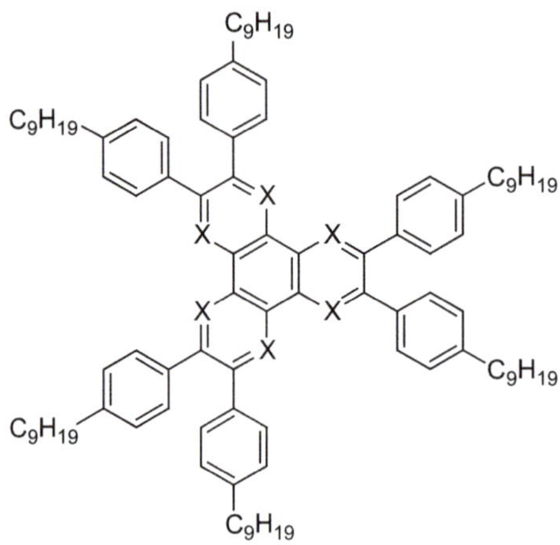

3a: X=N and **3b**: X=CH

Figure 15.6 The structure of hexakis(4-nonylphenyl)dipyrazino[2,3-f:2′,3′-h]-quinoxalene (**3a**) and 2,3,6,7,10,11-hexakis(4-nonylphenyl)triphenylene (**3b**).

Boden *et al.*[57] designed hydrophilic and hydrophobic side-chain-containing triphenylenes (TPs). TP with five hydrophobic and one hydrophilic alkyl chains showed a columnar phase. The stability of the columnar phase is enhanced by mixing TP derivatives **3a** and **3b** *via* CPI (Figure 15.6). Moreover, mesophases can also be induced in non-mesogenic materials *via* CPI. Boden *et al.*[58] again reported some advantages of CPI in discotic systems such as charge mobilization, phase separation of block copolymers, improved alignment characteristics and increased mesomorphic range.[59] Along with these beneficial effects of the CPI phenomenon, photo-conductivity and electrical conductivity[60] are also enhanced in the composites.[61]

Bushby *et al.*[62] conducted cyclic voltammetric studies on the TP system and a high charge mobilization is determined in CPI stacks; a large difference in the oxidation potential indicates the high hole transporting nature of the TP compounds. The columnar mesophase behaviour of TP derivatives could be drastically improved by introducing α-substitution and it is considered as a great tool.[63] But, α-substitution of bulky groups in the TP molecule does not favour an improvement in the mesomorphic properties. Bulky groups such as cyano and nitro induce steric hindrance and destroy the planarity of the discotics, which leads to problems in forming the columnar phases. However, the introduction of small α-fluoro substituents induces a lateral dipole without any complications.[63] Also, substitution of small functional groups facilitates the stabilization and widening of the columnar phase *via* CPI processes. The columnar phase is more stable at room temperature than at high temperatures and the molecules do not crystallize since these compounds have higher clearing points than their individual counterparts because of the CPI.

McLaren *et al.*[64] designed CPI columnar LC phases using the monomers **4** and **5** (Figure 15.7). CPI interactions are possible which stabilize the columnar mesophase by the combination of various **4** and **5** derivatives.[55–57,65,66] Component **4** could be placed in either the main-chain or the side-chain in the case of polymer matrices.[58] Here, McLaren *et al.* reported main-chain polymer components of **4** and **5** derivatives with low molecular weight used in the fabrication method. The resultant columnar phase from this mixture is stable down to room temperature and no crystallization occurs since the mixture has a higher clearing temperature than the individual components. For the long chain **4** derivatives, higher order columnar rectangular phases are obtained, and these species are very good hole transporters.[67] Formation of large domains is quite unusual for these mixtures with respect to typical main-chain DLC polymers. However, spontaneous formation of large aligned domains and highly ordered columnar phases could be obtained using a CPI polymer mixture. In line with the earlier reports, the CPI mixture is useful in the study of self-healing/self-organising semiconductors.[68,69]

15.4 Comparison Between Complementary Polytropic Interaction and Charge Transfer Concepts

Boden *et al.*[70] described the comparison of CPI and CT in a TP DLC system. A binary mixture is prepared using a 1 : 1 mixture of DLC TP derivatives (Figure 15.8). This mixture has a high clearing temperature and a wider range of mesophase than its individual components, which is determined using the phase diagram. These compounds showing CPI characteristics are further compared to CT compounds, by the addition of **7** to TP discogens. The mesophases formed by CPI and CT are different from one another, but stabilization and local molecular packing are essentially the same. To justify these effects, the extended electron distribution (XED) method is employed.

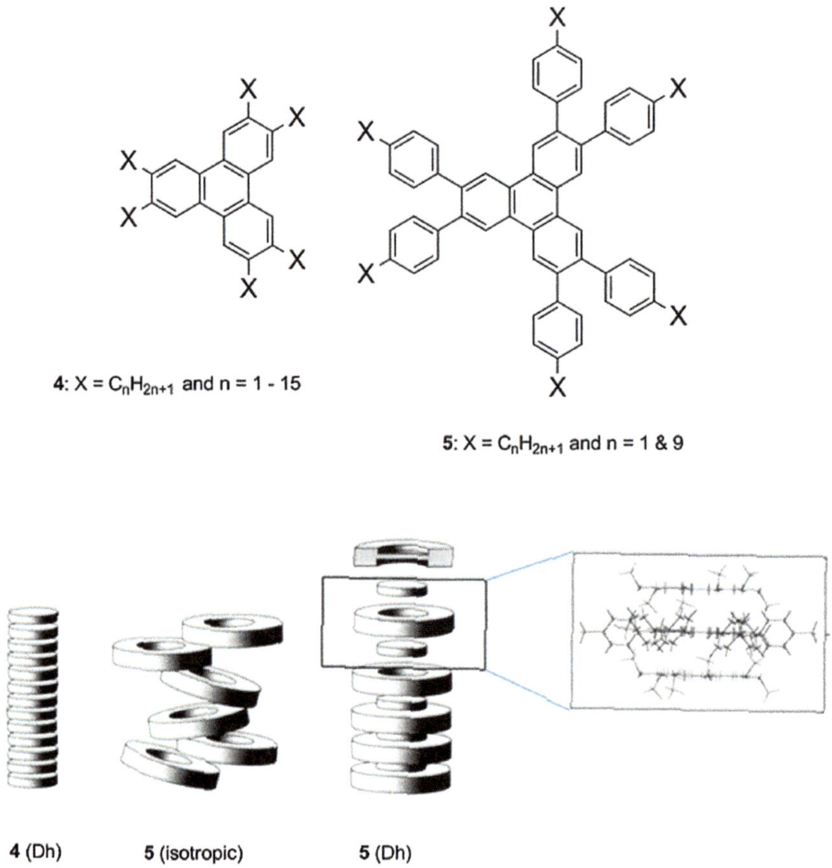

Figure 15.7 The structure of 2,3,6,7,10,11-hexakis(hexyloxy)triphenylene (**4**) and 2,3,6,7,10,11-hexakis-(4′-nonylphenyl)triphenylene (**5**).
Reproduced from ref. 64 with permission from the Royal Society of Chemistry.

It is assumed that van der Waals and dispersed coulombic interactions dominate as compared to other forces of attraction.

The theoretical calculations are obtained from XED software. The short chain analogues **6a** and **8a** were used for the modelling in order to minimize the computation time. The molecular geometry is optimised by adding extended electrons to the system. To locate global minima, a conformational search is performed. A convenient minimum global energy is applied to run a series of docking experiments. Here, a molecule is docked starting from 250 points on a spherical surface and the position of the neighbouring molecule is fixed. Thus, one can get the most probable structure of a dimer, energies of individual monomers and van der Waals and coulombic interactions in the system.

The columnar hexagonal mesophase range for **6c** is 70–100 °C[71] and for the binary mixture of **6c** and **8b** it is from below room temperature to

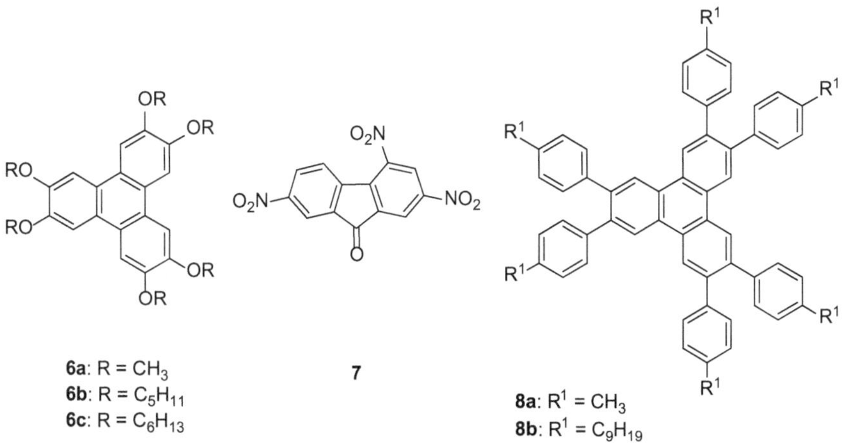

Figure 15.8 The compounds used to study the CPI and CT of two-component columnar phases.

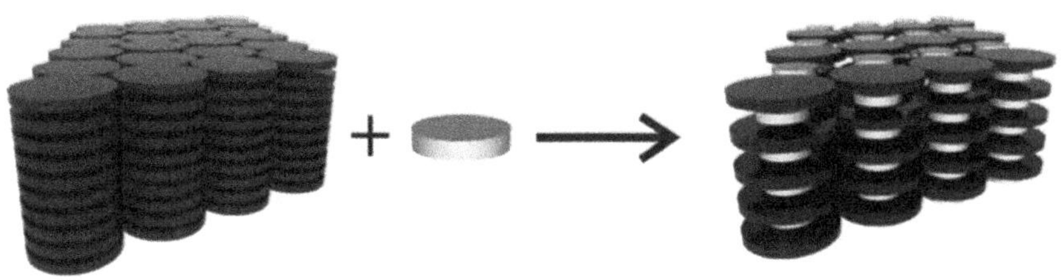

Figure 15.9 Hexagonal columnar phase obtained by the intercalation of 7 or **8a,b** into the discogen **6a–c**.
Reproduced from ref. 70 with permission from Taylor and Francis.

154 °C.[55] A broad maximum at the centre of the graph indicates the clearing temperature for the binary composition. In a binary mixture, excess **6a–c** or **8a,b** has limited solubility unlike **6a,c** and 7 composites. Low angle XRD confirms the columnar hexagonal phase for the binary mixture of **6c** and **8b** (Figure 15.9).

The phase obtained from the intercalation is much more ordered than the hexagonal columnar phase of **6c**. The partial three-dimensional ordering of the plastic phase can also be visualized from reflection patterns.[72] A homeotropic alignment is generated after the annealing process. There is no charge transfer and no colour change in the binary mixture of **6c** and **8b** in heptane as observed by UV–Vis spectroscopy.[55] The difference between the first reduction potential of **8a,b** and first oxidation potential of **6a–c** is ~1.69 V.[61] The CT interaction and attractive quadrupole moment are not responsible for the stability of the discogen **6a–c** and **8a,b** pairs. This is due to the presence of the same sign of the total quadrupole of **6a–c** and **8a,b** discogens.

A series of binary mixtures have been studied to understand their stability. The CPI technique is used to rationalize the stability of the discotic mesomorphic phases and evaluated by the XED model. Here, the sum of atomic centred coulombic and van der Waals terms are used to express the π-stacking interactions. The total energy ΔE of the system is given in eqn (15.1).

$$\Delta E = U_{AB} - \left[\frac{U_{AA} + U_{BB}}{2}\right] \tag{15.1}$$

In eqn (15.1), U_{AA} and U_{BB} are the energies of homodimers **6a–c**, 7 and **8a,b** and U_{AB} is the energy of the heterodimers. Apart from the information about the formation of these systems, analysis of the CPI model also helps to explain CT, structure and stability of the LC system.[73] The homodimer molecules of 7 are more likely stay parallel to the aromatic cores with the opposite dipolar interactions. In 7 and **6a** heterodimer molecular system, 7 molecules are arranged over the TP nucleus. It is difficult to get the parallel arrangement of TP nucleii, due to the non-planar nature of molecules in the **8a** homodimer system. However, this is possible in case of a 7 and **6a** heterodimer system. In **6a–c** and **8a,b** heterodimer systems, **8a,b** nucleii are fitted nicely into the gaps of the **6a–c** nucleii.

15.5 Polymer Dispersed Discotic Liquid Crystals

In 1997, Kitzerow *et al.*[74] introduced the concept of polymer dispersed discotic liquid crystals (PDDLC) which form when a small quantity of DLC is embedded in the polymeric cavities. Alignment of DLCs in PDDLC composites could be achieved by shearing the substrates with respect to each other. Electro-optical properties of PDLCs play an important role in applications. Good contrast in the device occurs due to the high tilt angle of the chiral DLC dibenzopyrene derivative (Figure 15.10). Two modified low field phases appear over the range of field strengths as compared with the values obtained for pure DLC, when investigated by electro-optic analysis. The larger tilt angle is obtained either by reducing the sample thickness or by increasing the applied voltage. However, the long response time for electro-optical switching and high transition temperatures of the studied chiral DLC (Figure 15.10) are major disadvantages for the fabrication of light shutters.

Interestingly, this PDDLC matrix showed appreciable storage capabilities, since after cooling the sample from mesophase temperature to room temperature, the field induced orientation of the DLC is still preserved.

The orientation order of the PDDLC matrix can further be improved as pointed out by Chenard *et al.*,[75] with a DLC rufigallol derivative showing a columnar mesophase as shown in Figure 15.11.

In the PDDLC composite, the phase separation of DLC, as well as the confined geometry and resulting morphologies of the columnar phase, are the essential concepts for study. As far as LC orientation is concerned, there

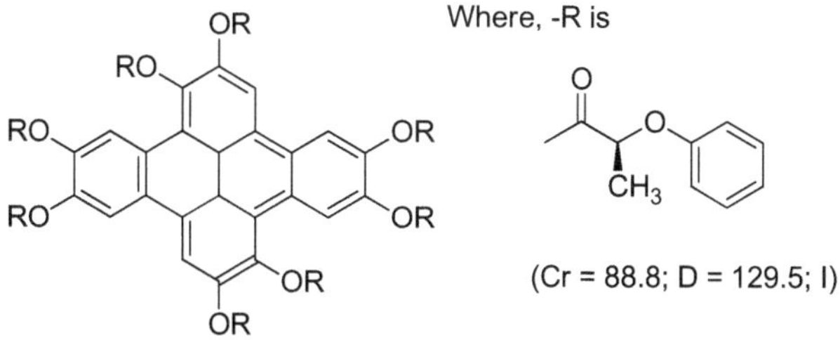

Figure 15.10 The molecular structure of a chiral DLC dibenzopyrene derivative with the transition temperatures.

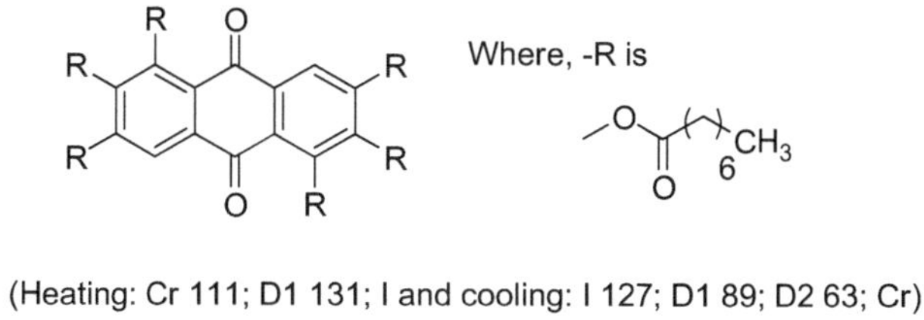

Figure 15.11 Chemical structure and heating/cooling transition temperatures of a DLC rufigallol derivative.

are three major components that play an important role. They are (a) uniform molecular orientation; (b) columnar orientation by mechanical stretching; and (c) anchoring effects of discogens. There are three types of polymer matrices that are used to prepare the composite mixtures. They are poly(ethyl methacrylate) (PEMA), poly(methyl methacrylate) (PMMA) and polystyrene (PS). Effective alignment of columnar axes of the matrix is achieved by stretching the rigid sample. However, the orientation of the rufigallol-based DLC is preserved only when the stretched film is cooled under strain. Otherwise, the columnar orientation is not stable if cooling without any external aid. However, the excellent orientational order of rufigallol-based DLCs can be demonstrated by infrared dichroism.

Bayer *et al.*[76] evaluated the significant change in the dynamic and structural properties in the PDDLC system as compared with pure TP-based DLC samples. Due to geometrical confinement effects in the polymer dispersion, the emission and absorption properties could be modified in the PDDLC layer. The transparent layers could be coated using spin coating techniques. This transparent layer plays a significant role in the application of organic light emitting diodes.

15.6 Summary

Polymer-stabilized DLCs may have a significant impact in the field of discotic liquid crystals. From the application point of view, polymer stabilization is very important. Unfortunately, there are no reports in the literature regarding polymer-stabilized DLCs available to date. However, this chapter discusses stabilization of DLCs using various disciplines. The chemical structures and physical properties of the stabilized DLCs are given briefly for convenience. Apart from the standard polymer stabilization as used in nematics, cholesterics, blue phases or ferroelectric liquid crystals, the stabilization of columnar phases could be achieved by two techniques. These are charge transfer stabilization (CT) and complimentary polytropic interaction (CPI). The concept of polymer dispersed DLCs was further discussed. There are possible methods of achieving polymer stabilization in DLCs. These may contribute valuable ideas to future display technologies and research on other applications of DLCs.

References

1. T. Geelhaar, K. Griesar and B. Reckmann, *Angew. Chem., Int. Ed.*, 2013, **52**, 8798.
2. M. Bremer, P. Kirsch, M. Klasen-Memmer and K. Tarumi, *Angew. Chem., Int. Ed.*, 2013, **52**, 8880.
3. K. Arakawa, *Liquid Crystal Display Having Positive and Negative Birefringent Compensator Films*, 1993, US005189538A.
4. H. Mori and Y. Ito, *Liquid Crystal Display with Compensators Having Minimum Retardations in the Inclined Direction*, 1998, US005805253A.
5. N. Boden, J. Clements and B. Movaghar, *Fluid Sensing Device Using Discotic Liquid Crystals*, 2002, US006423272B1.
6. H. Haick, *Electronic Nose Device with Sensors Composed of Nanowires of Columnar Discotic Liquid Crystals with Low Sensitive to Humidity*, 2008, US20100191474A1.
7. S. Kumar, Design Concepts and Synthesis of Discotic Liquid Crystals, in *Handbook of Liquid Crystals*, ed. J. W. Goodby, P. J. Collings, T. Kato, C. Tschierske, H. F. Gleeson and P. Raynes, Wiley-VCH, Weinheim, 2014, vol. 4, p. 467; S. Kumar, *Chem. Soc. Rev.*, 2006, **35**, 83; S. Kumar, *Chemistry of Discotic Liquid Crystals: From Monomers to Polymers*, CRC Press, Boca Raton, FL, 2011.
8. P. A. Heiney, Structure and Physical Properties of Columnar Liquid Crystals, in *Handbook of Liquid Crystals*, ed. J. W. Goodby, P. J. Collings, T. Kato, C. Tschierske, H. F. Gleeson and P. Raynes, Wiley-VCH, Weinheim, 2014, vol. 4, p. 521.
9. R. J. Bushby and N. Boden, Responses of Discotic Liquid Crystals to Mechanical, Magnetic and Electrical Fields, in *Handbook of Liquid Crystals*, ed. J. W. Goodby, P. J. Collings, T. Kato, C. Tschierske, H. F. Gleeson and P. Raynes, Wiley-VCH, Weinheim, 2014, vol. 5, p. 569.
10. M. Kumar and S. Kumar, *Poly. J.*, 2017, **49**, 85.
11. W. Pisula and K. Müllen, Discotic Liquid Crystals as Organic Semiconductors, in *Handbook of Liquid Crystals*, ed. J. W. Goodby, P. J. Collings, T. Kato, C. Tschierske, H. F. Gleeson and P. Raynes, Wiley-VCH,

Weinheim, 2014, vol. 8, p. 627; S. Sergeyev, W. Pisula and Y. H. Geerts, *Chem. Soc. Rev.*, 2007, **vol. 36**, 1902.
12. R. Blinc, I. Musevic, J. Pirs, M. Skarabot and B. Zeks, in *Liquid Crystals in Complex Geometries*, ed. G. P. Crawford and S. Zumer, Taylor and Francis, London, 1996.
13. D. J. Broer, G. P. Crawford and S. Zumer, *Cross-linked Liquid Crystalline Systems*, CRC Press, Boca Raton, FL, 2011.
14. I. Dierking, *Polym. Chem.*, 2010, **1**, 1153.
15. E. H. Kwok, S. Naemura and H. L. Ong, *Progress in Liquid Crystal Science and Technology*, World Scientific, Singapore, 2013.
16. I. Dierking, *Materials*, 2014, **7**, 3568.
17. H. Ringsdorf, R. Wüstefeld, E. Zerta, M. Ebert and J. H. Wendorff, *Angew. Chem., Int. Ed. Engl.*, 1989, **28**, 914.
18. K. Praefcke, D. Singer, B. Kohne, M. Ebert, A. Liebmann and J. H. Wendorff, *Liq. Cryst.*, 1991, **10**, 147.
19. B. Sabaschus, D. Singer, G. Heppke and K. Praefcke, *Liq. Cryst.*, 1992, **12**, 863.
20. H. Bengs, R. Renkel and H. Ringsdor, *Makromol. Chem., Rapid Commun.*, 1991, **12**, 439.
21. M. Möller, J. H. Wendorff, M. Wert, H. W. Spiess, H. Bengs, O. Karthaus and H. Ringsdorf, *Liq. Cryst.*, 1994, **17**, 381.
22. R. Andreu, J. Garin, J. Orduna, J. Barbera, J. L. Serrano, T. Sierra, M. Salle and A. Gorgues, *Tetrahedron*, 1998, **54**, 3895.
23. D. Suzuki and N. Koide, *Mol. Cryst. Liq. Cryst.*, 2001, **364**, 635.
24. M. Manickam, S. Kumar, J. A. Preece and N. Spencer, *Liq. Cryst.*, 2000, **27**, 703.
25. M. Manickam, M. Belloni, S. Kumar, S. Varshney, D. S. S. Rao, P. R. Ashton, J. A. Preece and N. Spencer, *J. Mater. Chem.*, 2001, **11**, 2790.
26. M. Manickam, G. Cooke, S. Kumar, P. R. Ashton, J. A. Preece and N. Spencer, *Mol. Cryst. Liq. Cryst.*, 2003, **397**, 99[399].
27. S. Kumar and S. Varshney, *Mol. Cryst. Liq. Cryst.*, 2002, **378**, 59.
28. P. H. J. Kouwer, O. van den Berg, W. F. Jager, W. J. Mijs and S. J. Picken, *Macromolecules*, 2002, **35**, 2576.
29. P. H. J. Kouwer, W. F. Jager, W. J. Mijs and S. J. Picken, *Macromolecules*, 2002, **35**, 4322.
30. P. H. J. Kouwer, W. F. Jager, W. J. Mijs and S. J. Picken, *J. Mater. Chem.*, 2003, **13**, 458.
31. K. Hatsusaka and K. Ohta, *J. Porphyrins Phthalocyanines*, 2003, 7, 105.
32. (a) O. Kruglova, E. Mendes, Z. Yildirim, M. Wbbenhorst, F. M. Mulder, J. A. Stride, S. J. Picken and G. J. Kearley, *ChemPhysChem*, 2007, **8**, 1338; (b) L. A. Haverkate, M. Zbiri, M. R. Johnson, E. Carter, A. Kotlewski, S. Picken, F. M. Mulder and G. J. Kearley, *J. Chem. Phys.*, 2014, **140**, 014903.
33. P. S. Kumar, S. Kumar and V. Lakshminarayanan, *J. Phys. Chem. B*, 2008, **112**, 4865.

34. (a) C. Jun-Rong, C. Jing, X. Bu-Yi, L. Quan and Z. Ke-Qing, *Chin. J. Chem.*, 2008, **26**, 2292; (b) X. Zarate, E. Schott, L. Alvarado-Soto and T. C. Sutherland, *Int. J. Quantum Chem.*, 2013, **113**, 2287; (c) X. Zarate and E. Schott, *RSC Adv.*, 2014, **4**, 15642.
35. (a) S. K. Gupta, V. A. Raghunathan, V. Lakshminarayanan and S. Kumar, *J. Phys. Chem. B*, 2009, **113**, 12887; (b) J. Li, Z. He1, H. Zhao1, H. Gopee, X. Kong, M. Xu1, X. An, X. Jing and A. N. Cammidge, *Pure Appl. Chem.*, 2010, **82**, 1993; (c) W. Yu, C. Chen, P. Hu, B. Wang, C. Redshawab and K. Zhao, *RSC Adv.*, 2013, **3**, 14099.
36. T. Hirose, T. Yumoto, K. Matsumoto, S. Mitsushio, O. Kawakami and M. Yasutake, *Mol. Cryst. Liq. Cryst.*, 2010, **524**, 68.
37. L. M. Klivansky, D. Hanifi, G. Koshkakaryan, D. R. Holycross, E. K. Gorski, Q. Wu, M. Chai and Y. Liu, *Chem. Sci.*, 2012, **3**, 2009.
38. L. A. Haverkate, M. Zbiri, M. R. Johnson, B. Deme, H. J. M. de-Groot, F. Lefeber, A. Kotlewski, S. J. Picken, F. M. Mulder and G. J. Kearley, *J. Phys. Chem. B*, 2012, **116**, 13098.
39. M. Gupta, I. Bala and S. K. Pal, *Tetrahedron Let.*, 2014, **55**, 5836.
40. W. G. Zhu, Y. G. Zhen, H. L. Dong, H. B. Fu and W. P. Hu, *Prog. Chem.*, 2014, **26**, 1292.
41. K. R. Leight, B. E. Esarey, A. E. Murray and J. J. Reczek, *Chem. Mater.*, 2012, **24**, 3318.
42. W. Pisula, M. Kastler, D. Wasserfallen, J. W. F. Robertson, F. Nolde, C. Kohl and K. Müllen, *Angew. Chem., Int. Ed.*, 2006, **45**, 819.
43. G. Schweicher, G. Gbabode, F. Quist, O. Debever, N. Dumont, S. Sergeyev and Y. H. Geerts, *Chem. Mater.*, 2009, **21**, 5867.
44. Y. Su, Y. Li, J. Liu, R. Xing and Y. Han, *Nanoscale*, 2015, 7, 1944.
45. S. Horiuchi, F. Ishii, R. Kumai, Y. Okimoto, H. Tachibana, N. Nagaosa and Y. Tokura, *Nat. Mater.*, 2005, **4**, 163.
46. A. Das and S. Ghosh, *Angew. Chem., Int. Ed.*, 2014, **53**, 2038.
47. L. Y. Park, D. G. Hamilton, E. A. McGehee and K. A. McMenimen, *J. Am. Chem. Soc.*, 2003, **125**, 10586.
48. N. Boden, R. Bushby and O. Lozman, *Mol. Cryst. Liq. Cryst.*, 2004, **411**, 345.
49. J. J. Reczek, K. R. Villazor, V. Lynch, T. M. Swager and B. L. Iverson, *J. Am. Chem. Soc.*, 2006, **128**, 7995.
50. L. M. Klivansky, D. Hanifi, G. Koshkakaryan, D. R. Holycross, E. K. Gorski, Q. Wu, M. Chai and Y. Liu, *Chem. Sci.*, 2012, **3**, 2009.
51. J. Y. Wang, J. Yan, L. Ding, Y. G. Ma and J. Pei, *Adv. Funct. Mater.*, 2009, **19**, 1746.
52. G. Zucchi, P. Viville, B. Donnio, A. Vlad, S. Melinte, M. Mondeshki, R. Graf, H. W. Spiess, Y. H. Geerts and R. Lazzaroni, *J. Phys. Chem. B*, 2009, **113**, 5448.
53. Y. Li, M. Li, Y. Su, J. Liu, Y. Han and S. Zheng, *J. Mol. Liq.*, 2016, **224**, 721.
54. M. Gupta, S. P. Gupta and S. K. Pal, *J. Phys. Chem. B*, 2017, **121**, 8593.
55. E. O. Arikainen, N. Boden, R. J. Bushby, O. R. Lozman, J. R. Vinter and A. Wood, *Angew. Chem., Int. Ed. Engl.*, 2000, **39**, 2333.

56. R. J. Bushby, J. Fisher, O. R. Lozman, S. Lange, J. E. Lydon and S. R. McLaren, *Liq. Cryst.*, 2006, **33**, 653.
57. N. Boden, R. J. Bushby, Z. Lu and O. R. Lozman, *Liq. Cryst.*, 2001, **28**, 657.
58. N. Boden, R. J. Bushby, G. Cooke, O. R. Lozman and Z. Lu, *J. Am. Chem. Soc.*, 2001, **123**, 7915.
59. T. Kreouzis, K. Scott, K. J. Donovan, N. Boden, R. J. Bushby, O. R. Lozman and Q. Liu, *Chem. Phys.*, 2000, **262**, 489.
60. N. Boden, R. J. Bushby, O. Lozman, Z. Lu, A. Mcneill, B. Movaghar, K. Donovan and T. Kreouzis, *Mol. Cryst. Liq. Cryst.*, 2004, **410**, 13.
61. N. Boden, R. J. Bushby and O. R. Lozman, *Mol. Cryst. Liq. Cryst.*, 2003, **400**, 105.
62. R. J. Bushby, O. R. Lozman, L. A. Mason, N. Taylor and S. Kumar, *Mol. Cryst. Liq. Cryst.*, 2004, **410**, 171.
63. R. J. Bushby, K. J. Donovan, T. Kreouzis and O. R. Lozman, *Opto-Electron. Rev.*, 2005, **13**, 269.
64. S. R. McLaren, D. J. Tate, O. R. Lozman, G. H. Mehlc and R. J. Bushby, *J. Mater. Chem. C*, 2015, **3**, 5754.
65. R. J. Bushby, I. W. Hamley, Q. Y. Liu, O. R. Lozman and J. E. Lydon, *J. Mater. Chem.*, 2005, **15**, 4429.
66. N. Boden, R. J. Bushby, Q. Y. Liu and O. R. Lozman, *J. Mater. Chem.*, 2001, **11**, 1612.
67. B. R. Wegewijs, L. D. A. Siebbeles, N. Boden, R. J. Bushby, B. Movaghar, O. R. Lozman, Q. Liu, A. Pecchia and L. A. Mason, *Phys. Rev. B: Condens. Matter Mater. Phys.*, 2002, **65**, 245112.
68. Q. Li, *Self-Organized Organic Semiconductors: From Materials to Device Applications*, Wiley, Weiheim, 2011.
69. R. J. Bushby, S. M. Kelly and M. O'Neill, *Liquid Crystalline Semiconductors Materials, Properties and Applications*, Springer, Dordrecht, 2013.
70. N. Boden, R. J. Bushby and O. Lozman, *Mol. Cryst. Liq. Cryst.*, 2004, **411**, 345.
71. (a) C. Destrade, M. C. Mondon and J. Malthete, *J. Phys., Colloq.*, 1979, **3**, 17; (b) C. Destrade, N. H. Tinh, H. Gasparoux, J. Malthete and A. M. Levelut, *Mol. Cryst. Liq. Cryst.*, 1981, **71**, 111.
72. (a) J. Simmerer, B. Glusen, W. Paulis, A. Kettner, P. Schuhmacher, D. Adam, K. E. Etzbach, K. Siemensmeyer and D. Haarer, *Adv. Mater.*, 1996, **8**, 815; (b) B. Glusen, W. Heitz, A. Kettner and J. H. Wendorff, *Liq. Cryst.*, 1996, **20**, 627.
73. O. R. Lozman, R. J. Bushby and J. G. Vinter, *J. Chem. Soc., Perkin. Trans. 1*, 2001, **2**, 1446.
74. H. S. Kitzerow and H. Bock, *Mol. Cryst. Liq. Cryst.*, 1997, **299**, 117.
75. Y. Chenard, N. Paiement and Y. Zhao, *Liq. Cryst.*, 2000, **27**, 459.
76. A. Bayer, J. Kopitzke, F. Noll, A. Seifert and J. H. Wendorff, *Macromolecules*, 2001, **34**, 3600.

CHAPTER 16

Polymer Modified Nanoparticle Laden Liquid Crystals

INGO DIERKING

School of Physics & Astronomy, University of Manchester, Oxford Road, Manchester M13 9PL, UK
Email: ingo.dierking@manchester.ac.uk

16.1 Introduction

Many of the previous chapters of this book were concerned with different phases of polymer stabilized liquid crystals, and the respective authors have given overviews and references of the most important techniques, properties and applications. In recent years there has been another topic of liquid crystal research which has gained increasing attention and interest, and that is related to the different possibilities of producing suspensions and dispersions of nanosized particles with liquid crystal phases. A very recent two volume book edited by Lagerwall and Scalia[1] gives an up to date summary of the state-of-the-art of this topic. The doping of nanoparticles into liquid crystal phases has three main aims (i) to modify and tune the liquid crystal properties, (ii) to add a specific functionality to a liquid crystal, and (iii) to exploit the liquid crystal self-organization to template order and transfer orientation to dispersed nanoparticles. Looking a bit further than thermotropic liquid crystals, one could add another aim (iv) the formation of liquid crystal phases by adding a certain volume fraction of shape-anisotropic nanoparticles to an isotropic solvent.[2]

Soft Matter Series No. 8
Polymer-modified Liquid Crystals
Edited by Ingo Dierking

Published by the Royal Society of Chemistry, www.rsc.org

Aim (i) is often achieved by simply suspending polydisperse and shape-irregular or spherical nanoparticles in the liquid crystal, which will change the properties such as threshold- and switching voltage, response times, viscosity, elastic constants, birefringence, or the helical pitch for chiral phases like cholesterics or SmC*.[3–11] Likewise, aim (ii) can be achieved by adding nanoparticles with certain properties to liquid crystal phases. Often these nanoparticles are anisotropic in shape. It is especially popular to use ferroelectric nanoparticles, like $BaTiO_3$,[12–14] but also magnetic rods and plates.[15–19] Quantum dots,[20,21] gold nanoparticles and nanorods[22–24] are used to change the plasmonic properties of the liquid crystal. There have also been reports of semiconducting nanoparticles.[25] A lot of attention has recently been devoted to the suspension of carbon-based materials, such as fullerenes,[26,27] graphene and graphene oxide,[28–34] and especially carbon nanotubes,[35–41] due to their large aspect ratio and directional conductivity. To achieve aim (iii), a whole range of materials have been employed, not only nematic liquid crystals. For thermotropic systems, due to their potential as a new generation of displays (see chapters 12 and 13), the Blue Phases (BP) were quite extensively studied, not only with polymer stabilization, but also simply with particle addition. Nanotubes were dispersed in nematics to achieve alignment and reorientation, but also in ferroelectric liquid crystals (FLC), discotics for one-dimensional conductors, and lyotropic systems. A slightly different aspect is (iv) which can involve the formation of lyotropic liquid crystal phases from anisotropic particles, ranging from inorganic and mineral rod-like dopants[42] to disc-like clays,[43] tobacco mosaic viruses (TMV),[44] DNA,[45] and cellulose nanocrystals,[46] all the way to nanotubes, various nanorods and nanowires, and graphene oxide.

At this point it should be noted that despite a large body and variety of literature on polymer stabilization, as well as on particle loaded liquid crystals, only few attempts have been made to date to combine both fields of study in an effort to produce polymer stabilized particle suspended liquid crystals. The available literature seems to reach back for about a decade, but is very scarce. In the following we will summarize the main findings on such systems, which may in fact be advantageous for future applications and improvements in device performance.

16.2 Investigated Systems and Electro-optic Performance

16.2.1 Polymer Modified Nanoparticle Doped Nematic Liquid Crystals

One of the first investigations to combine effects of nanoparticle doping and polymer stabilization on liquid crystalline properties was reported by Yaroshchuk *et al.*,[47] who studied the two limiting cases of polymer modification, (i) polymer dispersed liquid crystals, PDLCs,[48] with nanoparticle doped liquid crystals at large polymer contents, and (ii) the other side of the phase diagram involving low polymer concentrations, leading to polymer network structures, PSLCs.

Several different nanoparticles (Sb_2O_5, SiO_2 and TiO_2), with sizes of the order of 10–20 nm, were employed at concentrations between approximately 1–10vol%, together with the standard commercial liquid crystal E7 and a photo-curable material No65, which was varied in concentration between 0 to 50vol%, thus covering the PSLC and the PDLC regimes.

16.2.1.1 *Nematic PDLC system*

For the PDLCs it was found that the nanoparticles are located mainly in the polymer matrix, but do not substantially alter the polymer morphology as demonstrated in Figure 16.1 for (a) the pure PDLC as compared to (b) the particle doped system. Since the nanoparticles in the polymer matrix change the refractive index, stronger light scattering, and thus a larger electrooptic contrast, was observed for increasing particle concentration. The driving voltage, $U_{0.9}$, increases strongly with increasing polymer concentration above approximately 30vol% and is higher for the nanoparticle filled system than for the non-filled system (Figure 16.2(a)). The turn-off response times, τ_{off}, decrease with increasing polymer concentration, as well as with increasing particle concentration (Figure 16.2(b)). Further, the authors report an improved angular characteristic for the transmission, with a reduction of off-axis haze. Similar results were obtained by Li *et al.*[49] for SiO_2 particles, Zhu *et al.*[50] for silver particles and Hinojosa *et al.*[51] for gold particles. Particle doped PDLC systems were also employed for improving the electrooptic characteristics of paper-like reflective displays.[52] A relatively recent review with further literature references can be found in ref. 53.

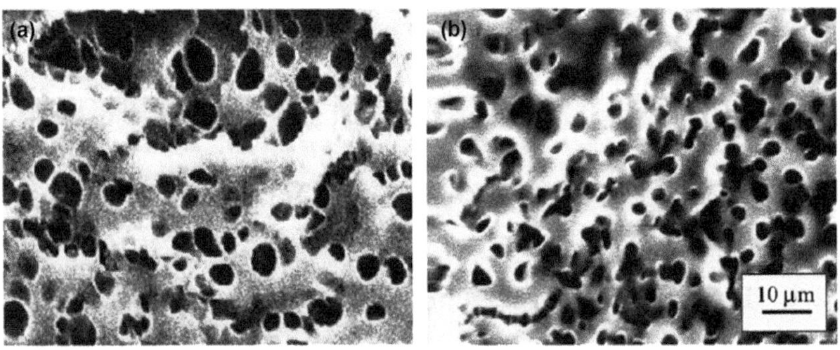

Figure 16.1 Scanning electron micrographs of the PDLC polymer morphology for (a) the pure and (b) the particle laden composite. The morphology is not substantially altered by introducing inorganic Sb_2O_5 nanoparticles. Reproduced from ref. 48 with permission from Elsevier, Copyright 2007.

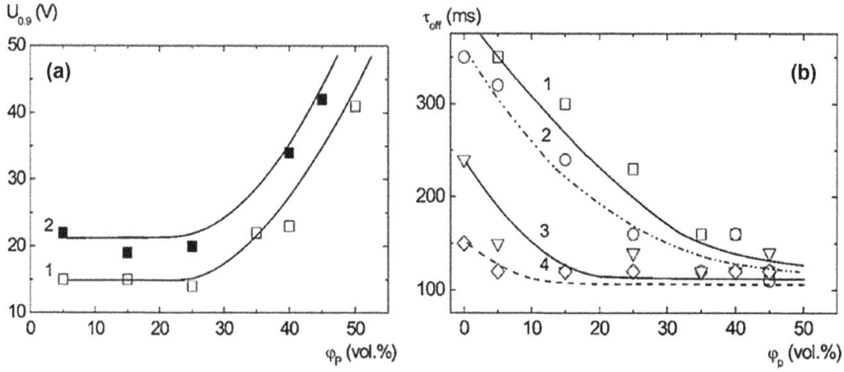

Figure 16.2 (a) Driving voltage, $U_{0.9}$, and (b) turn-off time τ_{off} for the Sb_2O_5 particle laden PDLC composite. The driving voltage increases for increasing polymer content, and is larger for the particle doped system (curve 2 in (a)) than the non-filled PDLC (curve 1 in (a)). The response time decreases with increasing polymer concentration and also decreases for increasing particle volume fraction (curves 2–4 in (b)). Reproduced from ref. 47 with permission from the American Physical Society, Copyright 2005.

16.2.1.2 Nematic PSLC system

At the low polymer concentration end of the polymer stabilized liquid crystals, the situation is somewhat different, because here we have a continuous polymer network dispersed throughout the continuous phase of the liquid crystal. In this case, the nanoparticles aggregate within the liquid crystal, which leads to a refractive index mismatch and light scattering, affecting contrast and viewing angle characteristics of the system.[47] The performance of the particle laden PSLC becomes similar to that of LC-aerosil mixtures.[54,55] Yaroshchuk and co-workers found that the driving voltage and hysteresis effects are practically constant for polymer concentrations below ~10vol% (see Figure 16.2(a)). At constant polymer concentration of the PSLC the driving voltage, $U_{0.9}$, as well as hysteresis effects, $\Delta U_{0.5}$, only slowly increase for increasing particle concentrations up to approximately 8vol% (Figure 16.3(a)), while the response times for the turn-on, τ_{on}, as well as the turn-off switching process, τ_{off}, decrease substantially, as illustrated in Figure 16.3(b).

For both polymer dispersed liquid crystals as well as polymer stabilized liquid crystals, it appears that additional doping with inorganic nanoparticles is beneficial for the overall electrooptic performance of devices. Light scattering can be increased, off-axis haze can be reduced, while at the same time the performance speed can be increased. Naturally, as with all switchable devices based on polymer or particle inclusions, these benefits come at a cost in the form of increased driving voltages. The overall performance of particle laden PDLCs and PSLCs thus needs to be optimized for the desired application with respect to applied voltage *versus* speed.

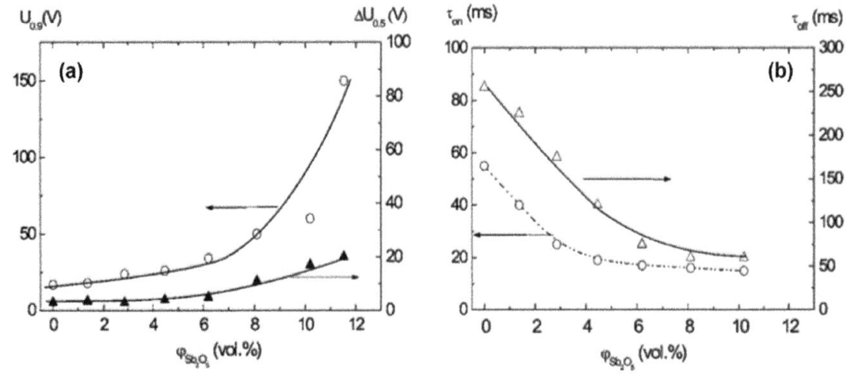

Figure 16.3 (a) Driving voltage $U_{0.9}$ and hysteresis $\Delta U_{0.5}$ slowly increase for increasing volume fraction of Sb_2O_5 nanoparticles in the PSLC composite at 5vol% polymer. (b) The response times τ_{on} and τ_{off} of the polymer-stabilized system both decrease substantially as the particle concentration is increased.
Reproduced from ref. 47 with permission from the American Physical Society, Copyright 2005.

16.2.2 Polymer Stabilized Cholesteric Liquid Crystal-aerosil Particle Composites

Polymer stabilized cholesteric textures, PSCTs, have promise in reflective displays and displays without active backlighting. They can be driven in two different ways and geometries, the so called "normal mode" and the "reverse mode". In the normal mode an electric field is applied to unwind the helical superstructure of the short pitch chiral nematic, or cholesteric phase, so that a homeotropic director configuration is obtained. In this state the polymerization is carried out. Turning the field off, the liquid crystal adopts a scattering focal conic texture at $E=0$ with low transmission, or the "white state". Application of an electric field E switches the liquid crystal to the homeotropic state, which is non-scattering, transparent, and thus represents the "black state". In the reverse mode on the other hand, a long pitch cholesteric is oriented in planar Grandjean texture and stabilized *via* polymerization of the monomers at zero voltage. This is the "black state", because no selective reflection occurs in the visible range of the spectrum and the transmission is high. Field application breaks up the helical superstructure into the focal conic, scattering state, which is the "white state".

Liang *et al.*[56] have investigated the electrooptic properties of a tertiary system composed of cholesteric liquid crystal, monomer and aerosil particles in the geometry of a normal mode PSCT. They found that the aerosil particles greatly influence the polymer network morphology, and thus also the electrooptic properties. As the aerosil particle concentration was increased from 0–1% by weight, the polymer network strands became substantially finer and the void sizes drastically increase, as shown for selected concentrations in Figure 16.4.

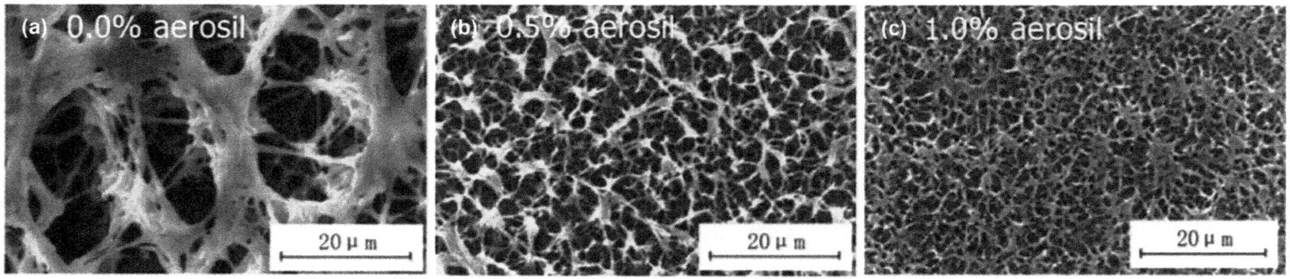

Figure 16.4 Network morphology for 5wt% polymer formed in a cholesteric liquid crystal with (a) 0.0%, (b) 0.5% and (c) 1.0% aerosil particles by weight. The addition of the particles has a drastic effect on the polymer network morphology, which results in a strongly changing electrooptic performance as the small particle concentration is increased.
Reproduced from ref. 56 with permission from Taylor and Francis.

The variation in morphology alone would suggest a strong change of the electrooptic properties.

Two independent experimental series were carried out, (i) a variation of polymer concentration at constant aerosil content, and (ii) a change of aerosil concentration at constant polymer concentration. The first series yielded the well known and expected results: the threshold as well as the saturation voltage increased with increasing polymer content. The turn-on times increased while the turn-off times became faster, and the contrast increased with increasing amount of polymer. This behaviour had been observed by many groups in the past. The more interesting aspect was through the simultaneous addition of aerosil particles (see Figure 16.5). Here it was found that with the particle addition at varying concentration, the threshold voltage and saturation voltage actually decreased, the turn-on time also decreased, while the turn-off response was practically independent of aerosil concentration. The contrast, on the other hand, was drastically decreased. The observed behaviour may be understood with the following interpretation. As the particle concentration is increased, the polymer network becomes finer, and voids become smaller, because the presence of more particles decreases the diffusion of reactive monomers. The tighter network represents an increasing amount of polymer surface and thus larger elastic interactions between the liquid crystal and the network, which in turn decreases the driving voltage into the templated favourable homeotropic state. This interpretation can also explain the response times, where the helix unwinding turn-on time becomes faster due to increased elastic interactions caused by the increased particle concentration leading to tighter voids, while the return to the helical state becomes slower.

Also for this application of polymer stabilized particle laden liquid crystals, different electrooptic parameters are a trade off between different performance aspects. Driving voltages and switching times can be reduced through the addition of particles, unfortunately at the cost of a reduced contrast. The simultaneous variation of polymer and particle concentrations on the other hand allows for an additional parameter to tune the electrooptic performance to that desired for a specific application.

16.2.3 Polymer Stabilized Nanoparticle Doped Blue Phase Liquid Crystals

Polymer stabilized Blue Phases (BP) were first introduced by Kikuchi *et al.*,[57] when a novel display mode on the basis of the Kerr effect was discussed. The properties of these materials are discussed in detail within the framework of this book. We simply would like to point out the significant advantages of this technology. The electrooptic response is significantly faster than that of a common nematic liquid crystal. But probably the most important aspect is that no alignment layer is needed so a complete step in the production process can be left out, specifically the step which causes the most dust to be generated, and thus the most rejects in display production.

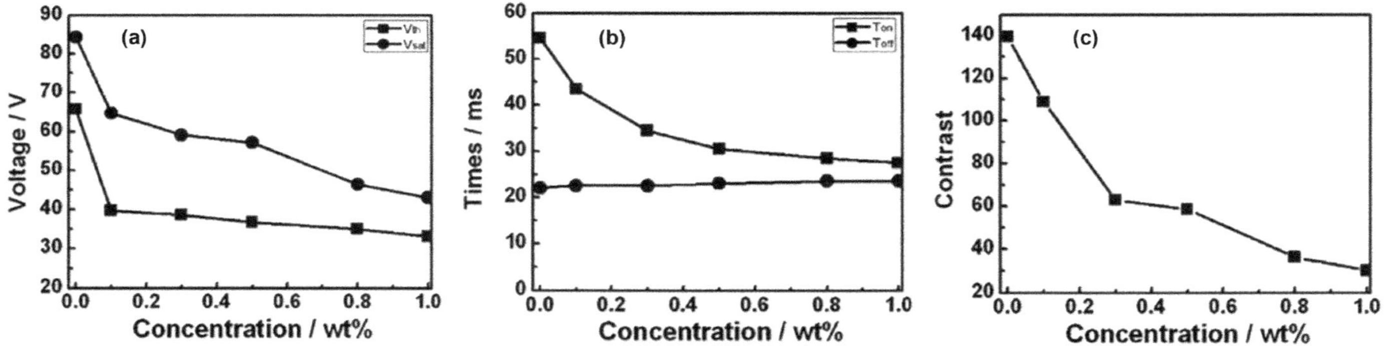

Figure 16.5 Electrooptic performance parameters for a 5wt% polymer stabilized normal mode cholesteric liquid crystal in dependence on the added aerosil particle concentration. (a) Threshold voltage (circles) and saturation voltage (squares), (b) turn-on (squares) and turn-off response times (circles), and (c) the obtained contrast.
Reproduced from ref. 56 with permission from Taylor and Francis.

The stabilization of the otherwise very narrow Blue Phases by a polymer network was attributed to the polymer network forming within the defects and disclinations of the Blue Phase. Defects cost elastic energy and thus increase the overall free energy density of the liquid crystal. By filling them with a polymer, the free energy density is lowered, and the phase stabilized. Therefore, the temperature regime of Blue Phase existence is widened, like in the case of Kikuchi's work, from a very few degrees to tens of degrees, and the Blue Phase becomes attractive for display applications.

A similar approach was followed *via* nanoparticle stabilization of the Blue Phase a few years later by Yoshida *et al.*[58] Gold nanoparticles of size approximately 4 nm were doped into a Blue Phase. It was anticipated that the nanoparticles would agglomerate within the defects, lower the free energy density and thus widen the existence regime of the BP. This was indeed observed with the temperature regime increasing from about 0.5 K to 5 K, while at the same time the clearing temperature is strongly decreased. The mechanism of the stabilization should thus be equivalent to that of polymer stabilization.

Besides several other mechanisms of BP stabilization, Dierking *et al.*[59] combined the methods of doping and polymer stabilization. This mechanism was explored *via* two systems, (i) polymer stabilized-bent-core doped Blue Phases, and (ii) polymer stabilized-nanoparticle doped Blue Phases, at varying concentrations of polymer. The results of these investigations are shown in Figure 16.6, together with a predicted behaviour, if one assumes simple additivity of both effects. Figure 16.6(a) depicts the width of the Blue Phase for pure polymer stabilization. The temperature range of the Blue Phase linearly increases from 1 K to approximately 8 K with increasing polymer content in the region of small concentrations (smaller than 5wt%), before phase separation occurs. In part (b) of the figure, the pure effect of adding monodispersed nanoparticles of diameter 40 nm is shown, as the concentration is increased along the estimated occupation of defects. It was demonstrated that at first the Blue Phase temperature range increases with increasing nanoparticle concentration from approximately 1 K to a saturation value of about 4 K, which is reached at approximately 100% defect occupancy. In Figure 16.6(c) the results of a combined investigation of polymer stabilized-particle laden Blue Phases are shown for a defect occupancy with nanoparticles of 100%. The contributions of the polymer as well as the nanoparticles at this single concentration are shown in black and red symbols, respectively. The blue symbols represent the combined effect on Blue Phase stability predicted from assuming simple additivity of the individual effects. From these investigations, which were carried out on systems that were in no way optimized, it could be concluded that the combination of two stabilization methods is beneficial, but nevertheless less effective than what could be expected by full additivity of the two individual effects of nanoparticles and polymer stabilization, as demonstrated by the green symbols from the respective experiment.

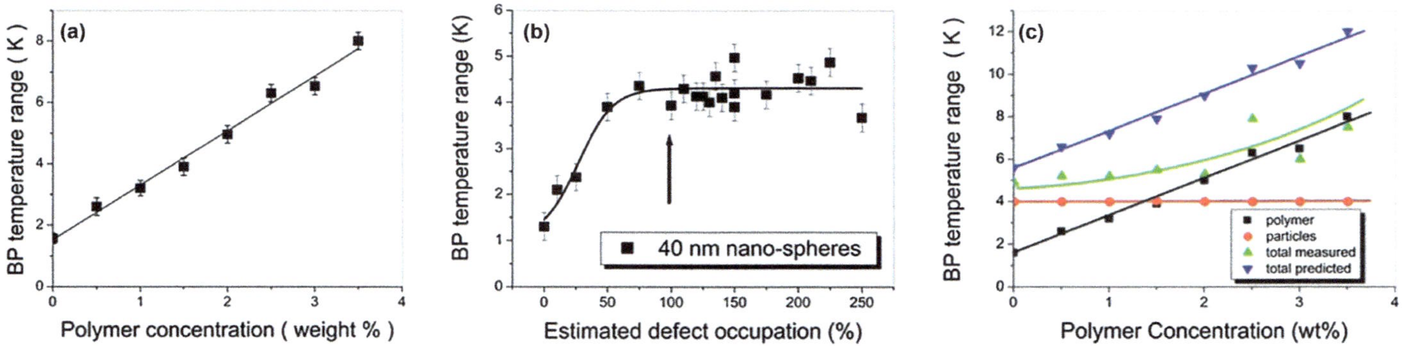

Figure 16.6 Temperature stability of the Blue Phase with (a) pure polymer stabilization, (b) pure particle addition, and (c) for the combined polymer-particle-Blue Phase composite. For a detailed discussion, see text.
Reproduced from ref. 59 with permission from the Royal Society of Chemistry.

But it can be seen that at very low polymer concentrations, smaller than 2wt%, the combination of two methods is more beneficial than the use of only an individual mechanism. This is probably the case because neither the nanoparticles nor the polymer fully fills all the defects. As the concentration of the polymer is increased above 2wt%, the experimentally observed stabilization of the Blue Phase for the combined composite approaches that of pure polymer stabilization. Nevertheless, in all cases the experimentally observed stabilization is smaller than the one predicted from additivity of individual composite components.

Wang *et al.*[60] used nanoparticles of very similar size, 30 nm, but came to a different conclusion. They found that their system responded with fast switching times, at low voltages, and with no hysteresis observed. These are all the characteristics of the Kerr effect. They also concluded that the benefits of combining nanoparticles with polymer stabilization were larger than a simple addition of individual contributions. On the other hand, their system also behaved differently to the behaviour commonly observed. The isotropic to BP transition temperature was reported to strongly increase for ferroelectric nanoparticles of $BaTiO_3$, significantly more than for ZnS. This was attributed to the spontaneous polarization of barium titanate. On the other hand, it has been shown for nanoparticles of $BaTiO_3$ that ferroelectricity is lost for particle sizes below approximately 100 nm.[61] The authors report an increase of the BP stability range for increasing particle concentration, as it was also observed before. The switching voltage decreased with increasing particle concentration, as did the Kerr coefficient until saturation at the same concentration as the switching voltage saturates. Also, Xu *et al.*[62] investigated polymer stabilized Blue Phases doped with $BaTiO_3$ in the size range of 30–100 nm. Here the reflectance was measured, and in general a trend towards lower driving voltages was recorded for polymer stabilized BPs with nanoparticles. This means that the same reflectance is achieved at much smaller applied voltages for the $BaTiO_3$ nanoparticle doped polymer stabilized BP systems than without nanoparticles. All in all, it appears that further systematic investigations with respect to phase stability and electrooptic behaviour would be of benefit to further elucidate the delicate interplay between Blue Phase, polymer network and nanoparticle addition.

16.2.4 Polymer Stabilized Nanotube Reinforced Liquid Crystals

An interesting aspect of polymer stabilized liquid crystals that was very recently reported by Prasad *et al.*[63] is the reinforcement by carbon nanotubes, thus using quasi one-dimensional nanoparticles, instead of the quasi zero-dimensional ones discussed above. In the standard polymer stabilized liquid crystal devices, a polymer network is dispersed in the continuous phase of the liquid crystal, which acts like an additional interface, besides the bounding substrates. This in turn results in a modification and often enhancement of the electrooptic properties, as demonstrated in the chapters above. The novel aspect that Prasad introduced was a reinforcement of the polymer network by polymer-capped single wall nanotubes, which elastically strengthens the network. The device is schematically shown in Figure 16.7.

This resulted in attractive properties, such as a temperature independent threshold voltage, in comparison to the strongly dependent threshold for devices without nanotubes. Furthermore, the value of the threshold voltage is reduced for the nanotube reinforced network structures, and the switching dynamics accelerated, due to an increased effective splay elastic constant. A two-stage switching process, as it has also been observed for standard polymer stabilized liquid crystals at low polymer concentrations,[64] could be avoided by adding polymer-capped nanotubes, without having to increase the polymer concentration. Some of the enhanced device properties are summarized in Figure 16.8.

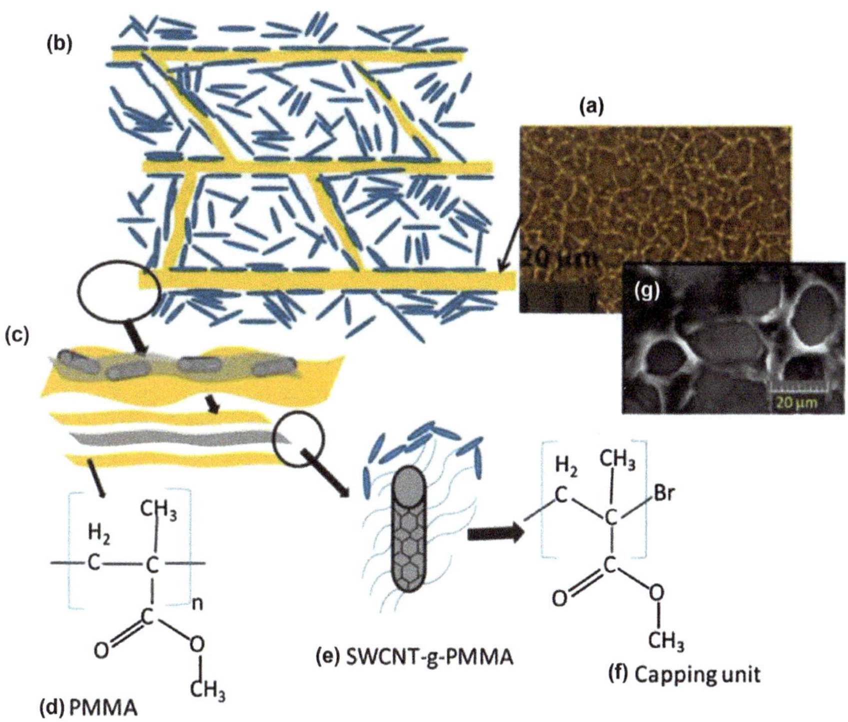

Figure 16.7 (a) Optical photomicrograph of the nanotube reinforced polymer network after removal of the liquid crystal; (b) schematic diagram of the polymer stabilized structure and (c) closeup of a single polymer strand, with (d) indicating the polymethyl methacrylate, PMMA, polymer, (e) the polymer modified nanotube with (f) the capping unit; (g) shows a scanning electron micrograph, SEM, closeup of the polymer network structure after removal of the liquid crystal.
Reproduced from ref. 63 with permission from American Chemical Society, Copyright 2017.

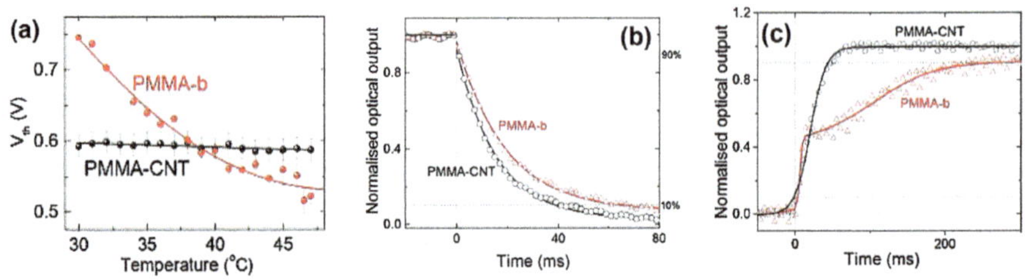

Figure 16.8 Comparison of the electrooptic properties of the PMMA polymer stabilized liquid crystal and the carbon nanotube reinforced polymer stabilized device, PMMA-CNT. (a) The threshold voltage becomes temperature independent, in contrast to the strong temperature dependence observed for the non-reinforced network device. (b) The electrooptic response becomes faster when the polymer is reinforced with capped nanotubes, and (c) a two-stage switching process can be avoided by adding nanotubes to the polymer network phase. Reproduced from ref. 63 with permission from American Chemical Society, Copyright 2017.

The sum of these improved features makes a nanotube reinforced device attractive when driving it through the addressing electronics, which in general is due to the good compatibility of the polymer matrix and polymer modified carbon nanotubes.

16.2.5 Nanoparticles in LC Elastomers

Elastomers have long been researched as materials for nonlinear optics, storage devices, actuators or sensors.[65] For applications of these systems it is on most occasions necessary to employ samples which are uniformly oriented over macroscopic sizes. This has been achieved by Küpfer and Finkelmann.[66] Materials were obtained that exhibited the order and anisotropy of a liquid crystal in the nematic, SmA or SmC phase, and at the same time showed the elastic behaviour of a crosslinked rubber. Reversible shape changes were observed as the material underwent temperature changes and phase transitions, while a high birefringence was also obtained. Similar to rubber elasticity, temperature changes occur when subjecting the liquid crystal elastomer to mechanical stimuli. Elastomers have been shown to exhibit potential for a wealth of applications, from artificial muscles[67] to actuators and sensors,[65] microfluidic valves,[68] propulsion and motors,[69] grippers in microsystems,[70] tuneable lasing materials[71] or bifocal contact lenses.[72] Although not being strictly polymer modified liquid crystals, because elastomers are rubber-like crosslinked polymers which exhibit liquid crystalline order, the embedding of nanoparticles can add a novel functionality to the system, which may enhance properties and response, especially in the area of attenuators and sensors. In the following we will present some selected examples. A recent review from the group of Terentjev about the topic of nanoparticle-liquid crystalline elastomer composites can be found in ref. 73.

Thermally stimulated shape changes of liquid crystalline elastomers are quite well understood, and incorporation of nanoparticles may create materials that are more sensitive to external stimuli. One example is swollen liquid crystalline elastomer networks within a carbon nanoparticle dispersion, which is subsequently deswollen to create a conductive surface layer, without loss of the macroscopic elasticity[74] (see Figure 16.9). These could be of interest for sensors,[75] interconnects,[76] or flexible heating elements.[77]

Another system has been demonstrated to produce a rapid and reversible actuation *via* laser irradiation of inclusions of gold nanospheres and nanorods in liquid crystalline polyacrylate based elastomer microactuators.[78] This was achieved at low loading fractions of less than 1wt% and without changing the thermal actuation ability of the non-doped elastomer. Surface plasmon resonance enhanced optical absorption can further be exploited to photothermally actuate the microactuators. This allows precise targeting of actuation *via* laser beams (Figure 16.10).

To provide a different type of stimulus, liquid crystalline elastomers can also be magnetically sensitized by embedding iron oxide nanoparticles at low concentration.[79] Reversible deformations can then be induced *via* the nanoparticles, due to the application of alternating magnetic fields, which cause a nematic to isotropic phase transition (see Figure 16.11). The transition temperature was found to decrease with increasing particle loading.

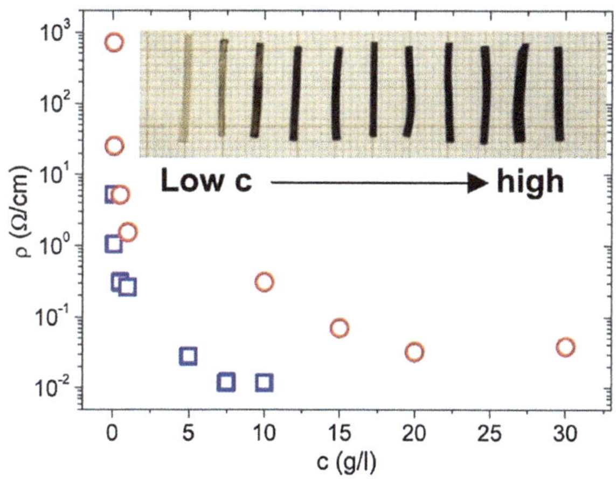

Figure 16.9 Resistivity of the deswollen liquid crystalline elastomer with carbon black (squares) and carbon horns (circles) nanoparticle surface layers, indicating an increase and saturation of the conductivity at large nanoparticle concentrations. The inset illustrates the change in opaqueness as the carbon black layer becomes thicker for increasing concentration.
Reproduced from ref. 74 with permission from IOP Publishing.

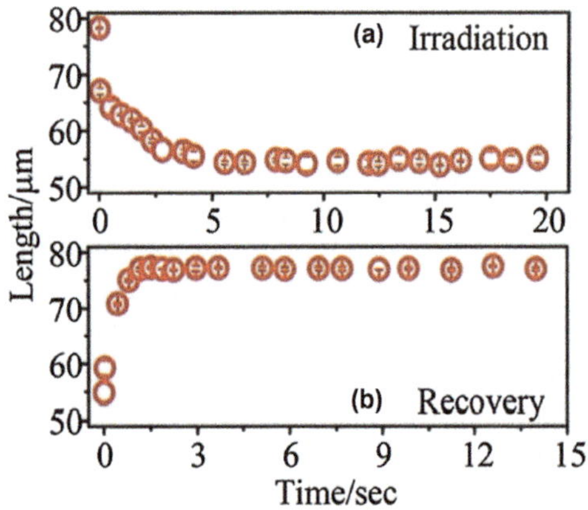

Figure 16.10 Photothermal actuation by laser irradiation of a liquid crystalline gold nanorod doped elastomer. Length change of the elastomer when the laser is (a) on and (b) off.
Reproduced from ref. 78 with permission from John Wiley and Sons, © 2015 WILEY-VCH Verlag GmbH & Co. KGaA, Weinheim.

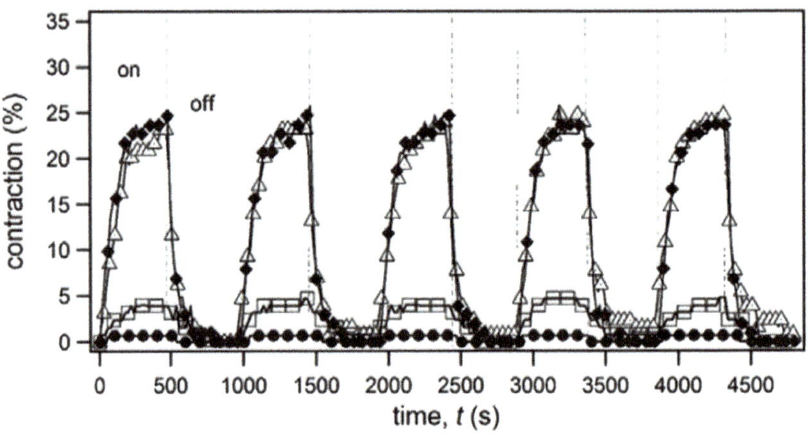

Figure 16.11 Magneto-stimulated actuation by an alternating magnetic field of approximately 50 kA m^{-1} at a frequency of about 300 kHz, switching the field on and off every 500 s. The contraction response increases with increasing iron oxide concentration.
Reproduced from ref. 79 with permission from Elsevier, Copyright 2016.

Nanoparticles do not need to be dispersed physically within the liquid crystalline elastomer, as uniform dispersions at high loading can also be achieved by incorporating them into the crosslinkers, as demonstrated very recently for gold nanoparticles of a diameter of approximately 2 nm.[80] The principle is schematically depicted in Figure 16.12.

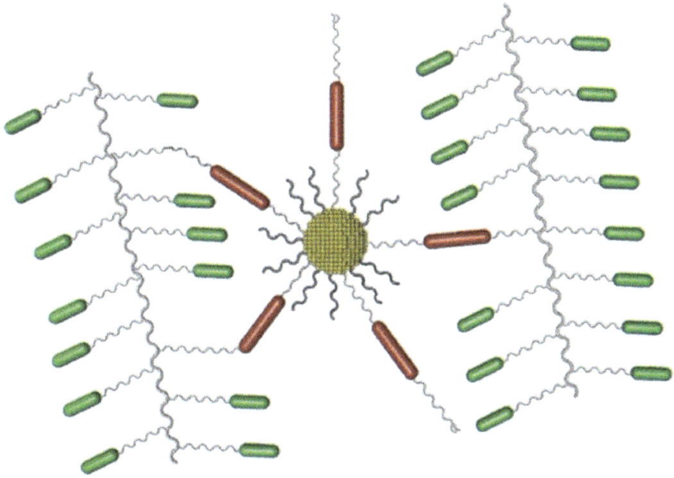

Figure 16.12 Schematic illustration of the embedding of gold nanoparticles into a liquid crystalline elastomer, by cross-linking *via* the nanoparticle. This allows the uniform incorporation of particles at relatively large concentrations.
Reproduced from ref. 80 with permission from John Wiley and Sons, © 2017 Wiley-VCH Verlag GmbH & Co. KGaA, Weinheim.

16.3 Concluding Remarks

It appears that the combination of nanoparticle dispersion within polymer-modified liquid crystals is only in its infancy, with experimental data sometimes inconclusive and controversial. It can thus be anticipated that many more investigations will be carried out, and innovative and novel ideas for devices will be proposed in the future, which add the mechanical, electric, magnetic or optic functionality of nanoparticles to that of the mechanical stabilization of liquid crystal phases and their physical properties.

References

1. *Liquid Crystals with Nano and Microparticles*, ed. J. P. F. Lagerwall and G. Scalia, World Scientific, Singapore, 2017.
2. I. Dierking and S. Al-Zangana, *Nanomaterials*, 2017, 7, 305.
3. Y. Reznikov, O. Buchnev, O. Tereshchenko, V. Reshetnyak, A. Glushchenko and J. West, *Appl. Phys. Lett.*, 2003, **82**, 1917.
4. F. H. Li, J. West, A. Glushenko, C. Il Cheon and Y. Reznikov, *J. Soc. Inf. Displ.*, 2006, **14**, 523.
5. T. Hegmann, Hao Qi and V. M. Marx, *J. Inorg. Organomet. Polym. Mater.*, 2007, **17**, 483.
6. O. Stamatoiu, J. Mirzaei, X. Feng and T. Hegmann, *Top. Curr. Chem.*, 2012, **318**, 331.

7. Y. S. Ha, H. J. Kim, H. G. Park and D. S. Seo, *Opt. Express*, 2012, **20**, 6448.
8. T. Zhang, C. Zhong and J. Xu, *Jpn. J. Appl. Phys.*, 2009, **48**, 055002.
9. A. Chaudhary, P. Malik, R. Mehra and K. K. Raina, *J. Mol. Liq.*, 2013, **188**, 230.
10. H. K. Chung, H. G. Park, Y. S. Ha, J. M. Han, J. W. Lee and D. S. Seo, *Liq. Cryst.*, 2013, **40**, 632.
11. U. B. Singh, R. Dhar, R. Dabrowski and M. B. Pandey, *Liq. Cryst.*, 2013, **40**, 774.
12. S. Al-Zangana, M. Turner and I. Dierking, *J. Appl. Phys.*, 2017, **121**, 085105.
13. P. Kumar, A. Kishore and A. Sinha, *Appl. Phys. Lett.*, 2016, **108**, 262903.
14. R. Basu and A. Garvey, *Appl. Phys. Lett.*, 2014, **105**, 151905.
15. T. Toth-Katona, P. Salamon, N. Eber, N. Tomasovicova, Z. Mitroova and P. Kopcansky, *J. Magn. Magn. Mater.*, 2014, **372**, 117.
16. A. Mertelj, N. Osterman, D. Lisjak and M. Copic, *Soft Matter*, 2014, **10**, 9065.
17. N. Podoliak, O. Buchnev, D. V. Bavykin, A. N. Kulak, M. Kaczmarek and T. J. Sluckin, *J. Colloid Interface Sci.*, 2012, **386**, 158.
18. A. Mertelj, D. Lisjak, M. Drofenik and M. Copic, *Nature*, 2013, **504**, 237.
19. M. Shuai, A. Klittnick, Y. Shen, G. P. Smith, M. R. Tuchband, C. Zhu, R. G. Petschek, A. Mertelj, D. Lisjak, M. Copic, J. E. Maclennan, M. A. Glaser and N. A. Clark, *Nat. Commun.*, 2015, 7, 10394.
20. R. Basu and G. S. Iannacchione, *Phys. Rev. E*, 2009, **80**, 010701.
21. J. Mirzaei, M. Reznikov and T. Hegmann, *J. Mater. Chem.*, 2012, **22**, 22350.
22. S. Khatua, P. Manna, W. S. Chang, A. Tcherniak, E. Friedlander, E. R. Zubarev and S. Link, *J. Phys. Chem. C*, 2010, **114**, 7251.
23. A. Choudhary, G. Singh and A. M. Biradar, *Nanoscale*, 2014, **6**, 7743.
24. S. Sridevi, S. K. Prasad, G. G. Nair, V. D'Britto and B. L. V. Prasad, *Appl. Phys. Lett.*, 2010, **97**, 151913.
25. M. V. Mukhina, V. V. Danilov, A. O. Orlova, M. V. Fedorov, M. V. Artemyev and A. V. Baranov, *Nanotechnology*, 2012, **23**, 325201.
26. N. Kamanina and I. Denisyuk, *Opt. Commun.*, 2004, **235**, 361.
27. J. Szydlowska, K. Trzcinska, P. Bilkova, J. Mieczkowski, D. Pociecha and E. Gorecka, *Liq. Cryst.*, 2006, **33**, 335.
28. N. Behabtu, J. R. Lomeda, M. J. Green, A. L. Higginbothan, A. Sinitskii, D. V. Kosynkin, D. Tsentalovich, A. N. G. Parra-Vasquez, J. Schmidt, E. Kesselman, Y. Cohen, Y. Talmon, J. M. Tour and M. Pasquali, *Nat. Nanotechnol.*, 2010, **5**, 406.
29. J. E. Kim, T. H. Han, S. H. Lee, J. Y. Kim, C. W. Ahn, J. M. Yun and S. O. Kim, *Angew. Chem., Int. Ed.*, 2011, **50**, 3043.
30. Z. Xu and C. Gao, *Nat. Commun.*, 2011, **2**, 571.
31. T. Z. Shen, S. H. Hong and J. K. Song, *Nat. Mater.*, 2014, **13**, 394.
32. S. Al-Zangana, M. Iliut, M. Turner, A. Vijayaraghavan and I. Dierking, *Adv. Optical Mater.*, 2016, DOI: 10.1002/adom.201600351.

33. S. Al-Zangana, M. Iliut, Gökçen Boran, M. Turner, A. Vijayaraghavan and I. Dierking, *Sci. Rep.*, 2016, **6**, 31885.
34. S. Al-Zangana, M. Iliut, M. Turner, A. Vijayaraghavan and I. Dierking, *2D Mater.*, 2017, **4**, 041004.
35. M. D. Lynch and D. L. Patrick, *Nano Lett.*, 2002, **2**, 1197.
36. I. Dierking, G. Scalia, P. Morales and D. LeClere, *Adv. Mater.*, 2004, **16**, 865.
37. I. Dierking, G. Scalia and P. Morales, *J. Appl. Phys.*, 2005, **97**, 044309.
38. J. Lagerwall, G. Scalia, M. Haluska, U. Dettlaf-Weglikowska, S. Roth and F. Giesselmann, *Adv. Mater.*, 2007, **19**, 359.
39. S. Kumar and H. K. Bisoyi, *Angew. Chem., Int. Ed.*, 2007, **46**, 1501.
40. J. P. F. Lagerwall and G. Scalia, *J. Mater. Chem.*, 2008, **18**, 2890.
41. S. P. Yadav and S. Singh, *Prog. Mater. Sci.*, 2016, **80**, 38.
42. A. S. Sonin, *J. Mater. Chem.*, 1998, **8**, 2557.
43. J. C. P. Gabriel and P. Davidson, *Colloid Chem. 1 Top. Curr. Chem.*, 2003, **226**, 119.
44. S. Fraden, G. Maret and D. L. D. Caspar, *Phys. Rev. E*, 1993, **48**, 2816.
45. F. Livolant and A. Leforestier, *Prog. Polym. Sci.*, 1996, **21**, 1115.
46. J. P. F. Lagerwall, C. Schütz, M. Solajkova, J. H. Noh, J. H. Park, G. Scalia and L. Bergström, *NPG Asia Mater.*, 2014, **6**, DOI: 10.1038/am.2013.69.
47. O. V. Yaroshchuk, L. O. Dolgov and A. D. Kiselev, *Phys. Rev. E*, 2005, **72**, 051715.
48. O. V. Yaroshchuk and L. O. Dolgov, *Opt. Mater.*, 2007, **29**, 1097.
49. W. Li, M. Zhu, X. Ding, B. Li, W. Huang, H. Cao, Z. Yang and H. Yang, *J. Appl. Polym. Sci.*, 2009, **111**, 1449.
50. T. Zhu, B. Tan, X. Pan, W. Tao and J. Xu, *Proc. Photonics Optoelectron.*, Wuhan, China, IEEE, NY, (2009), 716.
51. A. Hinojosa and S. C. Sharma, *Appl. Phys. Lett.*, 2010, **97**, 081114.
52. A. Masutari, T. Roberts, B. Schüller, N. Hollfelder, P. Kilickiran, A. Sakaigawa, G. Nelles and A. Yasuda, *J. Soc. Info. Display*, 2008, **16**, 137.
53. M. Jamil, Farzana Ahmad, J. T. Rhee and Y. J. Jeon, *Curr. Sci.*, 2011, **101**, 1544.
54. A. Glushchenko, H. Kresse, V. Reshetnyak, Y. Reznikov and O. Yaroshchuk, *Liq. Cryst.*, 1997, **23**, 241.
55. O. Kovalchuk, S. Zakrevska, O. Yaroshchuk and U. Maschke, *Mol. Cryst. Liq. Cryst.*, 2001, **368**, 129.
56. X. Liang, H. Cao, G. Pan, X. Cui, F. Li, G. Niu, D. Zhang, Z. Yang, H. Yang and S. Zhu, *Liq. Cryst.*, 2009, **36**, 93.
57. H. Kikuchi, M. Yokota, Y. Hisakado, H. Yang and T. Kajiyama, *Nat. Mater.*, 2002, **1**, 64.
58. H. Yoshida, Y. Tanaka, K. Kawamoto, H. Kubo, T. Tsuda, A. Fujii, S. Kuwanata, H. Kikuchi and M. Ozaki, *Appl. Phys. Express*, 2009, **2**, 121501.
59. I. Dierking, W. Blenkhorn, E. Credland, W. Drake, R. Kociuruba, B. Kayser and T. Michael, *Soft Matter*, 2012, **8**, 4355.

60. L. Wang, W. He, Q. Wang, M. Yu, X. Xiao, Y. Zhang, M. Ellahi, D. Zhao, H. Yang and L. Guo, *J. Mater. Chem. C*, 2013, **1**, 6526.
61. H. M. Atkuri, K. Zhang and J. L. West, *Mol. Cryst. Liq. Cryst.*, 2009, **508**, 183.
62. X. W. Xu, X. W. Zhang, D. Lun and H. T. Dai, *Opt. Express*, 2015, **23**, 32267.
63. S. Krishna Prasad, M. Baral, A. Murali and S. N. Jaisankar, *ACS Appl. Mater. Interfaces*, 2017, **9**, 26622.
64. I. Dierking, L. L. Kosbar, A. Afzali-Ardakani, A. C. Lowe and G. A. Held, *J. Appl. Phys.*, 1997, **81**, 3007.
65. C. Ohm, M. Brehmer and R. Zentel, *Adv. Mater.*, 2010, **22**, 3366.
66. J. Küpfer and H. Finkelmann, *Makromol. Chem., Rapid Commun.*, 1991, **12**, 717.
67. T. Mirfakhrai, J. D. W. Madden and R. H. Baughman, *Mater. Today*, 2007, **10**, 30.
68. A. Sanchez-Ferrer, T. Fischl, M. Stubenrauch, A. Albrecht, H. Wurmus, M. Hoffmann and H. Finkelmann, *Adv. Mater.*, 2011, **23**, 4526.
69. M. Yamada, M. Kondo, J. I. Mamiya, Y. L. Yu, M. Kinoshita, C. J. Barrett and T. Ikeda, *Angew. Chem., Int. Ed.*, 2008, **47**, 4986.
70. A. Sanchez-Ferrer, T. Fischl, M. Stubenrauch, H. Wurmus, M. Hoffmann and H. Finkelmann, *Macromol. Chem. Phys.*, 2009, **210**, 1671.
71. H. Finkelmann, S. T. Kim, A. Muæoz, P. Palffy-Muhoray and B. Taheri, *Adv. Mater.*, 2001, **13**, 1069.
72. A. Amigo-Melchior and H. Finkelmann, *Polym. Adv. Technol.*, 2002, **13**, 363.
73. Y. Li, J. E. Marshall and E. M. Terentjev, *Polymers*, 2012, **4**, 316.
74. M. Chambers, B. Zalar, M. Remskar, J. Kovac, H. Finkelmann and S. Zumer, *Nanotechnology*, 2007, **18**, 415706.
75. M. Chambers, H. Finkelmann, M. Remskar, A. Sanchez-Ferrer, B. Zalar and S. Zumer, *J. Mater. Chem.*, 2009, **19**, 1524.
76. S. P. Lacour, S. Wagner, Z. Huang and Z. Suo, *Appl. Phys. Lett.*, 2003, **82**, 2404.
77. F. El-Tantawy, K. Kamada and H. Ohnabe, *Mater. Lett.*, 2002, **56**, 112.
78. X. Liu, R. Wei, P. T. Hoang, X. Wang, T. Liu and P. Keller, *Adv. Funct. Mater.*, 2015, **25**, 3022.
79. S. Herrera-Posada, C. Mora-Navarro, P. Ortiz-Bermudez, M. Torres-Lugo, K. M. McElhinny, P. G. Evans, B. O. Calcagno and A. Acevedo, *Mater. Sci. Eng., C*, 2016, **65**, 369.
80. M. M. Wojcik, J. Wrobel, Z. Z. Janczuk, J. Mieczkowski, E. Gorecka, J. Choi, M. Cho and D. Pociecha, *Chem. – Eur. J.*, 2017, **23**, 8912.

Subject Index